T0251412

Molecular Modeling Techniques
in
Material Sciences

Molecular Modeling Techniques
in
Material Sciences

Jörg-Rüdiger Hill
Lalitha Subramanian
Amitesh Maiti

Taylor & Francis
Taylor & Francis Group

Boca Raton   London   New York   Singapore

A CRC title, part of the Taylor & Francis imprint, a member of the
Taylor & Francis Group, the academic division of T&F Informa plc.

Published in 2005 by
CRC Press
Taylor & Francis Group
6000 Broken Sound Parkway NW, Suite 300
Boca Raton, FL 33487-2742

10 9 8 7 6 5 4 3 2

International Standard Book Number-10: 0-8247-2419-4 (Hardcover)
International Standard Book Number-13: 978-0-8247-2419-1 (Hardcover)

This book contains information obtained from authentic and highly regarded sources. Reprinted material is quoted with permission, and sources are indicated. A wide variety of references are listed. Reasonable efforts have been made to publish reliable data and information, but the author and the publisher cannot assume responsibility for the validity of all materials or for the consequences of their use.

Library of Congress Cataloging-in-Publication Data

Catalog record is available from the Library of Congress

Taylor & Francis Group
is the Academic Division of T&F Informa plc.

Visit the Taylor & Francis Web site at
http://www.taylorandfrancis.com

and the CRC Press Web site at
http://www.crcpress.com

Foreword

It continues to be an exciting time for those scientists and engineers applying computational techniques in the materials sciences. Today's practitioners are the beneficiaries of dramatic and sustained improvements in hardware tools, in theory and algorithms, and in software implementations. Today's practitioners are presented, on a regular basis, with high-profile systems to consider; superionic conductors, high temperature superconductors, microporous or hierarchical solids, heterogeneous and homogeneous catalysts, self-assembling systems, biomaterials, complex fluids, nanotechnology are some that spring immediately to mind. Today's practitioners have a diverse set of computational approaches to such systems from which to select. And practitioners today are recognized and appreciated by the broader scientific and technical community to a pleasingly substantial degree. Still, there remain several areas deserving of further attention. One might, somewhat parochially, highlight four.

Firstly, there is real potential for a much more substantial industrial impact. One can point to cases of clear industrial benefit from the application of molecular modeling. Materials can be designed computationally to have targeted properties and with such properties then evidenced in the materials subsequently prepared; composition of matter patents for which reduction to practice was solely computational; materials for which known structural details derived from computational studies; chemical processes for which computed chemical properties have been key design parameters, and so on. However, the level of industrial impact remains far below potential. What is needed is a larger population of industrial simulation specialists. Such individuals need to be well versed in the various tools of molecular modeling, able to appreciate the dimensions of a given industrial problem, and, critically, adept at analyzing the problem in such a way that the most effective modeling approach can be both chosen and applied effectively. Industrial simulation specialists are still rare, suggesting interesting career opportunities for today's graduates.

Secondly, there will be benefits to a closer integration between the many classes of molecular modeling methods: quantum, molecular, geometrical, analytical, crystallographic, correlative, mesoscopic, etc., and in the communities that exploit each. A combination of methods might often be most effective in progressing towards a problem solution, or towards new insights into materials behavior.

Thirdly, while reference to experimental data is universally valuable, or even necessary, for validation, it will be interesting to see more in the way of exploitations of one of the prime strengths of simulation. Namely, that one can study systems or configurations that are not accessible experimentally. Not only can one address a range of "what if" questions, but models can be crafted or distorted so as to probe the basis for behaviors.

Fourthly, larger-scale collaborative or collective programs are anticipated. There are possibilities for integrating of molecular modeling work across different groups, with complementary experimental efforts, and with archival work already in the literature. As of the morning of drafting this foreword, performing a search on Google based on the terms "molecular modeling" and "materials" yields (in 0.27 seconds) approximately 48,000 hits; a similar search on "molecular simulation" and "materials" yielded some 12,600 hits. It might be argued that of all of the computational developments that have affected the materials modeling field over the past decade, none has been more impactful than the emergence of the internet.

In this context, then, an overview of molecular modeling techniques, one that connects well with applications in the materials sciences, is certainly timely. The authors, in addition to being former colleagues, have all dedicated many years of their careers to the business of molecular modeling in the materials realm. There is a large, global community of users of commercially supported molecular modeling codes. These users, in universities, governmental institutions, and corporations, address a broad range of topics, reflective of the diversity of the materials sciences themselves. There is an inevitability to an increased reliance on molecular modeling in materials research and development. A prime question is at what pace this reliance will develop.

John M Newsam
San Diego, CA
August 2004

Contents

Preface

Molecular modeling is a method which combines computational chemistry techniques with graphics visualization for simulating and predicting the three-dimensional structures, chemical processes, and physiochemical properties of molecules and solids. One of the strengths of molecular modeling is to be able to predict properties before experiments are performed. Another strength is the ability to generate data from which one may gain insight, and thereby rationalize the behavior of a large class of molecules. These strengths are being used extensively by major pharmaceutical companies for screening drug candidates, thus reducing the cost of experimental research. However, chemical and materials based companies use molecular modeling to a much lesser extent because of lack of knowledge.

Computers do not solve problems, people do. Therefore, this book has been designed to provide information about the various methods employed to study a wide variety of materials such as oxides, superconductors, semiconductors, zeolites, glass and nanomaterials. Polymers have been excluded since there are other publications devoted to them. Just as computers do not solve problems, mathematical formulas by themselves do not provide insight. It is in this spirit that we have tried to write this book. The necessary mathematical background has been provided only in the last chapter, the aim being that the reader should be able to understand the premises and limitations of different methods, and follow the main steps in running a calculation.

The goal of this book is to provide an overview over commonly used methods in atomistic simulation of a broad range of materials, giving enough theoretical background to understand why, for example, a certain type of forcefield may be applied to a zeolite while the same may not work well for glasses. Or why the local density approximation may be used for geometry optimization of a semiconductor structure while it is not used to predict the band gap of the semiconductor.

This book provides information about how to handle different materials, how to calculate their properties, what method is appropriate for a certain problem, etc. We have tried to include most methods which already are extensively used, together with some that we expect to become generally available in the near future. The amount of detailing in the description of the methods depends partly on how practical and commonly used the methods are (both in terms of computational resources and software), and partly reflects our own limitations in terms of knowledge.

We have assumed that the reader may or may not have prior knowledge of concepts specific to computational chemistry, but has a working understanding of introductory quantum mechanics and classical mechanics. This book is an introduction for all chemists, materials scientists, and researchers working in, and interested in, the field of molecular modeling applications to industrial problems.

Jörg-Rüdiger Hill would like to thank Prof. Joachim Sauer and Prof. Reinhart Ahlrichs for raising and directing his interest in theoretical chemistry and teaching him the basics of the underlying methodologies. A lot of the practical experience in industrial application of theoretical chemistry which forms the basis for this book would not have been gained without the support and encouragement from Dr. John Newsam and Dr. Erich Wimmer and the entire former contract research team at Molecular Simulations, Inc. Special thanks go to Dr. Clive Freeman for countless discussions and his help in preparing this book.

Lalitha Subramanian would like to thank her mentors, Prof. P. T. Manoharan, for instilling the spirit of cutting edge research, Prof. Roald Hoffmann for teaching the language of bonding in solids and surfaces, and Prof. Dennis Lichtenberger for bridging the gap between experiments and theory. Special thanks go to Dr. John Newsam and Dr. Erich Wimmer for giving her ample opportunity to successfully apply molecular modeling to solve industrial problems. Lalitha would also like to thank her husband and parents for their encouragement and son for his patience during various stages of writing this book.

Amitesh Maiti is greatly indebted to his mentors from Berkeley, North Carolina, and Oak Ridge for guiding him into the exciting fields of computational materials modeling and nanoscale research. Special thanks go to Profs. Leo Falicov, Jerry Bernholc, Sokrates Pantelides, and Steve Pennycook. Much of his recent work, presented in various chapters of the book, benefited significantly from scientific collaborations both within and outside Accelrys. He would especially like to mention the contributions of M. P. Anantram, Jan Andzelm, Matt Chisholm, Joe Golab, Niranjan Govind, Paul Kung, Simon McGrother, Marc in het Panhuis, Alessandra Ricca, Jose Rodriguez, Joachim Sauer, Marek Sierka, Alexei Svizhenko, James Wescott, and Peidong Yang. Stimulating discussions with Profs. Hongjie Dai and Ray Baughman, and with Accelrys colleagues Dominic King-Smith and John Newsam are also gratefully acknowledged. Finally, Amitesh would like to thank his parents for their constant encouragement, and his wife and daughter for their patience and support.

Last but not least we would like to thank all the colleagues at Biosym/Molecular Simulations/Accelrys, Inc. which we had the pleasure to work with over the years. Special thanks to Sam Kaminsky for providing literature hard copies promptly and meticulously. Finally, we would like to thank the publishing staff for their patience with us while preparing this book.

Chapter 1

Scope of Materials Modeling

1.1 Introduction

Materials are chemical compounds with numerous technological uses and applications. For instance, in a typical electronic device, materials form important components of the active device, interconnects, substrates, as well as the associated structural support. Materials can occur both naturally or be synthesized artificially, and come in a wide variety of shapes and sizes, which can range from liquid to solid, organic to inorganic, and crystalline to amorphous. Examples of some of the most widely used materials include polymers, ceramics, minerals, zeolites, semiconductors, superconductors, glasses, and nanomaterials. Materials can be analyzed at the microscopic level in terms of the different types of elements they contain, the spatial arrangement of constituent atoms and molecules, and the nature of interactions between them. The microstructure of materials governs important properties such as thermal and electrical conductivity, mechanical strength, ductility, electronic, optical, and magnetic properties.

For example, polystyrene consists of large numbers of carbon and hydrogen atoms bonded in a certain manner which renders low thermal conductivity and a low softening point. Steel consists of at least two main components, iron and carbon, and depending on the ratio of the components, properties such as mechanical strength and ductility vary. Glass is an amorphous mixture of silicon and oxygen and a number of other elements. Depending on the composition, process temperature and cooling rate properties of glasses can vary drastically.

Traditionally one relies on a variety of experimental approaches to characterize, measure, and manipulate materials properties in order to tailor them for new applications. However, in many situations it may be difficult or even impossible to perform controlled experiments that mimic the real process conditions. Examples include extremely hot temperatures in a nuclear reactor or exceedingly high pressures in the earth's core. In such situations modeling on a computer could provide an easier alternative. Another example is in heterogeneous catalysis where the details of molecular motion as well as the atomistic structure of the surface are difficult to monitor experimentally, but can be modeled on a computer. Computer simulation provides useful insight into a system and its properties either as a supplement to existing experimental data, or as the main knowledge source in the absence of reliable experimental measurements. Using simple atoms or molecules as building blocks in a simulation environment, chemists and physicists can rapidly solve the complicated puzzle of structure-property relationships. This enables the synthesis of systems with desired chemical behavior.

The wide range of macroscopic properties associated with different types of materials makes it difficult to have a universal theory that will explain the behavior of all systems. Calculations on different systems typically require different scales of simulation times and length-scales depending on the properties to be modeled. Figure 1.1 is a schematic representation of the distance and time units involved in macro-, meso-, and micro-level modeling. Macro-scale modeling is usually carried out using finite element analysis methods. Micro-scale simulations involve molecular modeling techniques. Mesoscale simulations can take various forms, but usually start by lumping a group of atoms into single entities or "beads". The dynamics of the beads is then investigated either by representing the beads as Newtonian particles [1] or as density fields [2]. Bead-bead interaction that governs bead-dynamics (and therefore properties at the mesoscale) can be determined either from fitting to experimental data or from micro (i. e., atomistic) level modeling. Simulations at the mesoscale, in its turn, can provide input parameters to macro-scale modeling. This book almost exclusively deals with modeling at the micro-scale, with only a few references to meso-scale modeling. Macro-level modeling is beyond the scope of this book.

With the advent of advanced graphics, faster and cheaper workstations and PCs, new advances in theoretical techniques, and clever numerical algorithms, scientists are now able to perform micro-scale experiments on computers and predict realistic materials properties at the macro-scale. They can predict if a polymer and a copolymer will mix, or if the pore of a zeolite is suitable as a molecular traffic controller. Molecular modeling enables one to calculate the rates of diffusion of gas molecules within a matrix and visually follow the path of diffusion. It is possible to see the charge density distribution on a molecule or surface and determine the point of electrophilic attack. One can compute the morphology of a crystal at a high temperature and predict which face would be stable. Electronic band-structure calculations can predict whether a material would be a metal, insulator, or a semiconductor. Modern quantum chemical methods allow one to visualize electronic charge redistribution during a chemical reaction and compute reaction barriers, or to predict how the mechanical deformation of a carbon nanotube manifests itself in altered electrical properties. These are just a few of nu-

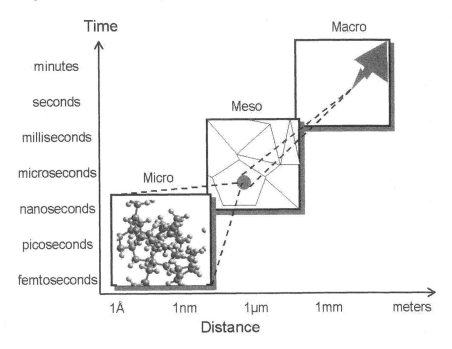

Figure 1.1: Length and time scales involved in modeling macro-, meso-, and micro-levels.

merous applications of molecular modeling. Its endless potential is limited only by the imagination of the user and available hardware resources.

Even at the molecular scale there are different theoretical approaches corresponding to the level of detail at which the electrons in the system are to be described. It is beyond the scope of this book to go into the detail of every theory. However, we have attempted to include the important ideas so that the reader can gauge the approximations and errors implicit in a given simulation method.

1.2 Theoretical Methods

The goal of materials modeling is to develop an understanding of the relation between a material's properties and the underlying atomic structure. Thus, simulating properties of materials requires us to determine the location of each atom in space and their motion as a function of time.

The one fundamental property which determines the structure of any molecular system is the energy of the system. A system at zero temperature will always try to reach a state where its energy is at a minimum. Determining the structure, therefore, involves calculating the total energy of the system and searching for that arrangement of atoms where the energy is at its minimum. Calculating the structure of a molecular system by minimizing its energy is commonly called geometry optimization or geometry minimization. The process of geometry optimization requires computing the net force on each atom due to all other atoms. The force on a given atom is obtained either from an analytical or numerical first-derivative of the system energy with respect to the coordinates of the atom in question. Once the forces on each atom are computed, the atoms can be moved down the energy gradient in small steps using various numerical schemes. The energy and forces are re-computed for the new position and the atoms are moved again. This process is continued until the force on each atom becomes smaller than a specified tolerance, which is usually a small number. The system is then said to have reached its equilibrium geometry, corresponding to a local minimum in total energy. The computed equilibrium geometry can be compared directly with an experimental structure, although small differences are to be expected due to neglect of atomic vibrations at experimental temperatures, intrinsic quantum effects like zero-point vibrations, or approximations in the computed total energy.

On the other hand, if one is interested in the behavior of a material at a finite temperature, one needs to perform a so-called "molecular dynamics" simulation. This approach entails solving the Newtonian equations of motion for each atom under the forces from other atoms (as mentioned above in the context of geometry optimization), as well as an effective force arising from energy exchange with a temperature reservoir. This yields the evolution of atomic positions and velocities as a function of time starting from an initial configuration as specified by the user. A large number of dynamical properties and correlation functions can be computed from such a time evolution, commonly known as a "trajectory".

One drawback of molecular dynamics is that it is typically constrained to a total simulation time of a few tens of nanoseconds or some microseconds at best. The main reason for this limitation is that the fastest processes, typically involving atomic vibrations, occur on time scales of 0.01-0.1 picoseconds, which forces one to use an integrating time step of 1 femtosecond or smaller. However, if one is not interested in such fast motions, but rather in the evolution and averages over a much longer time scale, then molecular dynamics alone may not be very helpful. The short simulation time in molecular dynamics does not allow the system to explore a large part of the accessible phase space that is essential to generate equilibrium statistical ensembles.

For this purpose Monte Carlo simulations are suited better. Monte Carlo simulations are not limited by the small time steps inherent to molecular dynamics and, therefore, allow a much more accurate averaging over the whole configurational phase-space. This method can be applied to problems where the average property for a larger number of particles is required. For example, the adsorption energy of a substance in a zeolite is a property which not only depends on a single molecule, but on the average of a larger number of molecules. In a typical adsorption experiment many molecules are adsorbed on the zeolite and the energy is measured for this process. Since not all molecules will be adsorbed at the same site in the zeolite, each molecule will have a different adsorption energy and the measured value is the average of all the different adsorption energies. To simulate such a property the simulation has to follow the experiment. Instead of studying the adsorption of a single molecule at a single site, the molecule is put into the structure a large number of times (hundred thousands to millions of times in practice) at randomly chosen locations in a randomly chosen orientation and the energy is calculated for all these adsorption sites and averaged.

In the foregoing discussion we have mentioned energy as an important quantity to compute without specifying how to compute it. Since we are dealing with atomic nuclei and electrons, physics tells us that we have to use quantum mechanics for this purpose. The basis of quantum mechanics is the wave function. De Broglie discovered in 1924 that a planar wave can be assigned to every massive particle. Energy and momentum of the particle can then be described by means of this wave. The square of the wave function describes the probability to find the particle at a given location at a given time. An equation of motion has to be found for this wave function which fulfills the following constraints:

1. In a force free state planar waves have to be obtained

2. The superposition principle has to be fulfilled (it has to be a linear and homogeneous equation)

3. It must fulfill $E = \vec{p}/2m$ (where $\vec{p}$ is the momentum and m is the mass of the particle)

Such an equation was found in 1926 by Schrödinger and forms the basis of quantum chemistry [3]. The Schrödinger equation describes all the chemistry as Dirac pointed out in 1929:

> "The underlying physical laws necessary for the mathematical theory of a large part of physics and the whole of chemistry are thus completely known, and the difficulty is only that the exact application of these laws leads to equations much too complicated to be soluble."

Dirac, Proc. Roy. Soc (London) 123:714 (1929)

Unfortunately, the Schrödinger equation can become very complex and can be solved in general only approximately (the only molecular system where an analytical solution is possible is the hydrogen atom). In its time-independent form the Schrödinger equation describes the location of atomic nuclei and electrons of a molecular system. To solve the Schrödinger equation a number of approximations have to be used. The first simplification applied is to assume a homogeneous and isotropic space. This allows for the separation of the external degrees of freedom (e. g., for a molecule there are three translations and three rotations which do not change the relative positions of the atomic nuclei and electrons) from the internal degrees of freedom. This separation can still be accomplished mathematically exact. The first approximation usually introduced is to assume that the movement of the atomic nuclei is much slower than that of the electrons since the mass of an atomic nucleus is much larger than the mass of an electron (even in the most unfavorable case of the hydrogen atom the nucleus is 1836 times heavier than an electron). This is the Born–Oppenheimer approximation [4] which enables us to separate the movement of the electrons from that of the nuclei. The Schrödinger equation is therefore in general solved for a given geometry of the nuclei. The location of the electrons is now only determined by the electron-nuclei attraction and the electron-electron repulsion. To make the Schrödinger equation mathematically tractable, the electron-electron repulsion is not treated explicitly. Rather, an electron experiences an average repulsion by all the other electrons. This is known as the model of independent particles. In this case the wavefunction for the whole molecular system can be set up as a combination of one-electron wavefunctions. The one-electron wavefunctions are assumed to be similar to the wavefunction of the hydrogen atom where the Schrödinger equation can be solved analytically. The solutions of the Schrödinger equation for the hydrogen atom are called orbitals and the wavefunction of the whole molecular system is obtained as a linear combination of atomic orbitals (LCAO). Unfortunately, the orbitals are required to set up the equations which are to be solved to obtain the energy and the wave function. Therefore the Schrödinger equation can only be solved iteratively. Since the solutions become self-consistent, the whole process is known as self-consistent field (SCF) or Hartree–Fock calculation.

The atomic orbitals are described using basis functions. Since the solutions of the Schrödinger equation for the hydrogen atom yield Slater-type functions (STF) these were used earlier on in quantum chemical calculations. Later Boys introduced Gaussian-type functions (GTF) which have computational advantages over Slater type functions and have replaced them nearly completely. Unfortunately, a single GTF is not a very good approximation for a STF. Therefore, a linear combination of more than one GTF is used for a single basis function. The combination coefficients are usually determined from calculations on atoms and are kept fixed. This combination of GTF's is known as contracted GTF (CGTF). Based on the number of basis functions used to describe a single atomic orbital, basis sets are categorized as minimal (one basis function per orbital), double-ζ (two basis functions per orbital, DZ), triple-ζ (three basis functions per orbital, TZ), etc. Since the core electrons of an atom do not play such a significant role as the valence electrons do, split-valence basis sets (SV) are in use which consist of one basis function per orbital for the core electrons and two for the valence electrons. To account for subtle differences in the electron distribution, polarization functions are often necessary. These are basis functions with an azimuthal quantum number larger by at least one than the maximum azimuthal quantum number an atom has occupied in its ground state. A polarization function for hydrogen, e. g., would be a p function while a polarization function for carbon would be a d function.

Since the solution of the Schrödinger equation using the above mentioned approach does not include any empirical data, but is based on a sound theoretical basis it is called an ab initio method (ab initio: Latin, from the beginning). As has been mentioned earlier, the electron-electron repulsion is treated only approximately. This approximate treatment of the electron-electron repulsion can lead to serious problems if the molecule under study cannot be correctly described without explicit electron-electron repulsion. This is the case for molecules which have two or more nearly degenerated electronic states. This is usually referred to as static correlation. In such a case the ansatz used to construct the wave function from one combination of atomic orbitals breaks down. It is necessary to use more than one combination of atomic orbitals to correctly describe the electronic state of the system. A standard method to do so is Multi-Reference SCF (MRSCF).

In addition to the static correlation the error introduced in the Hartree–Fock solution by averaging the electron-electron interactions has to be corrected. Since electron-electron repulsion will cause electrons to avoid each other during their motion this is also known as dynamic correlation. To account for dynamic correlation, methods have been developed which improve the Hartree–Fock solution (post Hartree–Fock methods). The most commonly used is perturbation theory where the electron-electron repulsion is considered to be a (small) perturbation to the Hartree–Fock solution. We will not go into the details of this method here. Most often perturbation theory is applied by Møller–Plesset calculations (MP2, MP3, MP4, etc.). Other methods to include electron correlation are Coupled Cluster (CC) and Configuration Interaction (CI). We will not explain here how these methods work, but it should be noted that Full CI yields the exact solution of the Schrödinger equation for a given basis set. If static correlation also plays a role for a given system, the corrections for both static and dynamic correlation have to be applied.

As we have mentioned before the square of the wave function yields the probability to find an electron at a certain location. The square of the wave function is therefore the electron density. Hohenberg and Kohn showed that the energy can be written as a unique functional of the electron density and Kohn and Sham were able to determine the energy by using an effective exchange-correlation potential. This led to density functional theory which has been used by physicists for quite some time, but has found general application in chemistry only in recent years. The use of the electron density simplifies the solution of the Schrödinger equation and can even include electron correlation, but requires some empirical knowledge about the so-called density functional. The simplest functional is based on a homogeneous electron gas, where an analytical relationship between electron density and potential can be derived. This is an example of a density functional which depends only on the electron density itself. Since these functionals account only for the local environment of an atom they are known as local density approximations (LDA). In recent years it has become clear that much better results can be obtained when the functional does not only depend on the electron density itself, but also on its first derivative. These functionals are called gradient-corrected functionals. Both local and gradient-corrected functionals are in use today.

While density functional theory is often considered to be an ab initio method, this is, strictly speaking, not true. The equations which are to be solved have been derived from theory, but the density functional used in these equations is based on empirical knowledge. It is therefore better justified to count density functional methods as a separate group of methods besides ab initio and semi-empirical methods. Some researchers refer to density functional theory therefore also as first principles methods [5].

The two significant advantages of density functional theory over ab initio methods are the possibility to include electron correlation and the speed at which density functional calculations can be performed. The Hartree–Fock method scales theoretically with $N^4/8$ where N is the number of basis functions. Density functional methods usually scale with N^3 which allows to treat much larger systems with the same computational resources. Even more important is that with basically the same computational effort electron correlation effects can be accounted for. To correct the Hartree–Fock solution by, e. g., perturbation theory requires significant more resources (the most demanding part of MP2 calculations scales with N^5).

If additional approximations are introduced to solve the Schrödinger equation, we arrive at semi-empirical methods such as the extended Hückel method (EHT) and methods [6, 7] based on neglect of differential overlap (NDO) such as the complete neglect of differential overlap (CNDO), intermediate neglect of differential overlap (INDO), and their variants such as MNDO and MINDO/3, Austin Model 1 (AM1), Parametric Method Number 3 (PM3), etc. These are based on experimental data used to fit certain difficult to compute terms in the energy expression and on the neglect of other terms. In contrast to ab initio methods, semi-empirical methods are fast and can be applied with ease to large systems. While MINDO/3, AM1, PM3 are in use for mainly organic systems and their successful application usually requires the user to stay within the groups of systems they were fitted for, the extended Hückel method has been used successfully for both inorganic and organic molecules as well as for periodic systems with a fair amount of qualitative success. This has been demonstrated extensively by Hoffmann et al. [8–12] using Extended Hückel tight binding methods [13].

Figure 1.2 sums up the range of available methods in terms of the maximum system size that can be handled and the compute time required for such calculations.

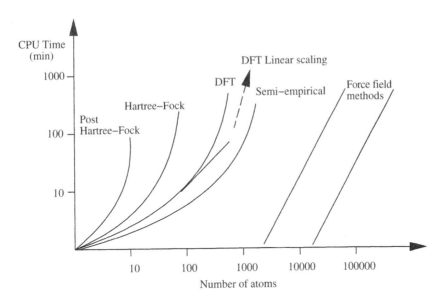

Figure 1.2: Differences between ab initio Hartree–Fock, post-Hartree–Fock, density functional, semi-empirical, and force field methods.

Most of the methods discussed so far are applied to molecules in the gas phase, typically containing a finite number of atoms. A macroscopic solid or liquid, on the other hand, contains such a large number of interacting atoms and electrons that it can be considered to be infinite for all practical purposes. However, if the solid is a perfect crystal, it essentially consists of a finite pattern repeated periodically in all three spatial dimensions. This finite pattern is called the "unit cell". The unit cell is typically described by a parallelepiped, whose edges are called "basis vectors" (typically denoted by $\vec{a}_1, \vec{a}_2, \vec{a}_3$). A (linear) integral combination of the basis vectors, i. e., $\vec{R} = n_1\vec{a}_1 + n_2\vec{a}_2 + n_3\vec{a}_3$ is called a "lattice vector". An infinite crystal is invariant under translation through any lattice vector. This invariance leads to the following theorem (due to Felix Bloch): in a pure infinite crystal the electronic wave function at two points separated by a lattice vector $\vec{R}$ are related to each other through a simple (multiplicative) phase factor of the form $\exp(i\vec{k} \cdot \vec{R})$. The reciprocal lattice vector $\vec{k}$ is a good quantum number describing an electronic state. Thus, Bloch's theorem recasts the original crystal problem of an infinite number of repeating unit cells to that of a single unit cell with electrons at an infinite number of k-points. The above method can be easily extended to study crystals with defects, surfaces, amorphous systems, liquids, or even gas molecules. Borrowing the concept of an "unit cell" from a crystal, one essentially constructs a suitable parallelepiped to be repeated periodically, called the computational "supercell". The repeat procedure is cleverly represented by what is known as the "periodic boundary condition" on the supercell, while the infinite number of k-points is suitably approximated by a finite sampling set determined by symmetry considerations.

If we consider a solid we have to deal with an infinite number of electrons and each basis function extends over the entire solid. It is possible to solve the Schrödinger equation for periodic systems using translational symmetry by employing symmetry-adapted basis functions (Bloch functions). This leads to so-called crystal orbital methods. Crystal orbital calculations are still very expensive in terms of resources and usually require supercomputer power, but they are increasingly being reported in the literature. The reported calculations are mostly so-called single point calculations which means that the energy and wave function have been evaluated for a single geometry only and no geometry optimization has been performed. Since crystal orbital methods require the ideal solid they cannot directly be used to study defects. Methods which can handle defects have been developed, but are quite demanding in terms of computer resources.

Density functional methods can be applied to solids in a similar way as ab initio methods. Since density functional methods usually require less computational effort they can be more easily applied to solids. Another approach using density functional methods on solids does not use the atom centered approximation of basis functions. Instead the wave function is expressed as a superposition of plane waves and the core electrons are treated by pseudopotentials. The pseudopotential approximation [14,15] thus allows the electronic wave functions to be expanded using a much smaller number of plane-waves. The great advantages of the plane wave basis set are that the Schrödinger equation can be solved using fast Fourier transformation and that the forces acting on a nucleus can be calculated from the Hamiltonian alone avoiding the difficult to evaluate dependency on the wave function. Due to these advantages ab initio programs for solids using plane waves are rather common.

As we have mentioned before, quantum chemical calculations of the energy are based on the Born–Oppenheimer approximation which assumes that the atomic nuclei are fixed in space and only the electrons move. If we want to perform geometry optimizations or molecular dynamics we have to move the atomic nuclei as well. It would of course be possible, albeit difficult, to solve a Schrödinger equation for the movement of the nuclei as well, but generally it is sufficient to assume that the nuclei behave like classical particles and to use the Newtonian equations of motion for them. But if it is sufficient to use classical mechanics for the movement of the nuclei, the question that arises is: Is it always necessary to treat electrons as quantum particles? Electrons are responsible for the bonds between atoms. If it would be possible to describe bonds by some classical (or semi-empirical) potential it would not be necessary to explicitly include electrons in the simulations. This would, of course, only allow us to study effects which are related to the arrangement of the atomic nuclei. Effects which are caused by the electrons, like, e. g., an optical spectrum, could not be simulated this way, but this is not always required.

It is indeed possible to use a set of classical potential functions to describe the interaction of the nuclei with each other. If the bonds are mainly covalent, as in organic systems, potential functions are used which describe the bond between two atoms explicitly. A carbon-carbon bond in a hydrocarbon, e. g., may be described by a Morse function which is a reasonable approximation to the energy/bond length relation one would obtain from quantum mechanical calculations. It is usually not enough to use only potential functions for bonds lengths, but such functions have also to be used for bond angles and for torsion angles. Atoms are simply assumed to be connected by springs of varying strengths in this picture. In addition to these terms, functions for so-called non-bond interactions are required. These non-bond interactions are electrostatic and Van der Waals interactions. Electrostatic interactions are caused by the charges on the atoms present in a molecule, while Van der Waals interactions model the electron-electron repulsion and dispersion which are required to avoid the collapse of a molecular system due to attractive interactions and the existence of intermolecular attractive forces which are responsible for the formation of intermolecular complexes. Electrostatic interactions usually use Coulomb's law as a functional form, while for Van der Waals interactions either a Buckingham-type (cf. p. 108) or a Lennard-Jones-type potential is used (cf. p. 235). Recent developments of classical potential functions have shown that a much better agreement between simulation and experiment can be obtained when so-called cross terms are included in the force field. Cross terms describe the coupling of changes of, e. g., the length of a bond, with the changes of the length of another bond attached to the same atom.

If the bonds between atoms are mainly ionic the use of explicit bond functions is not physically meaningful. In ionic systems bonding occurs mainly through electrostatic interactions which are not directed. The potential functions have to be selected accordingly. Therefore, for ionic systems so-called rigid-ion models are in use. These consist usually only of an electrostatic and a Van der Waals term. The electrostatic term provides the attractive force between two atoms and the Van der Waals term provides mainly the repulsive force. Since there are no explicit terms for each bond, ion-pair potentials can allow for varying coordinations of atoms in the course of a simulation and are therefore often used for inorganic systems. An extension to the ion-pair potentials is the shell-model potentials. Some ions (especially anions) are quite polarizable and a rigid-ion potential does not account for polarization which is often important to model properties of materials correctly. A shell-model potential [16–18] treats a polarizable ion as a combination of a core and a shell both of which have charges and which are linked by a harmonic spring. The electrostatic interaction between core and its shell is excluded. The ion now can become polarized by surrounding charges if core and shell are not located at the same position in space. The strength of the spring connecting core and shell corresponds to the polarizability of the ion. The shell model has, e. g., been successfully applied to alkali-metal halides. In alkali-metal halide melts, such as NaCl, the charge-ordering tendency is sufficiently strong and the local octahedral order is maintained on average. This means that the polarization effects are small. The polarization effects alter optical phonon frequencies and stabilize point defects. Therefore, the shell model which is a dipolar approximation is appropriate. However in transition-metal halides, polarization effects are more important and a quadrupolar approximation is needed. Moreover, due to the presence of quadrupole moments in transition metal ions, cross terms between the dipole and quadrupole moments are also possible. The polarizable-ion model (PIM) [19] is one of the models which takes the quadrupolar approximation into account. Polarizability of the oxygen anion O^{2-} is infinite and it is unstable in vacuum where O^{2-} is decomposed into O^{-} and one electron. However, in a crystal environment, O^{2-} is stabilized. This shows that the environment is very important in an oxide. The charge density of O^{2-} in a crystal, confined by the potential set up by its neighbors, is compressed and stabilized with respect to the free ion. The core and valence electron densities are more localized around the nucleus in the crystal compared to vacuum. This compression effect reduces the anion polarizabilities, often by 50 % or more, in the crystal relative to the free ion [20, 21]. The compressible ion model (CIM) [22] considers this compression effect for the interatomic potential functions.

If the bond is metallic, the itinerant behavior of the electrons is most important. There are two strategies to model metallic systems. For a simple metal, where the nearly free electron theory developed by Drude in 1900 [23] can be applied, the interatomic potential is derived from pseudopotential theory. However, for transition metals, this is inefficient and more recent methods such as the embedded-atom method (EAM) [24, 25] and effective-medium method (EMT) [26–28] (and other related methods) are better suited to model these systems. In general, these methods are called glue models. Usually, the interatomic potential is a function of the interatomic distance. However in the glue models, the attractive part is expressed as a functional of the electron density (or a similar variable which is calculated from the interatomic distances). The force is derived from the derivative of the interatomic potential with respect to the interatomic distance. Through this functional (e. g., in EAM this is called embedding function), the environmental many-body effects are realized. These methods have been applied to surfaces, defects, grain-boundaries, melts, etc. which are different from the bulk environment.

Another area which is fairly young and not yet well explored is that of interfaces. A fairly recent development in theory applied to the metal-ceramic interface is the Discrete Classical Model (DCM) [29].

All the methods based on classical (or semi-empirical) potential functions rely on the existence of parameters which describe the particular interaction, e. g., for the electrostatic interaction charges for each atom have to be known, while for a Morse function parameters like the position of its minimum and the depth of the minimum are required. These parameters can be obtained in two different ways. The first is to use available experimental data and to fit the parameters so that they reproduce the experimental data. This approach works reasonably well as long as the number of parameters required is small and there is enough experimental information to derive all the parameters. It becomes a problem if either the number of parameters required exceeds the experimental data available or the systems to be studied are not even accessible to experiments. In addition to these difficulties parameters obtained by fits to experimental data always introduce environmental effects of the experiment, e. g., experiments are performed at a certain temperature. If parameters are fitted to the data measured at this temperature, the simulation using these parameters will always reproduce the properties at this temperature. This will introduce errors if, e. g., these parameters are used in a molecular dynamics simulation at a different temperature.

Therefore a second approach to fitting parameters has won attention in recent years. Parameters are now fitted to the results of ab initio (i. e., Hartree–Fock) or first principle (i. e., density functional) calculations. This approach has the advantage that one can generate as much ab initio data as required to derive all parameters in a statistically meaningful way. A disadvantage to this approach is that it is not always possible to perform ab initio or first principle calculations for the system under study. Parameters can then only be derived by using model systems which are small enough to allow ab initio or density functional calculations and which represent typical structural units of the system to be studied. Another disadvantage is that ab initio and density functional calculations introduce systematical errors, e. g., Hartree–Fock calculations yield vibrational frequencies which are too large by around 10 % [30, 31]. A solution to this problem is to use ab initio or first principle calculations to derive raw parameters and then to apply a few scaling factors which are derived from a comparison between the outcome of the calculation with the raw parameters and experiment. Since there can be significantly less scaling factors than parameters, only a few experimental data are required.

This concludes our overview on the available methods to simulate materials. A more (mathematically) detailed description of many of the methods can be found in the last chapter.

1.3 Getting Started on a Modeling Project

Many molecular modeling programs are easily accessible from ftp and/or web sites [32–34]. It is often rather tempting to get started with modeling without actually planning. Similar to performing a laboratory experiment, it is necessary to define the goals of a modeling project and plan the experiment on the computer. Therefore, the following questions need to be answered before a simulation is started.

1. What is the objective of the simulation ?

   - to use as a tool to help understand experimental results or
   - to use as a graphics visualization tool or
   - to produce publication quality figures

2. What are the desired results?

   - to obtain structures
   - to obtain properties
   - to study structure-property relationships
   - to obtain spectral data
   - as a supplement to existing experimental data

3. How accurate do the results have to be?

   - level of theory required
   - kind of simulation engine
   - use of results as input to other software

4. How can the results of the simulation be validated ?

   - validate by experiment or
   - by agreement between different computational methods

5. What hardware is required to run the calculations?

   - memory required
   - storage (hard disk) space
   - speed of computing

6. What information is needed to set up the simulations?

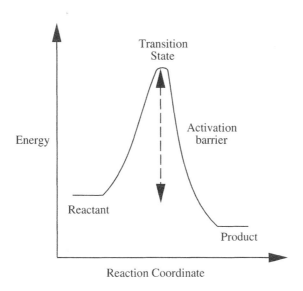

Figure 1.3: Energy versus reaction coordinate.

- molecular formula/composition of molecule or monomer type, molecular weight and linkage for polymers, or space group, composition and crystallographic coordinates for a crystal

- gas phase or solvated or solid-state

- and/or input experimental data if required

- and/or other parameters

An example of a molecular modeling study is to calculate the activation barrier for a reaction as shown in Figure 1.3. Knowing the geometry of the starting material and the product, it is possible to determine the structure of the transition state. Also, the energies of the reactant, product, and that of the transition state can be calculated and hence the activation barrier for the reaction can be deduced.

1.4   General Structure of Molecular Modeling Programs

A molecular modeling program consists of an user interface and one or more engines that calculate molecular structures and properties. In the most basic type

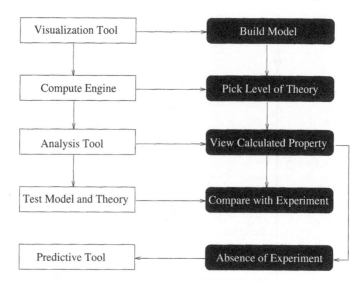

Figure 1.4: Flow chart of molecular modeling procedure.

of interface the user types commands into a file and the engine interprets this file. More complex interfaces use interactive commands issued while the program is running to control the program. The most user-friendly programs consist of graphical displays of the molecule and using a mouse the commands are selected from tool bars or pull-down menus. There are many programs that will accept a combination of the above mentioned ways to receive commands.

Molecules or periodic systems or polymers can be built and manipulated using molecular builders. Distance, angle, and torsion measurements can be performed on a three-dimensional molecular structure using the graphical interface. Graphics engines allow the user to visualize molecular structure, polymeric ribbons, solid-state surfaces as well as the calculated properties. In addition, some programs also allow the rendering of molecules with high-quality space-filling options that can be printed and used for publications or slides.

Coupling the graphics engine, the molecular builder, and various computational engines results in a software package which allows the user to interactively sketch a molecule on the graphics window, convert it to a 3D structure, optimize the geometry of the molecule interactively, and obtain a chemically reasonable structure displayed on the graphics screen within a short time. A simple flow diagram of a general molecular modeling procedure is shown in Figure 1.4.

1.5 Computer Hardware

Computational chemistry ranges from quantum mechanical calculations of the electronic structure of molecules to classical mechanical simulations of the dynamical properties of many-atom systems, from understanding and predicting structure-activity relationships and reaction intermediates to informatics. Although chemical theory and insight play crucial roles in this effort, the prediction of physical observables is controlled by the available computer capacity.

The evolution of theory, computer hardware, and software has reached the point where it is now practical to convert theory into quantitative information about industrially important chemicals and chemical processes via molecular simulations. Computed chemical data complement that obtained from experiments and observations, offering the opportunity to greatly accelerate the solution of research problems.

When one thinks of computational chemistry and hardware there are a few aspects to be considered: (1) computation, (2) visualization, and (3) computer architecture.

1) Computation: Depending on the type of computation and type of system studied as well as on the accuracy needed, hardware requirements differ. The hardware might have to deal with large number of atoms such as in a polymer or in a periodic system. It might have to handle enormous databases (gigabytes) in the case of informatics. As the system to be simulated increases in size or if the theory requires number-crunching or huge matrix solutions or if animation of simulation results is required, there is a need for greater computational power and more memory.

Unix-based workstations (such as Silicon Graphics, Sun, HP, and IBM) were and are still in use for high-performance computations. With the availability of faster processors the speed of personal computers has increased many-fold compared to a few years ago. For example, in the case of force field based calculations or semi-empirical quantum mechanical studies on small molecules a personal computer is sufficient.

Many popular computational chemistry applications have been in development for a substantial length of time (20 years or more) and are large and complex. The existing methods are being refined continually for greater accuracy. For computational efficiency, larger and more complex programs are optimized for use on supercomputers so that several computations are carried out simultaneously.

Massive parallel computers with hundreds of processors significantly outpace conventional supercomputers in both capacity and price/performance ratio. While increases in raw computing power alone will greatly expand the range of problems that can be treated by theoretical chemistry methods, the last few years saw an increased investment in new algorithms needed to fully exploit this potential. Though rewriting existing computational chemistry software to run in parallel by delegating tasks to numerous processors is a time consuming endeavor, this is necessary since merely porting available software to the parallel computers does not provide the increased efficiency. In fact, many existing parallel applications show a deterioration in performance with an increase in the number of processors used. The details of how the performance scales with changes in the number of processors depend on factors such as how well the various parts of the program can be parallelized, the size of the input data, and the performance of the parallel computer being used. New algorithms that exhibit linear increase in performance with increase in the number of processors are being developed. If the number of jobs to be run is greater than the number of processors in the computer, then the most efficient strategy is to run each job on a separate processor. This avoids the overheads inherent in parallel programs and guarantees that each processor is being used at maximum efficiency. The scalability of parallel programs has also been improved by using a superior network interconnect between the processors. Such improvements have resulted in substantial scientific and commercial gains by increasing the number and complexity of chemical systems that can be studied.

The other development in the last few years is cluster computing. Clusters are often created from existing networks of workstations or personal computers sitting on people's desks. A cluster management system is used to place jobs on machines that are idle otherwise.

Though the concept of cluster computing has been around for many years, high-end workstations (Unix based workstations from DEC, Sun, IBM, or Silicon Graphics) and high-end networks were initially used, both of which were very expensive. With the performance improvements in PCs, clusters have become very popular over the last few years to the point where they are now comparable to high-end Unix workstations for scientific computations. While low-end networks are still orders of magnitude slower than the proprietary high-speed networks used in commercial parallel computers, they have also improved greatly over the past few years with 100 Mbit/sec Fast Ethernet or Gigabit Ethernet being used. The other important driver for PC clusters has been the development of the Linux operating system, a freely available Unix implementation that was originally developed for PCs and has now been ported to a range of architectures, providing a reasonably standardized platform for program development. The development of Linux meant that the programming environment, compilers, cluster management systems, and parallel computing software used on Unix clusters could be used on clusters of low-cost PCs. Current trends in low end cluster computing technologies are that the processor clock speeds are continuing to improve. To illustrate the pace of growth of processor performance, Figure 1.5 shows the development of CPU speed over the last two decades for Intel and MIPS CPUs.

Extensive measures such as caching at different levels, architectural enhancements in the processor core, etc., are now standard and supply these extremely fast processors with an adequate amount of data in order that they are well uti-

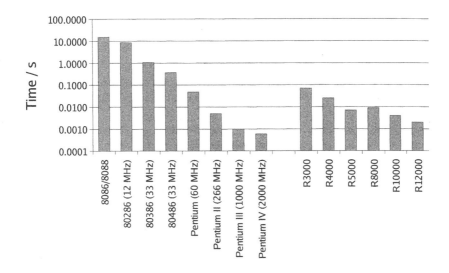

Figure 1.5: Time it takes various generations of Intel (left) and MIPS (right) processors to execute a Fortran benchmark program (note the logarithmic scaling of the time axis).

lized and not kept idle. All distributed-memory parallel computers need some message-passing software which is a collection of routines that allow communication among the processors. To get the most out of a computer with multiple processors on different nodes, it is vital to keep the processors occupied. The number of nodes involved in a calculation is constantly changing, so a piece of software known as a scheduler is called in to help manage various tasks more efficiently.

Chemistry-computing performance is also getting a boost from new advances in computer hardware. IBM has a number of new devices such as the series of accelerator chips known as MD-GRAPE. The chips operate on principles similar to graphics accelerators, except that they speed up dynamics calculations.

2) Visualization: Molecular modeling is interpreted as the interactive combination of visualization and computational techniques, with emphasis on the former. The power of scientific visualization is the ability to display all the data from a simulation or computer experiment. This allows for relationships among the data to be more easily determined. These could be 2D and 3D views of the molecular and crystal structure, molecular orbitals, electron density maps, anima-

tion of molecular motions, etc. There are numerous ways to display the data and uncover the relationships among it such as using color, texture, isosurfaces, slices through a data set, geometric representation of data, using multiple displays, or using time sequences data. Three dimensional displays provide true spatial information, relative positions, relative sizes, the ability to overlay information on maps, the ability to display grids that can give orientations, etc.

Traditionally Silicon Graphics has been the gold standard for visualization. However, the performance of 3D graphics hardware for personal computers has meanwhile outpaced the development in the workstation area. The computational chemist today can visualize results of computations on his cheap PC of the same or even better quality than on an expensive workstation a few years ago. The use of the OpenGL graphics standard has given the screen output on a PC a much more professional feel and appearance.

The standard visualization software used by computational chemists renders images which are flat. However, there are companies such as VRex [35] that provide solutions for a more complete 3D stereoscopic view. This allows to perceive depth or stereoscopic 3D which is natural to the human vision system. These images are extremely successful in commanding attention, presenting complex concepts, improving retention, and conveying a more complete information than 2D images.

3) Computer architecture: With the increased use of Linux servers and clusters, most of the new generation computational chemistry software makes use of a client-server architecture. This approach allows more efficient use of research dollars by buying several specialized machines; e. g., cost effective Linux servers (single or in a cluster). Also these servers can do without the state-of-the art graphics environment. The display component, for example a stereo-graphics card or a very high resolution/high performance graphics card, can now be on a client which does not need high end processing power. Having the client on a standard desktop environment also gives the end user an ability to use the programs in conjunction with other desktop software; e. g., results from the simulation or informatics program can be utilized directly in presentation software.

Computational chemists have emerged as major players in developing tools for new computer architectures such as computer clusters, new ways to visualize data, new hardware which allows lengthy calculations to be reasonably fast, and numerous tests of parallel versions of computational chemistry software.

1.6   Software Related to Materials Modeling

Since a wide range of chemicals with wide ranging properties are included under the broad umbrella of materials, there is not one software available that can fulfill all the requirements of a "materials modeler". However, some of the commercial and most of the academic codes are well tested, extremely robust and satisfy various demands of the user. In this section we have attempted to list the most commonly used codes for materials modeling.

Accelrys Inc., historically, had expertise in various materials modeling areas and this in turn grew to a wide range of products that cover the entire spectrum ranging from ab initio, density functional and semi-empirical calculations on molecular and extended systems to molecular mechanics, molecular dynamics, Monte Carlo simulations, powder and crystal structure diffraction simulations as well as informatics. Accelrys' software is powerful and has been used successfully in various industrial research projects. Accelrys has two software packages called Cerius^2 and InsightII, both of which run on Silicon Graphics IRIX workstations. A more recent development is Materials Studio which is based on a client-server architecture with a front end for Microsoft Windows PCs. The server end can run on Microsoft Windows, Red Hat Linux, Silicon Graphics IRIX, or HP Tru64. Materials Studio contains most of the functionality present in Cerius^2 and InsightII. All three suites of products, i. e., InsightII, Cerius^2, and Materials Studio, have sophisticated graphical user interface for data manipulation, structure drawing, scientific visualization of spectra, molecular and crystal structures, orbitals, electron density maps, isosurfaces, maps, slices, etc. Accelrys Inc. provides training courses that help users get a jump-start in using their software [36].

The Materials and Processes Simulation (MAPS) platform of Scienomics provides an extendable framework for using any simulation code in a unified environment. Users can use a whole range of simulation codes, both electronic structure and force field codes, provided by Scienomics or easily integrate their own codes into MAPS. MAPS is oriented towards team work since it supports sharing of simulation results between different users and contains data storage and management functionality. MAPS is operating system independent and can be used on Microsoft Windows or Linux workstations [37].

Pacific Northwest National Laboratory, funded by the U. S. Department of Energy, has developed and is distributing a modeling package called NWChem. It provides many methods to compute the properties of molecular and periodic systems using standard quantum mechanical descriptions of the electronic wave function. In addition, NWChem has the capability to perform classical molecular dynamics and free energy simulations. These approaches may be combined to perform mixed quantum-mechanics and molecular-mechanics simulations. The system is designed to run on high-performance parallel supercomputers (CRAY T3E, IBM SP2) as well as on conventional workstations (Silicon Graphics IRIX, Sun Solaris, HP RISC, Tru64, or Fujitsu). It also runs on PCs under Linux as well as on clusters of desktop platforms. The front end is the Extensible Computational Chemistry Environment (Ecce) which provides a graphical user interface and the underlying data management framework for efficient set up of calculations [38].

Materials Design's MedeA (Materials Exploration and Design Analysis) integrates experimental databases with computed data and runs on Microsoft's Windows operating system. It provides tools for fast all-electron density functional electronic structure calculations based on the augmented spherical wave method as well as tools for ab initio molecular dynamics [39].

Gaussian 03 is a program package which implements ab initio and semiempirical methods for molecules and periodic systems. It runs on HP Tru64, Sun Solaris, Red Hat Linux on Intel processors, and Silicon Graphics IRIX. Gaussian 03M and Gaussian 03W are complete implementations of Gaussian 03 for the Mac OS X and the Windows environments, respectively. GaussView is a graphical user interface for Gaussian 03. Gaussian 03 is based on ab initio quantum mechanics (including Hartree–Fock, Density Functional Theory, MP2, Coupled Cluster, and high accuracy methods) and can predict the energies, molecular structures, and vibrational frequencies of molecular and extended systems. Gaussian 03 can also compute a range of spectra such as IR and Raman, UV-Visible, Nuclear Magnetic Resonance, vibrational circular dichroism, electronic circular dichroism, optical rotary dispersion, harmonic vibration-rotation coupling, anharmonic vibration and vibration-rotation coupling and g tensors and other hyperfine spectra tensors [40].

Wavefunction's SPARTAN created by Professor Warren Hehre (University of California, Irvine) provides tools for molecular mechanics, semi-empirical, ab initio, and density functional theory quantum mechanics calculation on discrete molecules. Capabilities are accessed via a graphical user interface which includes molecule builders and interactive dialogs for input preparation and output analysis. SPARTAN displays structural models, orbitals, electron densities and electrostatic potentials as isosurfaces or slices. Computational components are fully parallelized. Additional SPARTAN features include five methods of conformational analysis, similarity/superpositioning, reaction coordinate simulation, calculation of quantitative structure activity relationship (QSAR) descriptors, and seamless interfaces with Gaussian, Allinger's MM3, IBM's Mulliken, and Schrodinger's Jaguar software. SPARTAN runs on Silicon Graphics, IBM, DEC Alpha, HP, and Cray as well as on Windows, Linux, and Macintosh.

Schrodinger's Jaguar runs only on Unix. A program called Maestro provides the graphical user interface. It is an ab initio (Hartree–Fock, Density Functional, MP2, GVB, and GVB-RCI methods) electronic structure software package. Jaguar's solvation modeling enables one to calculate geometries, transition states, and properties in a variety of solvents, based upon the standard self-consistent reaction field model, using the actual molecular surfaces for the dielectric boundary. Generalized Valence Bond theory (GVB) and GVB-RCI (restricted configuration interaction) offer a convenient means to treat spin couplings in metals, dissociate bonds correctly, obtain a very intuitive picture of metal-ligand bonding, or calculate accurate solvation energies. Schrodinger has another code which runs on Windows called MaterialsExplorer. This is a molecular dynamics package which they claim can work for polymers, metals, ceramics, and semiconductors [41].

Turbomole (Prof. Reinhart Ahlrichs, University of Karlsruhe) is an ab initio quantum mechanical code. It has been specifically designed for Unix workstations and Linux PCs and efficiently exploits the capabilities of this type of hardware [42]. It implements Hartree–Fock, MP2, and density functional methods, allows the calculation of vibrational and NMR spectra and of a large number of properties, and provides tools for ab initio molecular dynamics.

Molecular mechanics and dynamics calculations for inorganic systems can be performed using the General Utility Lattice Program (GULP) [43]. GULP allows to use ion pair, shell model, and molecular mechanics force fields and is quite flexible in the functional form of the force field. It allows to calculate structures, vibrational modes, dielectric and elastic properties of crystals, and a number of other properties.

Visualizations of the results of Turbomole or GULP calculations, as well as a number of other programs, are possible with the Open Source program Viewmol [44]. Viewmol allows the display of structures, of molecular orbitals and electron densities, of vibrations, and of molecular dynamics trajectories. It is scriptable using Python.

YAeHMOP (developed in Prof. Roald Hoffmann's group at Cornell University) is based on the extended Hückel method and can be used for calculations on isolated molecules as well as on solids and surfaces. It has parameters for all elements in the periodic table. The programs "bind" and "viewkel" form the core of the package and run on Unix and Windows [45].

Chapter 2

Metal Oxides

2.1 Introduction

Modeling of metal oxides has found the most interest in two areas: corrosion inhibition and catalysis. Metal oxides occur as products of corrosion, which is an unwanted process due to its economical consequences. Prohibiting corrosion is of great importance, e. g., in construction, and requires a thorough understanding of the underlying mechanisms. This understanding can be enhanced by modeling. Metal oxides also play important roles in the chemical industry as catalysts for a large variety of processes and as electrolytes. There are other application areas as well, including paints and lotions, microelectronic devices, gas sensors, and structural materials. They also play a role in nuclear energy production and in pollution control.

Compared to the experimental and computational study of metals and semi-conductors the study of metal oxides has started comparatively late. The reasons for this are manifold. Metal oxides show a large variety in structure and composition. It is often very difficult to prepare metal oxides in a reproducible way since they can be non-stoichiometric. This results in a large number of defects in the prepared oxides, which may have a profound effect on their properties. Transition metal oxides can have very complex electronic structures, which makes certain modeling approaches very hard to apply. Nevertheless, modeling efforts of metal oxides have seen a tremendous growth in the last decade mostly due to the increase in available computer power and theoretical advances in electronic structure methods.

Due to the nature of the O^{2-} anion, which is only stable in a crystal lattice, metal oxides occur exclusively as solids and are mostly ionic in nature. There-fore, modeling of metal oxides poses some serious challenges. Due to the ionic nature only 3D-periodic or large cluster calculations can produce reasonable re-

25

sults. The use of electronic structure methods is therefore limited. Furthermore, oxides of transition metals are generally difficult to handle in electronic structure methods due to the large number of close lying electronic states, which make the convergence of the electronic wave function a major problem. Suitable codes, almost exclusively based on density functional theory, have been used by solid state physicists for some time, but have become more common place among chemists and engineers only over the last decade. Some examples are discussed in Section 2.2.3.

Force field based methods, on the other hand, have seen significant application for metal oxides. One of the main reasons for the success of force fields in the area of metal oxides is that the limited number of elements involved leads to only a small number of parameters required. Compared to force field based methods for organic molecules, which usually require hundreds of parameters, metal oxides are usually modeled with maybe a dozen parameters at best. There is also a qualitative difference between force fields necessary to describe a metal oxide system from that developed for organics. This is because an important feature to correctly model oxides is the polarizability of the oxygen anions. Metal oxides are therefore often modeled using ion pair potentials, which are extended by shells on the oxygen anions to allow for polarizability.

In the following we will briefly discuss electronic structure methods and their application to metal oxides. Then we will take a look at various force fields, their derivation and their differences with respect to force fields for organics. Finally we will discuss applications of force field based methods to the modeling of metal oxides.

2.2   Electronic Structure Methods

Until the last decade, electronic structure calculations have seen rather little application in the modeling of metal oxides compared to force field based methods. However, increasing computational power and current developments in electronic structure methods have led to a growing number of studies using electronic structure calculations on metal oxide systems in recent years.

Density functional calculations are often the electronic structure method of choice for metal oxides. The reason for this is twofold. First, density functional calculations are computationally less demanding than Hartree–Fock and post-Hartree–Fock methods, thereby allowing computation on systems with a larger number of atoms. Second, many technologically important oxides are oxides of transition metals and density functional theory has proven to be more robust for modeling of transition metal compounds than Hartree–Fock methods [46, 47]. Furthermore, plane-wave or tight-binding methods, often used for the modeling of metal oxides, are based on density functional theory.

There are two fundamental approaches to modeling metal oxides using electronic structure methods. First, one can use cluster models suitably cut out from the periodic system. Cluster models can be handled like molecules and all electronic structure methods available for molecules are readily applicable. The second (and usually more accurate) approach involves modeling of the 3D-periodic structure through periodic boundary conditions and has employed almost exclusively tight-binding and plane-wave methods.

2.2.1 Cluster Models

Before the availability of density functional codes with periodic boundary conditions cluster models used to be the most widely used quantum mechanical approach for modeling metal oxides. Clusters can be thought of as aggregates of a rather small number of ion pairs. In such a sense a cluster is an intermediate between the single ion pair and the bulk structure. Metal oxide clusters can today be prepared chemically and are also used as catalysts. Therefore using clusters as models for the bulk of metal oxides seems quite natural, more so than in areas such as zeolites (cf. Chapter 3).

A cluster can be constructed in one of two ways. First, it can be made by cutting the appropriate number of atoms out of the bulk of the oxide. In this case some of the atoms of the cluster are usually restricted to stay at their bulk positions during calculations to model, e. g., a surface. The second way of making a cluster model consists of assembling the desired number of atoms more or less arbitrary followed by use of computational methods to relax the entire cluster. This way the structure of the cluster will represent minima on its potential energy surface and the behavior of a limited size metal oxide system can be studied. The second approach is better suited if the metal oxide to be studied is known to be present in nano-scale sized particles.

In general, the use of clusters to model solids requires that the clusters are terminated to avoid dangling bonds (cf., e. g., zeolites, page 56). Therefore, clusters are often terminated by hydrogen atoms which introduces problems related to the artificial formation of hydroxyl groups. Fortunately, for metal oxide clusters a termination is not always necessary, since metal oxides can occur in a number of stoichiometries. It is important, however, in particular for transition metal oxides, to make sure that the metal is in the correct oxidation state.

Brønsted acid sites on the surface of oxides of magnesium, aluminum, and titanium have been modeled using clusters of different size [48]. It was found that the acidity of the surface sites of the oxides is mainly governed by three structural parameters. The deprotonation energy decreases – and hence the acidity increases – with increasing coordination number of the acidic hydroxyl group and with the coordination number of the metal close to that OH group. The acidity also increases with the charge on the cation, which means that magnesium oxide exhibits less Brønsted acidity than aluminum or titanium oxide.

Cluster models have also been employed to model vanadium oxide. Vanadium oxides play an important role as catalyst in a number of industrial processes such as SO_2 oxidation or oxidative dehydrogenation of alkanes. Vyboishchikov and Sauer have used density functional methods to determine structural motifs, vibrational frequencies, and atomization and electronic excitation energies for several clusters and their anions [49].

2.2.2 Periodic Calculations

Since the process of creating clusters involves cutting off bonds to the rest of the solid, these cut bonds need to be artificially terminated (i. e., chemically "saturated"). For a cluster of organic material or a zeolite one would typically use hydrogen atoms for termination. However, for metal oxides artificial termination may lead to unwanted oxidation states of the metal centers. Using cluster models of strongly ionic systems has another shortcoming. Long range electrostatic interactions are not represented accurately. Because of the above reasons, calculations on metal oxides using periodic supercells are fast becoming the simulation method of choice. Periodic density functional calculations can be used to obtain information about the geometrical and electronic structure of metal oxides as well as to study chemical reactions, phase transformations, phonon dispersion, and many other properties. Since periodic calculations for metal oxides are very demanding the geometrical structure is often taken from experiment and only single point calculations of the electronic structure are carried out.

Gale, Catlow, and Mackrodt have used periodic ab initio calculations on α-alumina to derive force field parameters for a rigid ion pair potential [50]. They have derived different sets of parameters for different schemes of assigning the charges (formal, Mulliken, fitted charges) and found that the parameters fitted using formal charges are inferior to parameters derived using either Mulliken or fitted charges. The use of ab initio calculations in deriving parameters was found to result in a superior force field compared to fitting to empirical data.

The experimental data available about the structure of alumina has given rise to a long history of disagreement among scientists, although γ-alumina is widely used. The most widely held view is that γ-alumina is a stoichiometric oxide of aluminum with a defect spinel structure. The minority view is that γ-alumina actually contains hydrogen. Density functional studies employing periodic calculations have been able to solve this dispute [51]. Sohlberg and co-workers performed a series of computations fully relaxing structures for γ-alumina at various levels of dehydration. They also calculated vibrational frequencies for hydrogen atoms in nominal tetrahedral and nominal octahedral sites and the energy barrier to diffusion of hydrogen through the bulk.

The results from each of these calculations were in good accord with experimental data available through the literature. The calculations showed that certain spinel aluminas exist over a range of hydrogen content. In particular, γ-alumina behaves as a "reactive sponge" in that it can store and release water in a reactive way. This chemical activity offered a basis for understanding long-standing puzzles in the behavior of aluminas in catalytic systems. The structure of alumina leads to very unusual surface chemistry, which can explain some of the versatility and value of transition aluminas. When a water molecule reaches the surface of γ-alumina it breaks up. The hydrogen atoms enter the material, and the oxygen atom stays on the surface. Aluminum atoms migrate from the center of the material and combine with the oxygen atom, thus extending the crystal matrix. Valence requirements determine that for every three water molecules, six hydrogen atoms move into the material, two aluminum atoms move out, and the crystal extends by a stoichiometric Al_2O_3 unit. The reverse process is also possible. This behavior is a consequence of the fact that the hydrogen content in γ-alumina may fall anywhere within the range $0 < n < 0.6$ for $Al_2O_3 \cdot n(H_2O)$. The availability of hydrogen and oxygen at the surface of γ-alumina is likely to have significant implications for the understanding of the remarkable catalytic properties of this material [51].

Due to the increasing demand for light weight and compact rechargeable batteries considerable research and development effort has been put on modeling cathode materials for these batteries, in particular lithium manganese spinels. A key operational issue for lithium insertion batteries is the structural stability of the host electrode during repeated insertion and extraction of lithium. Ideally, lithium insertion/extraction reactions should be accompanied by a minimum distortion of the oxygen ion array and minimum displacement of the transition metal cations within the host [52].

It is known that for many nonmagnetic oxides such as TiO_2 [53] and slightly paramagnetic oxides such as RuO_2 [54], the ability of the local density approximation to predict the structural, electronic, and optical properties is comparable to the prototypical semiconductor results. However, for anti-ferromagnetic systems, such as $LiMn_2O_4$, the ability of spin-unpolarized calculations to accurately account for magnetic properties had not been demonstrated. Thackeray was able to show that periodic density functional calculations may be applied to the largely ionic MnO_2 material and that such calculations may be used to assess the quality of ionic potential model calculations for such systems [55]. For lithium doped systems, the calculation points to the importance of an accurate description of the spin polarization and increased metallic character in the system's bonding. For localized basis sets, single point calculations are in reasonable agreement with observed lattice parameters.

Periodic electronic structure calculations employing the Full Potential Linearized Augmented Plane Waves method (FLAPW) have been performed by Wang and co-workers to study the (0001) surface of hematite (α-Fe_2O_3) [56]. Hematite is probably the catalytically active material for the production of styrene, but very little is known about its surfaces. The calculations were able to identify the most stable iron and oxygen terminated surfaces of hematite and to show that they exhibit huge relaxations. Therefore, it can be assumed that under typical experimental conditions the surface of hematite should consist of these two domains. The surface reconstruction has a tremendous impact on the magnetic properties of the atoms close to and on the surface.

Band structure calculations on Fe_3O_4 (magnetite) have been performed by Anisimov et al. to study the charge-ordered insulating state of magnetite [57]. They have shown that the correct electronic description of a complex transition metal oxide is a very challenging task since only modifications to the local spin density approximation can account for this state.

Density functional calculations have also been used in the engineering of mixed metal oxides [58]. The introduction of a second metal into the lattice of an oxide can have a significant impact on the band structure of the oxide giving the material new properties. The tuning of electronic properties is required for the application of an oxide, e. g., as sensor material, microwave dielectricum [59], or catalyst. Figure 2.1 shows the density of states for β-MgMoO$_4$, β-FeMoO$_4$, and β-NiMoO$_4$ as obtained from gradient-corrected density functional calculations. The effect of the second metal on the band structure is clearly visible. The additional bands in NiMoO$_4$ cause a higher reactivity of this oxide compared to the other two oxides as has been shown experimentally [60–63].

2.2.3   Adsorption on Metal Oxide Surfaces

The use of metal oxides in catalysis is related to the fact that the oxides can bind molecules on their surface. Depending on the strength of the interaction between the molecule and the surface the molecule might dissociate and react either with other species adsorbed on the surface or directly with gas phase species. If the adsorption does not lead to dissociation it will at least activate the adsorbed molecule so it can undergo reactions more easily. Modeling catalysis on metal oxides therefore always requires the modeling of oxide surfaces and species interacting with this surface. While it is feasible to study oxide surfaces using force field based approaches (cf. p. 38) the interaction between surface and molecule usually requires electronic structure methods.

Metal oxides which are semiconductors can be used as sensor materials. The adsorption of a molecule on the metal oxide surface causes bending of the bands or changes in the surface conductivity of the metal oxides, which results in a measurable signal. Such a system is most often employed to detect gases. The modeling of these interactions necessitates the use of electronic structure methods as well.

Interactions between surfaces and molecules are often modeled by employing

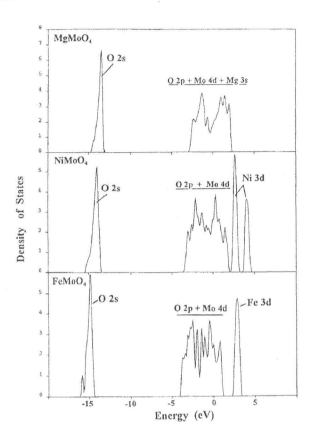

Figure 2.1: The density of states for molybdenum oxides with various other metals as obtained from gradient-corrected density functional calculations [58].

a cluster for the surface. The use of periodic calculations in such a case is problematic since applying periodicity also means to replicate the adsorbed molecule. Therefore periodic calculations would result in possibly strong interactions between different adsorbed molecules, which might not be of interest to the modeler in particular if in reality the occupation of the surface is low [58].

Comparing results obtained from modeling of species adsorbed on surfaces with experiment is particularly difficult. First, there is very little detailed experimental information about adsorbate structures. Second, modeling usually provides static configurations of atoms on a surface (relative stabilities), but has little to say about the kinetic routes by which these can be achieved. Therefore, adsorption processes which are kinetically controlled cannot be modeled easily. There

Table 2.1: Comparison between calculated and observed adsorption energies of CO on the MgO(100) and TiO$_2$(110) surfaces, [58]

Reference	Model	Method	Adsorption energy [kcal/mol]
		MgO(100)	
Nygren et al. [64]	Cluster	Ab initio model potential	1.6–2.1
Pacchioni et al. [65]	Cluster	DFT/LDA	11–27
Mejías et al. [66]	Cluster	Ab initio SCF	0.2–3.5
Yudanov et al. [67]	Cluster	DFT/GGA	0.5–1.5
Nygren and Pettersson [68]	Cluster	Ab initio model potential	1.8
Chen et al. [69]	Slab	DFT/LDA	6.5
Snyder et al. [70]	Slab	DFT/LDA	7.4
	Slab	DFT/GGA	1.9
Rodriguez et al. [71]	Slab	DFT/GGA	4.3
Observed [72, 73]			3.2–3.9
		TiO$_2$(110)	
Pacchioni et al. [74]	Cluster	Ab initio SCF	18.5–23.5
	Slab	Ab initio SCF	15.7
Casarin et al. [75]	Cluster	DFT/GGA	6.7
Sorescu and Yates [76]	Slab	DFT/GGA	11.1
Yang et al. [77]	Slab	DFT/LDA	18.3
		DFT/GGA	5.8
Rodriguez et al. [58]	Slab	DFT/GGA	12.2
Observed [76]			9.9

is a need in this area for directly comparable experiments and calculations.

A comparison of adsorption energies of small molecules on various metal oxide surfaces between experiment and calculation has been given by Rodriguez [58]. Tables 2.1 and 2.2 list the results of density functional calculations for the adsorption of CO and NO on MgO(100) and TiO$_2$(110). The calculated adsorption energies vary widely depending on the model and the method used. The best agreement with experimental results is obtained for slab models and gradient-corrected density functional calculations, which are also the most expensive cal-

Table 2.2: Comparison between calculated and observed adsorption energies of NO on the MgO(100) and TiO$_2$(110) surfaces, [58]

Reference	Model	Method	Adsorption energy [kcal/mol]
MgO(100)			
Rodriguez et al. [71]	Slab	DFT/GGA	5.9
Observed [73, 78]			5.1
TiO$_2$(110)			
Sorescu et al. [79]	Slab	DFT/GGA	10.5
Rodriguez et al. [80]	Slab	DFT/GGA	11.6
Observed [79]			8.4

culations performed. The differences observed for the adsorption on MgO and TiO$_2$ surfaces can be explained by band-orbital mixing [58].

The interaction of sulfur containing molecules with metal oxide surfaces plays a major role in the removal of these compounds from oil to prevent the release of SO$_2$ into the atmosphere on combustion of fuels and in the deposition of sulfur on oxide surfaces to protect or passivate electronic devices. Metal oxides are also used for the removal of SO$_2$ from feed streams in several petrochemical processes to avoid catalyst poisoning. Electronic structure calculations have shown that the reactivity of a metal oxide surface towards dissociative adsorption of H$_2$S is higher if the band gap of the metal oxide is smaller [58]. This trend can be explained by band-orbital mixing models as in the case of the CO adsorption energies shown above.

Chaturvedi et al. have studied the adsorption of SO$_2$ on ZnO using electronic structure calculations [81]. They have modeled ZnO by a cluster consisting of 13 zinc and 13 oxygen atoms arranged as in the geometry of bulk zinc oxide. Both the zinc terminated (0001) and the oxygen terminated (000$\bar{1}$) surfaces were used. Different adsorption geometries of SO$_2$ were considered.

Table 2.3 shows the geometries and adsorption energies obtained. SO$_2$ adsorbs on the surface forming an SO$_3$ species. The formation of the experimentally observed SO$_4$ species could not be modeled since the cluster used was too small to allow for the reconstruction of the surface necessary for this. However, these results show that the high affinity of SO$_2$ to oxygen and the stability of the species formed on metal oxide surfaces is responsible for the poisoning of oxide based catalysts by SO$_2$.

Table 2.3: Adsorption of SO_2 on ZnO [81]

	geometry				adsorption
	X–S [pm]	X–O [pm]	S–O [pm]	O–S–O [deg]	energy [kJ/mol]
on ZnO(0001)–Zn (X=Zn)					
η^1–S a-top	240		142	118	16.7
η^1–S bridge	242		144	118	20.9
η^2–O,O bridge		212	146	120	50.2
η^2–S,O bridge	243	214	145	118	29.3
on ZnO(000$\bar{1}$)–O (X=O)					
η^1–S a-top	154		141	119	121.4

The adsorption of H_2S on the MgO(100) and ZnO(0001) surfaces has been studied by Rodriguez and Maiti using a slab model [82]. It could be shown that the dissociative adsorption of H_2S on the MgO(100) surface is an energetically favorable process. The addition of alkali atoms to the ZnO(0001) surface enhances the reactivity of this surface towards sulfur containing species [83–85].

The same authors have employed density functional calculations to the problem of NO adsorption on magnesium and cerium oxide surfaces [86, 87]. Transition metals can form solid solutions in magnesium oxide. These doped oxides are active as catalysts, e. g., in the oxidative coupling of methane to C_2 hydrocarbons, the reforming of CH_4 with CO_2 or H_2O, or the DeNOx and DeSOx processes [58]. Density functional calculations using a slab model have been used to study the properties of these catalysts [88–91]. These calculations have shown that, e. g., in $Cr_xMg_{1-x}O$ (x=0.03–0.06), the chromium atoms are at a lower oxidation state than in Cr_2O_3 [88]. Consequently, $Cr_xMg_{1-x}O$ is a better catalyst than MgO and Cr_2O_3 for the Claus process and the reduction of SO_2 by CO. The energetical position of the electronic states of the dopants can also be correlated with the adsorption energy of NO on these doped magnesium oxides [58].

Additional information about the adsorption of small molecules on metal oxide nanomaterials can be found in Section 6.3.2.

2.3   Force Field Methods

Force field methods are often the method of choice for simulation of metal oxides. In developing a force field for any kind of chemical system it has to be decided how to best model interactions between atoms in a mathematically straightforward, but physically meaningful way. If one thinks of metal oxides it seems obvious to describe them as a collection of positively and negatively charged ions. Therefore, ion pair and shell model potentials are mostly used for modeling of metal oxides. They are described in Chapter 7. Buckingham potentials (see p. 212) are often used for the repulsive interactions, but they can lead to problems in practical calculations since at very short distances the sixth power dispersive term can overtake the exponential resulting in a collapse of the system modeled. These short distances might be sampled in molecular dynamics calculations with time steps too long.

The charges in the electrostatic term are, strictly speaking, parameters of the force field, but in metal oxide simulations formal charges are often used. In ion pair potentials for metal oxides the number of possible interactions is rather limited. For example, in MgO there can be only interactions between $Mg^{2+} \cdots Mg^{2+}$, $O^{2-} \cdots O^{2-}$, and $Mg^{2+} \cdots O^{2-}$. The interactions between like charged ions are mostly determined by electrostatics; therefore the use of Van der Waals parameters is not required for the like pairs $Mg^{2+} \cdots Mg^{2+}$ and $O^{2-} \cdots O^{2-}$. The only attractive electrostatic interaction is for the third pair $Mg^{2+} \cdots O^{2-}$ and a Van der Waals term is needed to prevent a collapse of the system. Adding another cation to the system would only require Van der Waals parameters for one more metal cation/oxygen anion pair. Since the number of these Van der Waals parameters therefore remains small, it is common to specify the parameters for each pair explicitly. This is in contrast to the practice in force field calculations on organic molecules, where Van der Waals parameters are usually derived from like pairs using combination rules. Explicitly specifying the Van der Waals parameters offers some additional flexibility compared to the use of combination rules, but not all software used for force field calculations is able to handle this approach.

The rigid ion model works well for systems where the ions are polarizable only to a small degree. Unfortunately, the oxide anion, O^{2-}, is quite polarizable and a better description of metal oxides is obtained when this polarization is accounted for in the force field. Considering that polarization is an electronic effect it might seem complicated to include it in a force field which does not treat electrons explicitly. However this can be accomplished with the shell model introduced by Dick and Overhauser [16].

For isostructural oxides one may assume that there is an approximately inverse dependence of the polarizability on the Madelung potential [92,93]. If the oxygen polarizability, α_0, and the corresponding Madelung potential, V_0, are known for a reference structure, α' and V' for a second isostructural oxide can be obtained by

$$\alpha' = \alpha_0 V_0 / V' \tag{2.1}$$

Therefore, in this approximation the shell model can be applied in a systematic way to oxides. The cation polarizability is considered to be less strongly influenced by the crystal environment [93].

When rigid ion or shell model potentials are to be applied to simulations of metal oxides the derivation of the parameters deserves significant attention. Parameters can be derived either from electronic structure calculations or by fitting to experimental data. Most of the potentials still in use today have been fitted to experimental data. In general, the number of experimental data points available for metal oxides is limited. In most cases only the structure and lattice constants are known. Therefore, there is insufficient data to determine all parameters even for binary oxides. In addition, potential parameters for oxides cannot be transferred between different systems since the O^{2-} ion is not stable as a free ion. The second electron in O^{2-} is only bound by the Madelung potential, which results in a strong variation of the electronic polarizability with crystalline environment for ionic oxides [92–94].

The problem of insufficient experimental data to derive all parameters can be reduced by using the following assumptions made by Catlow [95] and Sangster and Stoneham [94].

- The $O^{2-} \cdots O^{2-}$ interaction is taken to be the same for all crystals. At equilibrium separations of $O^{2-} \cdots O^{2-}$ this interaction is very small.

- Cation–cation interactions are assumed to be purely coulombic. This assumption is reasonable since cations are usually smaller than O^{2-} ions whose interaction is already small at equilibrium lattice spacings.

- The cation–anion interaction is considered to be purely repulsive. Therefore the attractive r^{-6} term is ignored.

For binary oxides these assumptions reduce the number of parameters to be derived to six (the cation–anion repulsive parameters and shell charges and spring constants of the cation and anion), but for many oxides even this is too much. Lewis and Catlow [93] have therefore derived parameters based on whole classes of oxides holding certain parameters constant.

For oxides where very little experimental data is available the hardness parameter, ρ_{ij}, was kept constant for all cations–anion interactions in one row of the periodic table. The parameter A_{ij} was then fitted to lattice parameters. The results of calculations employing these parameters are shown in Table 2.4.

For crystals of lower symmetry the structural data itself might contain enough information to determine ρ_{ij} as well. Finally, for those oxides where elastic constants, dielectric constants, cohesive energy, and structure are known, the short-range and polarizability parameters could be determined. Since the dielectric constants strongly depend upon the shell parameters, the cation-anion interaction parameters for a particular oxide are somewhat dependent on the way shell parameters are assigned. Starting from an oxide where the cations can be considered unpolarizable, such as MgO, the oxygen polarizability can be calculated

Table 2.4: Comparison of experimental and calculated cohesive energies (in eV) and elastic constants (in 10^{11} dyn cm^{-2}) for a number of metal oxides using a rigid ion potential [93]

Oxide	Experimental			Calculated			Exp. E_{lat}	Calc. E_{lat}
	c_{11}	c_{12}	c_{44}	c_{11}	$c_{12}(=c_{44})$			
NaCl structure								
CaO	22.4	6.0	8.1	22.35	9.87		−36.2	−36.1
ScO				24.89	13.78		−38.9	−38.5
TiO				26.59	18.30		−40.8	−40.6
VO				27.23	21.08		−41.1	−41.6
CrO				26.86	18.78		−41.8	−40.8
MnO	22.4	11.4	7.8	24.83	13.87		−39.5	−38.5
FeO	35.9	15.6	5.6	25.74	15.81		−40.7	−39.6
CoO	25.6	14.4	8.4	25.94	16.78		−41.4	−40.0
NiO	27.0	12.5	10.5	26.31	17.98		−42.3	−40.5
ZnO				25.77	16.40		−42.2	−39.8
CdO				20.71	11.06			−36.6
EuO				13.08	7.86			−33.2
Fluorite structure								
UO$_2$	39.6	12.1	6.4	45.1	9.8	6.4	−106.8	−102.0
ThO$_2$	36.7	10.6	7.97	42.2	9.9	7.2	−104.7	−99.9
CeO$_2$				46.8	10.1	6.4	−103.9	−102.7

for MgO and Eq. (2.1) can then be used to calculate the oxygen polarizability in other oxides. Table 2.5 compares experimental and calculated results for the longitudinal optic branches of phonon dispersion relations in the principal symmetry direction. MgO is the best, MnO the worst case [93].

Force field parameters for metal oxides have also been determined from ab initio calculations employing periodic codes [50]. A collection of shell model parameters for the simulation of metal oxides is shown in Table 2.6. Additional force field parameters for use in metal oxide simulations can be found on the Internet [96].

Woodley and co-workers have used the force field parameters listed in Table 2.6 to predict crystal structures for a number of metal oxides based on a genetic algorithm and energy minimization [97]. They are able to randomly place the constituent elements in a unit cell of known dimensions. By applying a genetic algorithm based on a simple cost function [97] they generate candidate structures

Table 2.5: Experimental and calculated longitudinal optic frequencies (10^{12} Hz) for MgO and MnO using shell model potentials [93]

		MgO				MnO	
	q	Exp.	Calc.		q	Exp.	Calc.
(q00)	0.40	19.82	19.78	(q00)	0.00	14.51	15.96
	0.60	18.07	17.91		0.20	13.82	15.72
	0.80	16.78	16.40		0.80	8.21	10.21
	1.00	16.61	16.05		1.00	8.80	10.39
(qq0)	1.00	13.29	13.67	(qq0)	0.20	12.21	13.23
	0.80	14.30	14.36				
	0.60	16.16	16.64				
	0.40	19.05	19.67				
	0.20	20.87	21.20				
(qqq)	0.10	21.73	21.55	(qqq)	0.30		15.81
	0.20	20.72	21.27		0.35	13.06	
	0.30	19.87	20.77		0.40		15.74
	0.40	18.40	20.00				

the best of which are undergoing a lattice energy minimization. The structures of 38 binary and ternary oxides could be predicted this way, including the difficult to model polymorphs of TiO_2.

We have already mentioned the importance of lithium manganese oxides in the previous section. Modeling using force fields has been applied to these oxides by Ammundsen et al. [98, 99]. Their simulation of infra-red and Raman spectra allowed for the first time to assign the bands observed in the spectra of these materials. They were also able to locate where protons are inserted by lithium–proton exchange.

2.3.1 Surfaces and Crystal Morphology

Many of the interesting properties and applications of metal oxides involve their surfaces. For example, in the case of heterogeneous catalysis the catalytic reaction occurs on the surface of the catalyst. The stability of certain surfaces has an effect on the morphology of the catalyst particle and highly reactive surfaces will only be useful in catalysis if they are present in a real crystal. Therefore, modeling of metal oxides in catalysis requires the modeling of surfaces and crystal morphology. Similarly, uses of metal oxides as sensors or as materials for electronic

Table 2.6: Shell model parameters for use in metal oxide simulations

	q(core) [e]	q(shell) [e]	A_{ij} [eV]	ρ_{ij} [Å]	B_{ij} [ev Å^{-6}]	K_{ij} [Å^{-2}]
Ag^+	1.000		962.197	0.3000	0.00	
Ag^{3+}	3.000		4534.200	0.2649	0.00	
Al^{3+}	0.043	2.957	2409.505	0.2649	0.00	403.98
Ba^{2+}	0.169	1.831	4818.416	0.3067	0.00	34.05
Ca^{2+}	0.719	1.281	2272.741	0.2986	0.00	34.05
Ce^{4+}	4.000		2409.505	0.3260	0.00	
Cu^+	1.000		585.747	0.3000	0.00	
Eu^{3+}	−0.991	3.991	847.868	0.3791	0.00	304.92
Fe^{2+}	2.000		2763.945	0.2641	0.00	
Fe^{3+}	1.971	1.029	3219.335	0.2641	0.00	179.58
Ga^{3+}	3.000		2339.776	0.2742	0.00	
Gd^{3+}	−0.973	3.973	866.339	0.3770	0.00	299.96
Ge^{4+}	4.000		3703.725	0.2610	0.00	
K^+	1.000		3587.570	0.3000	0.00	
La^{3+}	5.149	−2.149	5436.827	0.2939	0.00	173.90
Li^+	1.000		426.480	0.3000	0.00	
Mg^{2+}	1.580	0.420	2457.243	0.2610	0.00	349.95
Mn^{4+}	4.000		3329.388	0.2642	0.00	
Na^+	1.000		1271.504	0.3000	0.00	
Nb^{5+}	5.000		3023.184	0.3000	0.00	
Nd^{3+}	1.678	1.322	13084.217	0.2550	0.00	302.35
O^{2-}	0.513	−2.513	25.410	0.6937	32.32	20.53
Pb^+	1.000		5564.374	0.2610	0.00	
Po^{4+}	4.000		10984.535	0.3260	0.00	
Pr^{3+}	1.678	1.322	13431.118	0.2557	0.00	302.36
Rb^+	1.000		2565.507	0.3260	0.00	
Sn^{4+}	4.000		6327.497	0.2610	0.00	
Sr^{2+}	0.169	1.831	1956.702	0.3252	0.00	21.53
Ta^{2+}	2.000		2608.581	0.2610	0.00	
Tb^{3+}	−0.972	3.972	845.137	0.3750	0.00	299.98
Ti^{4+}	4.000		4545.823	0.2610	0.00	
Tl^{3+}	3.000		1753.153	0.3247	0.00	
U^{2+}	2.000		7784.072	0.2610	0.00	
V^{2+}	2.000		1946.254	0.2610	0.00	
V^{4+}	4.000		3228.972	0.2610	0.00	
Y^{3+}	3.000		1519.279	0.3291	0.00	
Yb^{3+}	−0.278	3.278	991.029	0.3515	0.00	308.91
Zr^{2+}	2.000		4796.164	0.2610	0.00	
Zr^{4+}	4.000		7290.347	0.2610	0.00	

packaging depend on their surface behavior. Also, control on sintering, mechanical strength, or the behavior of oxide films that may determine the corrosion rate of the corresponding metal all require knowledge of surfaces.

Today experimental techniques exists which allow studying of surfaces at atomic detail. Nevertheless, modeling of surfaces still plays a vital role in the interpretation of experiment, its extrapolation beyond the usual or accessible conditions and in the prediction of new effects for experimental validation. Modeling is particularly important in understanding the complex and competing effects in multi-component materials and is the only way to analyze the effects of certain components. Such analysis can lead to the prediction of which components in a material give rise to the desired behavior.

Great care has to be taken in interpreting modeling results. Much as measurements do not look at ideal surfaces, modeling has to be performed on reasonable model surfaces and not on surfaces which only exists in the imagination of theoreticians. A useful classification is the following [100]

Experiment

Uncleaned surface = Dirty surface

Cleaned surface = Probably dirty surface

Vacuum-cleaned surface = Possibly clean surface

Theory

Unrelaxed surface = Science fiction

Relaxed/polarized surface = Model of ideal surface

Relaxed, defective, impure surface = Model of real surface

Surfaces present some unique challenges to the modeler. In calculations of the bulk of a metal oxide the 3D periodic nature of the bulk can be utilized in that the simulation has to treat only one unit cell explicitly (unless, for example, defects are to be simulated where even in the bulk more than one unit cell needs to be used, see below). In simulating surfaces the best use of periodicity one can make is in 2D. For dealing with the third dimension two techniques are possible. First, one can create a pseudo-3D system by stacking sufficiently thick slabs separated by a sufficiently large empty space on top of each other and use 3D bulk methods. In this approach each slab possesses two surfaces and care has to be taken that the slabs are thick enough so that both surfaces do not interact with each other. Also, the empty space between the slabs has to be wide enough to prevent interactions between neighboring slabs. One advantage of this approach is that surface dipoles which can occur in some systems are easily dealt with, since the surface dipole of one surface is compensated by the surface dipole of the opposite surface. Another advantage is that such calculations can be performed using standard computer codes, which can handle 3D periodicity. A disadvantage is that the bulk area is reduced in size or completely missing, which might affect the surface structure.

The second approach to modeling surfaces consists of using one slab with 2D periodicity. In this approach the slab is usually constructed in a way that the atoms at the bottom are fixed at their bulk positions and only atoms at a certain distance from the surface are allowed to relax. The advantage of this approach is that bulk effects on the surface can be studied. A disadvantage is that in systems which possess a surface dipole, measures have to be taken to compensate for the surface dipole, since otherwise the surface will not be stable. Another disadvantage is that the computer code used for these simulations has to be able to deal with 2D periodicity. The atoms at the top of the slab which are allowed to relax are called region I atoms. The atoms at the bottom of the slab fixed at their bulk positions are called region II atoms (vide infra).

The stability of surfaces can be measured in two different ways. First, the so-called attachment energy can be calculated which is defined as the energy released when a further growth slice is brought from an infinite distance on to the growing crystal surface [101]

$$E_{attach} = E_{bulk} - E_{slice} \qquad (2.2)$$

In case of non-defective surfaces the growth slice is simply a flat slab of ions; however, when a defective surface is being modeled the growth slice has to contain troughs and steps to accurately reflect vacancies and interstitials [101]. The second way of obtaining a measure of the stability of a surface is the surface energy. It is defined as the energy required to split the crystal to produce the desired surface

$$E_{surface} = \frac{E_s - E_{bulk}}{A} \qquad (2.3)$$

where E_s is the energy of the ions at the surface, E_{bulk} the energy of the ions in the bulk, and A the surface area [101].

Ionic surfaces can be classified into three types [102, 103] (cf. Fig. 2.2)

- Type I
  Surfaces which consist of neutral planes, perhaps with a stoichiometric proportion of anions and cations on each plane.

- Type II
  Surfaces which consist of charged planes, for example just anions or cations, but where the repeat unit of planes has no net dipole moment.

- Type III
  Surfaces which consist of charged planes which also have a dipole moment.

Type I and II surfaces are stable; type III surfaces are not and will reconstruct in nature to remove the instability. Since the instability can be removed by removing the surface dipole, type III surfaces are often found where half of the cation or anion positions in the surface layer are vacant. In this case the surface energy is not much higher than that for other low energy surfaces.

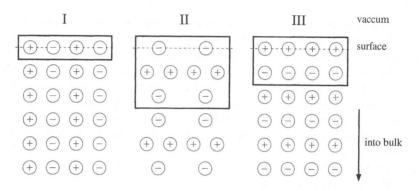

Figure 2.2: The three types of ideal surface terminations. The repeat units are shown in the boxes. Types I and II are stable, type III is not.

While setting up simulations of surfaces the surfaces are usually created by cleaving the bulk material. Care has to be taken as to where to place the cut. If the cut results in a type I surface no problems arise. If a type II surface is created the cut can only be placed at a position which leaves the repeat unit intact. Otherwise a surface dipole would be created. If a type III is generated there will always be a surface dipole. Since a surface dipole has to be avoided to get the electrostatic energy to converge, rearrangements are required to remove the surface dipole. The best approach is usually to remove the appropriate number of cations or anions from the top surface and place them on the bottom of the simulation cell as shown in Fig. 2.3. This approach also requires the inclusion of additional atoms from region II in region I to keep the number of atoms in both regions constant (cf. Fig. 2.3). Since it can make a difference whether cations or anions are moved both possibilities should be studied.

Surfaces can have different terminations depending on where the cut is placed. If this is the case only calculations can determine which cut results in the most stable surface. Therefore, if the possibility of different terminations exists all possible terminations need to be studied to obtain meaningful results.

Surfaces of metal oxides or other ionic materials can be modeled using the same ion pair or shell model potentials which are used for bulk simulations. This is a fortunate and not generally applicable property of ionic materials since the interactions in these materials are largely determined by electrostatics. It has been proven that potentials derived for the bulk can also predict surface energies, structure, vibrations, and impurity heats of segregation for oxides adequately [103].

To obtain meaningful results in simulations only relaxed surfaces have to be used. The amount of relaxation depends on the nature of the surface. Type I and II surfaces will show very little relaxation. For example, for the (100) surface of

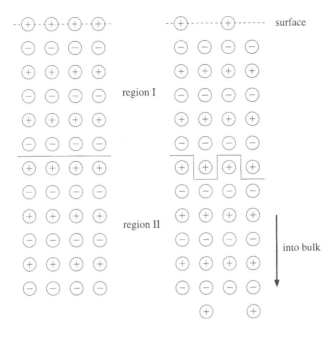

Figure 2.3: Surface dipoles on type III surfaces can be removed by moving the appropriate number of cations or anions from the top of the simulation cell to the bottom. Atoms in region I are allowed to relax; atoms in region II are fixed at their bulk positions.

MgO, a type I surface, the minuscule rumpling and relaxation of the surface are calculated to be -0.7% and 2.5%, respectively [104]. These values compare well with the observed ones of $0\pm0.75\%$ and $2\pm2\%$ [105]. Although these relaxations appear small, they have a profound effect on the surface energy. For alumina, for example, relaxation reduces the surface energy by one half [103]. The relaxation of even the simplest surface will produce a dipole layer in the surface. This does not produce a long-range field and has therefore no large effect on the energetics of most observable defect processes. But it is important in considering processes such as defect or impurity surface segregation, ionization, or ion sputtering.

The surface free energy of a material determines the crystal shape when equilibrium pertains and provides the thermodynamic opposition to fracture. Relative surface energies may determine the ability to cut or fracture a crystal in its various directions [103]. Once the attachment or surface energies for all low energy surfaces are known, it is possible to predict the morphology of a crystal.

There are two separate methods of determining the shape of a crystal. The

Table 2.7: Calculated and measured surface energies in Jm^{-2} [106]

Crystal surface	Calculated	Experimental (extrapolated to 0 K)
$UO_2(111)$	1.06	1.05, 1.4
ThO_2	1.02	
CeO_2	1.05	
$MgO(001)$	1.16	1.04, 1.2, 1.15
NiO	1.15	
$\alpha\text{-}Al_2O_3(01\bar{1}2)$	2.57	
$\alpha\text{-}Al_2O_3(0001)$	2.97	

growth morphology is obtained by using attachment energies. Therefore it is based upon the energetics associated with the deposition of blocks of materials upon all of the competing growth surfaces. In contrast, the equilibrium morphology is simply that which reduces the total surface energy to a minimum value and is based on the surface energies of each possible surface. The growth morphology is usually better suited to predict the shape of a crystal [101].

Fig. 2.4 shows as an example the calculated morphologies of $ZnCr_2O_4$ [101]. The top row shows the morphologies predicted based on unrelaxed surfaces while the bottom row shows the results obtained using relaxed surfaces. It is clear from this picture that the use of unrelaxed surfaces does result in a significant change to the predicted morphology and, when comparing with experimentally observed crystal shapes, to the wrong morphology. Therefore, unrelaxed surfaces are, as mentioned previously, pure fiction and the determination of morphologies through modeling requires the use of relaxed surfaces.

Aluminum is widely used as construction material. Its surface oxidizes to form a protective oxide film. Hydrocarbons are used as lubricants, paints, sealants, and adhesives for metal parts, and these organic materials must adhere well to the part to provide a good coating. Water often plays a crucial role in the quality of this coating, as it can act as both an adhesion promoter and an inhibitor. Thus a molecular level understanding of the interaction of hydrocarbons and water with the alumina (Al_2O_3) surface can be very beneficial in selecting the correct material to create a coating. De Sainte Claire et al. have employed modeling to examine the behavior of hydrocarbons and water on the (0001) surface of α-alumina using molecular dynamics calculations and a force field [107]. They could show that water is strongly attracted to the alumina surface and that alkane molecules, such as butane, octane, or dodecane, form ordered layers on the surface with

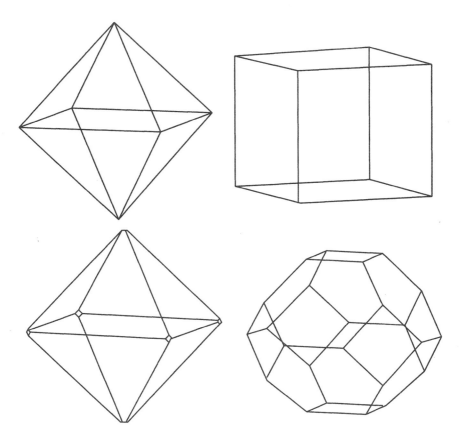

Figure 2.4: The calculated growth (left) and equilibrium (right) morphologies of $ZnCr_2O_4$ using unrelaxed (top row) and relaxed (bottom row) surfaces [101] (*Journal of Materials Science*, vol. 31, 1996, pp. 1151 – 1156, "Morphology and structure of $ZnCr_2O_4$ spinel crystallites", D. J. Binks, R. W. Grimes, A. L. Rohl, and D. H. Gay, Fig. 3 to 6, Copyright ©1996 Chapman & Hall with kind permission of Springer Science and Business Media)

a much higher density than in bulk hydrocarbons. If water and hydrocarbons are present together the water molecules are attracted by the alumina surface, but initially repelled by the hydrocarbons. Over longer time scales some water molecules penetrate the hydrocarbon layer breaking up the protective film. This can disrupt the adhesion of a coating such as paint.

Coatings, paints, and other materials are placed on surfaces of materials to protect them or to lubricate in mechanical operations. In automotive engines oil is required to reduce friction. These oils contain wear inhibitors as additives. The interaction of one such wear inhibitor, zinc dithiophosphate $(Zn((RO)_2PS_2)_2)$, with an iron oxide (hematite) surface, has been studied using a force field [108]. It has been found that zinc dithiophosphates form self-assembled monolayers on the hematite surface and that the cohesive energy of the self-assembled monolayers correlates well with the anti-wear performance of the zinc dithiophosphates. Therefore new wear inhibitors can now be evaluated on the computer prior to performing costly experiments.

Ceria, which plays a role in the catalytic decomposition of pollutants in car exhausts, has been studied by Vyas and co-workers employing a number of different rigid ion and shell model potentials [109]. Despite their finding that the different potentials result in significant differences in the absolute surface energies they show that the relative surface energies and therefore the calculated morphologies of ceria crystallites are largely independent of the potential used. Just one of the force fields used cannot produce meaningful results for high-indexed surfaces. In the derivation of this force field only bulk properties have been used. This shows how important the careful derivation of force field parameters is, in particular, the sampling of a wide range of interatomic distances.

The partial oxidation of hydrocarbons is an important industrial catalytic process. The catalysts used for this process are based on vanadium oxide on a titanium oxide (anatase) support. Vanadium oxide without the support is, however, much inferior as a catalyst. To understand the effect of the support Sayle and co-workers have performed a number of force field based studies of vanadium and titanium oxides [110–112]. The morphology of V_2O_5 can be correctly reproduced by force field calculations and adsorption sites for ethene molecules can be predicted [110]. Furthermore, it is found that vanadium oxide when it forms a thin film on the anatase support undergoes substantial relaxation and reconstruction compared to the bulk. The vanadium oxide film stabilizes oxygen vacancies. Both effects together can explain the remarkably different catalytic activity of supported vanadium oxide [111, 112].

The effect of defects and impurities on the morphology of NiO crystals has been investigated by Oliver, Parker, and Mackrodt [113] employing a shell model potential. They show that the morphology of NiO crystals strongly depends on the presence of nickel vacancies and Ni^{3+} holes. The good agreement between calculated and observed morphologies indicates that the surfaces of NiO crystals contain a large amount of defects.

The stability of different NiO surfaces as a function of temperature has been studied by Oliver, Watson, and Parker [114]. NiO has the same cubic structure as MgO, but the morphology of NiO crystals is different from the morphology of MgO crystals. One possible explanation would be the effect of temperature, which is usually not considered in calculations of morphologies. To account for possible temperature effects molecular dynamics simulations have been performed at different temperatures up to the melting point of NiO using a rigid ion potential. It has been found that different surfaces melt at different temperatures, the denser packed surfaces having the higher melting temperature.

Tin dioxide is a metal oxide with semiconducting properties. It is therefore used in sensor devices, in particular for O_2. To understand the sensing properties of SnO_2 Mulheran and Harding have performed a shell model study of the stability of SnO_2 surfaces [115]. Their results provide energetical reasons for different surface structures and their changes upon heating observed in thin films of SnO_2. Higher indexed surfaces might be present as steps which would provide sites for gas adsorption.

Another study of tin dioxide as a material for gas sensors has been performed by Skouras, Burganos, and Payatakes [116, 117]. They have used molecular dynamics and Monte Carlo calculations employing the UFF and COMPASS force fields to obtain mean residence times, Henry's constants, and the heat of adsorption of CO, CO_2, O_2, CH_4, and Xe on SnO_2, $BaTiO_3$, CuO, and MgO surfaces.

2.3.2 Defects

So far we have dealt with simulations of metal oxides where the crystals were considered to be ideal. In nature ideal crystals do not exist. All crystals have defects and only the number of defects varies. In most cases it is the presence of defects which gives a material its interesting properties. For example, the ionic conductivity of a number of metal oxides which gives rise to the use of these oxides as electrode materials in batteries or fuel cells is largely a result of defects. Defects allow diffusion through solids and thereby control corrosion and dissolution processes. The importance of such defects has resulted in a lengthy history of their simulation.

Defects can be classified as point or extended defects. Point defects are cation or anion vacancies (Schottky defects) or cation or anion interstitials (Frenkel defects). Extended defects are twin planes or shear planes, which might affect the entire crystal and are, therefore, difficult to simulate. The problem with all simulations of defects is that one has to deal with a large piece of crystal using only limited computational resources. In practice, a few thousands atoms can be handled, but this is insufficient to study the behavior of bulk material or defects in bulk material. Even with a cube of one thousands atoms, nearly half of them are on the surface. As mentioned above, one can impose periodic boundary conditions to eliminate the surfaces, but this results in another artificially imposed periodicity – the defects are now part of a regular array of images. In case of metal oxides, where the defects are often charged, the Coulomb summation of the electrostatic interactions will no longer converge. This can be corrected by introducing a uniform background of charge density and subtracting the Coulomb interaction between the defects [118]. It has been shown that this approach gives

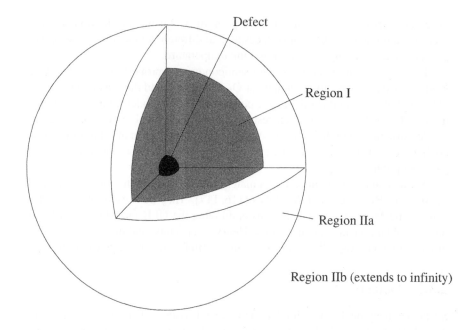

Figure 2.5: Mott–Littleton procedure of simulating defects.

the same values as the isolated defect calculation in the limit of large supercells.

Alternatively, Mott and Littleton have suggested an atomistic simulation methodology [119], which describes the area around a defect by regions of different complexity. In the region in the vicinity of the defect free atomic relaxation is allowed. This atomically described region is embedded in a dielectric continuum. A buffer region links these contrasting descriptions of the solid (cf. Fig. 2.5). The essential assumption of this approach is that the only important feature of the defect at long distance is its effective charge with respect to the lattice. This will give rise to a polarization field. This polarization is divided up among the sublattices in proportion to their polarizabilities [120, 121]. The resulting defect energies are generally in good accord with experimental results for ionic materials [122] (cf. Table 2.8).

It should be noted that these kinds of calculation are usually performed with a fixed lattice and as such yield constant volume thermodynamic properties. At higher temperatures there might be a significant difference between constant pressure and constant volume parameters [120].

Defects have been studied by Tomlinson and co-workers in La_2CuO_4, $LiNbO_3$, and TiO_2 (rutile) [124] using the Mott–Littleton method. They are able to contradict defect models proposed in the literature.

Table 2.8: Calculated and experimental defect energies for magnesia (in eV) using different force fields [123]

Defect	empirical	electron-gas	exp.
Cation vacancy energy	23.8	25.4	
Anion vacancy energy	24.7	22.9	
Schottky energy	7.7	7.5	
Cation vacancy migration	2.1	2.2	2.2 - 2.3
Anion vacancy formation	2.1	2.4	2.4 - 2.6

The Mott–Littleton approach is only useful for considering defects at infinite dissolution, i. e., defects which do not interact with each other. Defect concentrations might be much higher in reality and interactions between defects need to be accounted for. In this case the conventional methods of modeling described before need to be applied.

Defects also play an important role in the electrochemical properties of lithium manganese oxides. Islam and Ammundsen have used shell model calculations to study lithium insertion and associated defect properties [99]. They have been able to shed light on the migration pathway of lithium ions in manganese oxides and on the preferred location of protons, which are inserted into the material in aqueous acid environments.

Read and co-workers have studied the defect and surface properties of $LaCoO_3$ focusing on the doping with Sr and Ca [125]. Their results indicate in agreement with experiment that these alkaline earth metals are the most soluble dopants in this system. Oxygen vacancies and Co(IV) sites are formed for charge compensation. The oxygen vacancies give the material its catalytic activity. The energy of oxygen vacancy formation at the surface is shown to be smaller than in the bulk.

Manganese dioxide is produced in large quantities for the use in batteries. It is known that defects in the manganese dioxide lattice are essential for the electrochemical activity of the material. Defects are very difficult to study experimentally, e. g., the x-ray powder diffraction pattern of γ-MnO_2 shows only a number of broad and unstructured peaks (cf. Fig. 2.6). Force field calculations have been used to provide a more detailed picture of defects in this material [126]. Three types of defects play a role in γ-MnO_2, de Wolff disorder, microtwinning, and point defects. De Wolff disorder and microtwinning are extended defects, which require the use of large simulation cells. To predict which defects are important and have an effect on the x-ray powder diffraction pattern a systematic study of

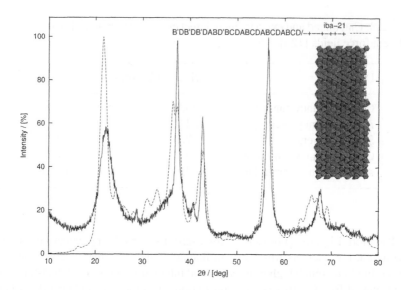

Figure 2.6: The x-ray powder diffraction pattern calculated for the MnO_2 defect structure shown in inset and compared with the observed pattern for a manganese dioxide (iba-21). Reprinted from *Journal of Solid State Chemistry*, vol. 177, J.-R. Hill, C. M. Freeman, and M. H. Rossouw, "Understanding γ-MnO_2 by molecular modeling", pp. 165 – 175, Copyright (2004), with permission from Elsevier.

various possible defect structures had to be performed. For this purpose structures with all possible positions of a defect had to be generated followed by a geometry optimization with a force field. The x-ray powder diffraction pattern was calculated and compared with observed ones. It was found that single defects could not explain the observed powder diffraction patterns. Only a combination of de Wolff disorder and microtwinning led to a calculated x-ray powder diffraction pattern which closely resembled the observed one (Fig. 2.6). Therefore, systematic studies of defect structures employing modeling techniques can be used to aid in the structure determination for experimentally difficult cases. In analogy to combinatorial experimental methods such systematic modeling studies of a large number of structures are called *virtual screening*.

2.3.3 Transport

For metal oxides, defects within the lattice are inextricably linked with transport properties [127, 128]. The diffusion of a cation in an oxide, for example, involves

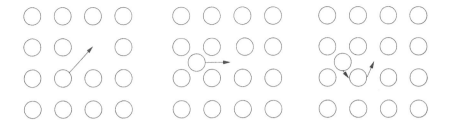

Figure 2.7: Migration mechanisms in solids: (a) vacancy migration, (b) direct interstitial migration, (c) concerted or interstitialcy migration.

the formation of vacancy or interstitial states within the crystal and the migration of these species leads to a net transport of material through the lattice [120]. At the macroscopic level transport phenomena can be observed as either diffusion or conductivity, i. e., matter or charge transport. At the microscopic level they are, however, linked since charge transport requires the movement of electrons or, more likely in metal oxides, ions. Although transport of matter and charge are linked their mechanisms might not. If matter transport is accomplished by ions jumping into vacant sites then charge transport can be considered in terms of the hopping of the effectively charged vacancy, which is a random process. The transport of the ions is, however, nonrandom, as after an ion has jumped into the vacancy there is a probability that the ion will jump back to the original site [120].

Under the assumption of transport being effected by discrete hops of ions random walk theory gives the following relation between the diffusion coefficient, D, the concentration of the hopping species, x, the hopping frequency, ν, and the hopping distance, d,

$$D = \frac{1}{6}x\nu d^2. \tag{2.4}$$

The hopping frequency is then given by

$$\nu = \nu_0 \exp(-G_{TS}/k_B T) \tag{2.5}$$

where G_{TS} is the difference in free energy of the hopping species between its ground state and the saddle point, i. e., the maximum in the free energy profile on the most favored route for the hopping process. k_B is the Boltzmann constant and T the temperature. Thus, the accurate calculation of G_{TS} is a priority for simulations methods.

Different types of migration mechanisms are shown in Fig. 2.7.

In most crystalline materials, the diffusing species can be correlated with some type of defect. The concentration of the defect then depends on intrinsic defect reactions (Frenkel or Schottky), impurities or non-stoichiometric compositions, or "non-equilibrium" defects induced by mechanical or radiation damage. The concentration of defects can be calculated by a number of methods, see, e. g., Lidiard [129] and Catlow [120, 127]. The most important source of defects in a crystal is impurities, which are deliberately or accidentally introduced. The majority of those impurities are aliovalent ions, impurities with a different valence from the host ions they replace. The charge imbalance they induce causes the formation of an oppositely charged defect population. Impurities attract each other owing both to Coulombic and elastic forces and form well-defined complexes (cf. [120]). Therefore, not all the defects created by impurities might be available to take part in transport and the concentration of defects needs to be corrected to account for this phenomenon.

Transport coefficients usually show Arrhenius behavior. For materials where this is the case transport is largely controlled by the values of the free energies of defect formation, interaction, and migration. Therefore, simulation of transport requires the calculation of these quantities. This can be achieved with either the Mott–Littleton defect approach [130], direct molecular dynamics techniques [131], or Monte Carlo methods to describe [132] overall transport on the basis of calculated individual process statistics.

Static calculations based on the Mott–Littleton defect approach have been discussed in the previous section. Molecular dynamics methods allow the direct calculation of the diffusion coefficient via the Einstein relationship (cf. p. 252)

$$6D_i t + B_i = \langle r_i^2(t) \rangle \tag{2.6}$$

where B is the thermal factor arising from vibrations of atoms and $\langle r^2(t) \rangle$ is the mean-square displacement of particles of type i at time t relative to their position at $t = 0$. The conductivity, σ, can be calculated via the Nernst-Einstein equation

$$\frac{\sigma}{D} = \frac{nq^2}{f k_B T} \tag{2.7}$$

where f is a correlation factor, which correlates charge and matter transport [120].

Molecular dynamics simulations are limited to systems where the diffusion processes are rapid since the time scale of such simulations are limited to a few nanoseconds. Also, the Arrhenius energy cannot be extracted from such simulations unless a number of simulations is run at different temperatures.

Monte Carlo calculations can be used to obtain the correlation factor f shown above by using

$$f_i = \frac{\langle r_i^2 \rangle}{na^2} \tag{2.8}$$

where $\langle r_i^2 \rangle$ is the mean-square displacement of atoms of type i after an average of n jumps, each of length a [120].

Table 2.9: Calculated and experimental energies for oxygen vacancy migration in $LaBO_3$ perovskites [133] (reproduced by permission of The Royal Society of Chemistry)

Compound	E_m/[eV] calc.	exp.
$LaGaO_3$	0.73	0.79, 0.66, 0.727
$LaMnO_3$	0.86	0.73
$LaCoO_3$	0.61	0.58, 0.78
$LaYO_3$	1.22	1.3

Transport properties of metal oxides play a role in the use of these oxides as solid electrolytes in high-temperature fuel cells, hydrogen sensors, and other electrochemical applications. In these applications the metal oxide acts as a proton conductor. Typical proton conductors are perovskite based structures, such as $AB_{1-x}M_xO_{3-\alpha}$ (A^{2+} = Ba, Sr, Ca; B^{4+} = Ce, Zr) where a trivalent dopant, such as Ln^{3+}, can replace either the tetravalent or the bivalent cation. In the former case, oxygen vacancies are formed to compensate for the lower charge and these vacancies can be occupied by OH^-. In the latter case, modeling has shown that the replacement of bivalent cations is compensated for by a bivalent cation vacancy [134, 135]. The same modeling study has also shown that the preferred replacement site depends on the size of the lanthanoid ion where larger ions prefer to replace a bivalent cation while smaller ones replace a tetravalent cation. These modeling results have been confirmed by EXAFS experiments [135].

Islam has reviewed simulations on ion transport in ABO_3 perovskite oxides [133]. Table 2.9 shows that calculated and observed oxygen vacancy migration energies agree well despite a large scatter in the observed data. The review also shows that simulations can give valuable insight into mechanistic details of ion transport in oxides, which are not easily accessible using experimental methods. Simulations have also stimulated further experimental work on these systems.

Ceria's catalytic properties can be enhanced by small amounts of zirconia. Solid solutions of ceria and zirconia have been studied by Balducci and co-workers [136]. The cell constants of the mixtures were found to be in very good agreement with experimental data. The ceria reduction energy is significantly reduced even by small amounts of zirconia and the activation energy for oxygen diffusion in the bulk is found to be low indicating an easy diffusion of oxygen through the bulk catalyst.

Chapter 3

Microporous Materials

3.1 Introduction

Microporous materials are all compounds which have pores of molecular dimensions, the best known are probably zeolites, which are mostly made up of silicon and oxygen. Glasses and quartz are also made up from silicon and oxygen (and from a number of other elements in the case of glasses), but form denser packed solids then microporous materials. Nevertheless, since they possess a similar chemical composition and share structural elements, e. g., SiO_4 tetrahedra, the same force fields are usually applied to both microporous materials and glasses. However, the amorphous nature of glasses requires the use of slightly different modeling approaches compared to the crystalline microporous materials. Therefore, we will focus on the latter in this chapter. Glasses will be dealt with in the next chapter.

Zeolites and other microporous materials, like silicon and aluminum phosphates (SAPOs and AlPOs), are mainly used as catalysts. Most of oil refinement today relies on zeolite catalysts. Zeolites have also taken over the function as water softeners in detergents from phosphates because they are more environmentally friendly. All these different uses together with the rather straightforward application of modeling techniques have made modeling of microporous materials one of the best studied areas of materials simulations. There have been all kinds of electronic structure calculations applied as well as force field based empirical methods. Current research includes cutting edge technologies such as first principles molecular dynamics to study, e. g., the interaction of sorbates with zeolite frameworks.

The following sections will show some examples of these techniques and provide insight into the results, which can be obtained from modeling for these systems. We will cover both electronic structure calculations and force field methods.

55

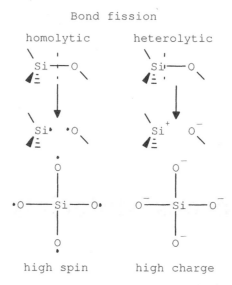

Figure 3.1: Two possibilities to cut bonds when creating clusters from solids.

3.2 Ab Initio and Density Functional Methods

3.2.1 Cluster Models

The application of ab initio methods to the study of microporous materials was long hampered by the lack of computers and algorithms powerful enough to treat the whole periodic system. Even for small zeolites the unit cells are rather large. Sodalite, one of the smallest and industrially unimportant zeolites, has, e. g., 36 atoms in its unit cell. ZSM-5, which is widely used in industry, already has 288 atoms in the unit cell. To make use of ab initio methods it was therefore necessary to find *models* for the real zeolite framework. On one hand these models have to be small enough to be treated by ab initio methods. On the other hand they have to represent structural features of the zeolite framework to allow conclusions from the ab initio calculations to be transferred back to the zeolite. We are faced here with one of the basic dilemmas of modeling in materials science. The real system is too large to be treated adequately. Therefore, smaller, representative structural units have to be found in the real material.

Zeolites are formed by linking SiO_4 tetrahedra together. It therefore seems reasonable to cut one of these tetrahedra out of the solid. To do this we have to cut some bonds and as Figure 3.1 shows there are two possibilities — we can either use a homolytic or a heterolytic bond fission. In the case of the homolytic

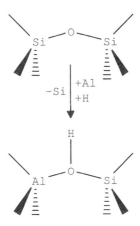

Figure 3.2: The formation of a bridging hydroxyl group in a zeolite.

bond fission the electrons of the bond are distributed to each atom and a cluster with a large number of unpaired electrons (and therefore a high spin) results. If we use a heterolytic bond fission the electrons are assigned to the more electronegative bond partner and a cluster with a high charge is obtained. Both clusters are obviously not very well suited to model the solid since the cut bonds have introduced artifacts, which are not present in the real system. A common way to avoid these problems is to use hydrogen atoms as "terminators". This results in a cluster which is neutral and has all electrons paired. Hydrogen atoms are preferably used because they are easiest to handle in ab initio calculations with respect to system resources.

If a single SiO_4 tetrahedron is used as a cluster orthosilicic acid, H_4SiO_4 is obtained after saturation with hydrogen. Orthosilicic acid can and has indeed been used as a very simple model for zeolites [137–140]. While some basic features of zeolites such as bond distances and some vibrational frequencies could be studied this way, orthosilicic acid is too small to account for structural differences found in different zeolites. For that purpose larger models have to be constructed, which represent larger building units of zeolites.

A very important feature of zeolites is their Brønsted acidity, which is caused by so-called bridging hydroxyl groups. These groups are formally formed by replacing a silicon atom with an aluminum atom. Since aluminum has a formal charge of +3 while silicon has +4 a cation has to be added to keep the whole system neutral. If a proton is used as the charge compensating cation it will bond to one of the four oxygen atoms surrounding the aluminum atom thus forming a bridging hydroxyl group as shown in Figure 3.2. Significant consideration has been given in the literature to the simulation of the behavior of this hydroxyl group and catalytic reactions involving it. The simplest model used for a bridging hydroxyl group is $H_3Si(OH)AlH_3$ [137], which cuts all but one of the SiO and AlO bonds. A more realistic, but significantly bigger model is $(HO)_3Si(OH)Al(OH)_3$ [141], which keeps the environment of the silicon and aluminum atoms the same as in the zeolite. As Table 3.1 shows the geometries obtained with the same basis set, but different model sizes are quite different.

With increasing computer power the use of larger models became feasible [143–146]. The enlargement of the models can be done in two different ways as shown in Figure 3.3. First, one can simply add more and more TO_4 tetrahedra (T=Al, Si) and form a branched structure. Second, one can add more tetrahedra, but fuse them to form rings and cages. The second approach builds models which are closer to structural units found in real zeolites. They also have the advantage that due to the formation of rings there are less degrees of freedom for the system. The branched structures have a rather large number of torsions parts of the system could rotate about. This makes geometry optimizations much harder to converge since torsions usually have very shallow potentials.

Another problem occurring with branched structures is the possibility to form intramolecular hydrogen bonds between different parts of the model. These hydrogen bonds are completely unphysical since they can only be formed due to the presence of the terminating hydrogen atoms, which are not part of the real zeolite. Even clusters as small as di- or trisilicic acid (two and three SiO_4 tetrahedra, respectively) will form intramolecular hydrogen bonds, which force the whole cluster into a ring-like shape (cf. Figure 3.4, [147–149]). The formation of these hydrogen bonds can be prevented by using symmetry constraints on the model. Of course, symmetry constraints might introduce other problems, which are not present in the modeled system. Therefore care has to be given to balance model and reality and models which contain rings are more appropriate here.

Table 3.1 lists the calculated structure of the bridging hydroxyl group as a function of model size. This table shows that the smallest possible model, $H_3Si(OH)AlH_3$, results in an AlO bond which is much longer than that obtained for larger models. Similarly the SiO bond is predicted too short. Only the OH bond is nearly independent of the model size since it can relax to its optimum bond length without much constraint from the rest of the system.

The results in Table 3.1 show that the local geometry of the bridging hydroxyl group undergoes large changes depending on the size of the model. The Al–O and Si–O bonds in all models are significantly longer than usual. The O–H bond is also elongated compared to a terminal OH group. Further studies, especially with larger models (cf. Table 3.1) and periodic calculations (cf. p. 62), confirmed this observation. The Al–O bond in a bridging hydroxyl group is especially sensitive to the environment and can undergo large changes. The SiOH angle is nearly

Table 3.1: Calculated bond lengths and bond angles for bridging hydroxyl groups

Model	Ref.	r(AlO)	r(SiO)	r(OH)	∠(SiOH)	∠(SiOAl)
6 − 31G* basis set						
H₃Si(OH)AlH₃	[137]	203	171	95.1	117.4	131
H₃Si(OH)Al(OH)₃ᵃ	[142]	196	170	95.6	120.4	137
H₃Si(OH)Al(OH)₃ᵇ	[142]	195	169	95.6	121.4	134
(HO)₃Si(OH)Al(OH)₃ᶜ	[141]	199	167	96.0		139
DZP/TZP Basis Set						
(HO)₃Si(OH)Al(OH)₃	[143]	198	166	95.3	114.8	133
(HO)₃Al(OH)Si(OH)₂OSi(OH)₃	[143]	192	168	95.5	117.7	137
(HO)₃Si(OH)Al(OH)₂OSi(OH)₃	[143]	194	167	95.3	115.0	133
(HO)₃Si(OH)Al(OH)₂[OSi(OH)₂]₂OH	[143]	199	166	95.5	114.7	141
(HO)₃Si(OH)Al(OH)[Si(OH)₃]₂	[143]	196	167	(96.4)ᵈ	112.9	144
Four-membered ring with 1 Al	[143]	195	168	95.5	118.3	135
Four-membered ring with 2 Al	[143]	190	172	95.4	111.6	140
Six-membered ring with 1 Al	[143]	194	169	95.5	109.8	146
Six-membered ring with 3 Al	[143]	190	171	95.6	109.5	145
Double four-membered ring with 2 Al	[143]	193	169	95.5	115.8	134
Double four-membered ring with 4 Al	[143]	188	173	95.3	115.4	134
					114.9	
Double six-membered ring with 2 Al	[143]	193	169	95.6	116.0	136

ᵃ eclipsed, ᵇ staggered, ᶜ effective core potential and valence part of the 6 − 31G* basis set, ᵈ bond affected by an intramolecular hydrogen bond

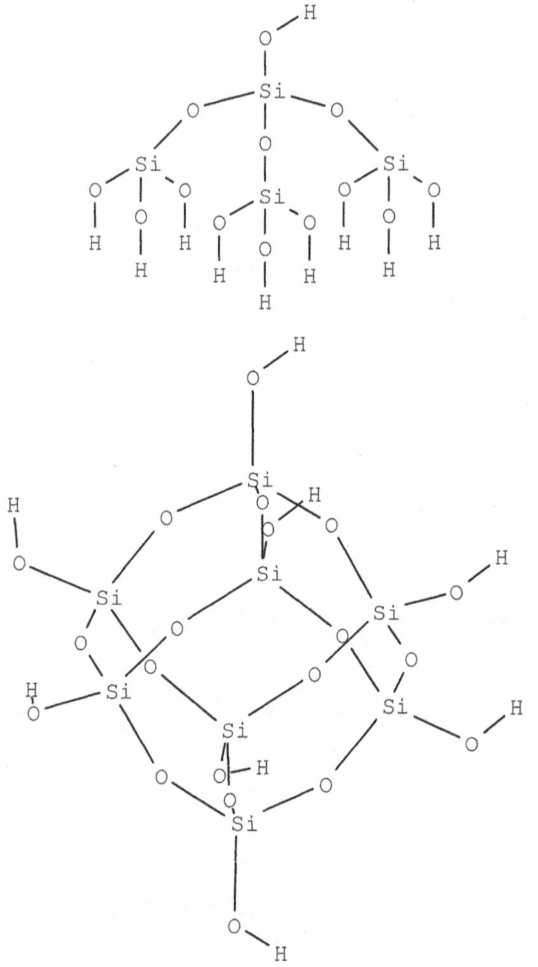

Figure 3.3: Adding more SiO$_4$ tetrahedra to a cluster can be done forming branched structures or rings and cages.

constant while the other two angles in the bridging hydroxyl group can also show large changes. Because of this flexibility the assumption of a bridging hydroxyl group is barely valid. The catalytic active site is rather an internal silanol group with a threefold coordinated aluminum atom in its vicinity.

But even larger models do not account for the effect of the crystal field, e. g.,

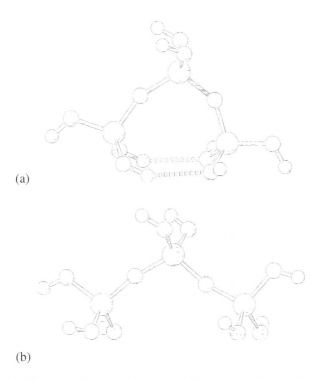

(a)

(b)

Figure 3.4: Cluster models may form unrealistic intramolecular hydrogen bonds as shown here with trisilicic acid. (a) the structure obtained with no constraints, (b) the structure obtained after imposing C_s symmetry and constraining two OSiOH torsion angles.

on a catalytically active center in the zeolite. One possibility to account for the crystal field is to embed the quantum mechanically treated cluster in its crystal environment. The crystal environment is then described by a force field to make the whole system computationally manageable. The force field used for embedding has to be carefully derived to reproduce the results of the actual quantum mechanical calculation. We therefore need a "Hartree–Fock" or a "density functional" force field to combine the corresponding quantum mechanical methods with a force field. These force fields can, of course, not be derived the usual way by fitting parameters to experimental data since this would result in an "experimental" force field. Fortunately, the quantum mechanical calculations on clusters can help here.

Cluster models have also been used to study the behavior of metal cations [150–156] and extra-framework aluminum species [157] in zeolites.

3.2.2   Periodic Calculations

In recent years it has become possible to perform periodic electronic structure calculations even for zeolites with their large unit cells. A number of studies have meanwhile been published dealing with structural aspects [158–168]. So far most of these studies have employed density functional theory.

Hill et al. have used periodic density functional calculations to determine the structure and vibrational properties of bridging hydroxyl groups in faujasite [158]. They obtain excellent agreement with observed OH stretching frequencies if anharmonic contributions are included. Comparisons are also made with calculations on cluster models and large differences are found for certain bond lengths and charges compared to the periodic calculations.

Periodic electronic structure calculations can also be helpful in structure determinations of zeolites. The structure of zeolite Na-MAP could not easily be resolved since the positions of the sodium cations were not clear. A periodic density functional calculation could provide a suitable starting model for the Rietveld refinement [159].

Demuth and co-workers have used density functional theory with plane waves to determine the structural effect of substituting silicon with aluminum in mordenite [160]. They have also studied the structure and vibrational properties of Brønsted acid sites in mordenite and of silanol groups [163] and various other defects [165] on the mordenite (001) surface. Substituting silicon with aluminum leads to large but localized distortions in the framework. The AlO bond is significantly longer than the SiO bond, but the lattice is able to accommodate these longer bonds by compressing SiO bonds in the vicinity. A quantum mechanical analysis of the OH vibration shows that it is sufficient to treat such vibrations classically. Bridging hydroxyl groups are unlikely to be stable on surfaces.

Catti, Civalleri, and Ugliengo have used periodic electronic structure calculations to predict the structures of α-quartz, α-cristobalite, α-tridymite, and coesite [161]. They have used density functional as well as Hartree–Fock calculations and are able to reproduce experimental structures very accurately (deviation of less than 1 % for the unit cell parameters in general). Hartree–Fock and LDA calculations give more accurate structures than the gradient-corrected B3LYP functional employed, but the relative stability of the polymorphs is predicted better by B3LYP. Defects in sodalite and chabazite have been studied using periodic Hartree–Fock and B3LYP calculations [164]. Significant differences between the results for Hartree–Fock and B3LYP are found here as well.

The location of Zn(II) cations in chabazite has been studied by Barbosa et al. using periodic density functional calculations with plane waves [162, 166].

Framework distortions play a large role in stabilizing the cation. Larger rings are preferred over smaller rings. The location of water molecules in the zeolites LiABW, NaNAT, and BaEDI was investigated by Larin and co-workers using periodic Hartree–Fock calculations [167]. Chains of water molecules present in the channels of the zeolites were found to have properties similar to ice or liquid water. The adsorption of water in Ti-chabazite was studied by Damin et al. [168]. They have also looked at the energetics of adsorption and the structure of the adsorption complexes of NH_3, H_2CO, and CH_3CN.

The isomerization reactions of toluene and xylene isomers in mordenite were investigated by Rozanska and co-workers using periodic plane wave density functional calculations [169, 170]. Transition states are found to be stabilized by zeolite framework oxygen atoms. Cluster and periodic calculations give in essence the same reaction mechanisms. The same authors have also studied the chemisorption of propene in chabazite [171] and of isobutene in chabazite, ZSM-22, and mordenite [172]. They find that the formation of an alkoxide is energetically favorable in the sequence tertiary > secondary > primary, although the shape of the zeolite's micropore plays an essential role. Relaxation of the zeolite framework is essential for obtaining correct energetics for the reaction path. The activation energy for the reaction is significantly lowered compared to a cluster calculation.

Recently, Fois et al. have even been able to perform an ab initio molecular dynamics simulation of water in zeolite Li-ABW [173]. They can show that one-dimensional chains of water are stable in this zeolite. The adsorption of water in mordenite has also been investigated by Demuth and co-workers using a plane wave density functional approach [174]. They were also able to perform a molecular dynamics simulation. The existence of a short-lived hydroxonium cation could be observed.

3.2.3 Combining Ab Initio Methods and Molecular Modeling

The derivation of parameters for potential functions based on available experimental data is problematic. On one hand there are, in general, only very few experimental data available which have to be used to determine a larger number of parameters. The available data are on the other hand often not directly computable by means of a potential so that detours have to be used to get parameters. If the potential is to be used in embedding calculations a derivation based on experimental data is generally not desirable as explained above.

A different, and in recent years more often used method, consists in the use of data from ab initio calculations to derive parameters [175]. Ab initio calculations offer the capability to provide a lot of data ("observables")[1] so that the derivation of parameters for mathematically sophisticated potentials becomes possible.

---

[1]"Observables" as used here does not mean observables in a strict quantum mechanical sense, but

Mathematically sophisticated potentials usually contain a large number of parameters and to get statistically meaningful fits the number of "observables" has to be even larger. The "observables" obtained in ab initio calculations (e. g., structures, forces, and force constants) can also be directly calculated with the potential, which simplifies the parameter derivation process.

The use of ab initio data for the derivation of potential parameters requires that ab initio calculations are feasible for the systems under study. In the modeling of solids that is in general not the case as outlined above. Because of this reason ab initio calculations for molecular models are the only possibility to get data for solids. The molecular models have to be selected in a suitable way. The procedure to derive potential parameters based on ab initio calculations can be summarized as follows:

- Selection of molecular models which are suitable for ab initio calculations

- Checking the suitability of these models with respect to a correct description of properties of the solid

- Use of ab initio data for these models to derive potential parameters and checking of the potentials using the models

- Checking the transferability of the parameters by calculations on larger, more complex models, which are still accessible to ab initio calculations using the newly derived parameters and finally

- Checking the parameters by calculations on solid structures and comparison of results with experimental data.

There are a few attempts to derive potential parameters based on ab initio calculations in the literature (cf. Tab. 3.2), but only a few of these attempts have been done in the way outlined above. The transferability and accuracy of the potentials which were not derived as outlined above is generally not very good.

The potentials for microporous solids available in the literature can be classified as ion pair potentials (cf. p. 210) and molecular mechanics potentials (cf. p. 218). Ion pair potentials use simple ion–ion interactions in the form of a Coulomb and a Van der Waals term occasionally supplemented by a harmonic core–shell term. In the latter case the ion's charge is distributed over two "particles" (the core and the shell) which are linked by a harmonic spring. Since core and shell can be at different positions in space this shell model potential allows the ion to be polarized. Shell model potentials have been introduced by Dick and Overhauser [16] in 1958 and have been proven valuable in the study of ionic crystals. Molecular mechanics potentials do not consider ion–ion interactions, but follow a covalent approach. The molecule is described by using harmonic or anharmonic

---

any quantity which can be calculated in a quantum mechanical calculation.

Table 3.2: Potentials derived based on ab initio calculations (SBU – secondary building unit)

Potential type	derived from	applicable to	Ref.
Ion pair potentials (partial charges)	SiO_4^{4-}	SiO_2 polymorphs	[176, 177] [178]
	H_4TO_4 (T=Al, Si, P)	SiO_2 polymorphs, AlPO's	[179]
	H_4TO_4 $(HO)_3Si(OH)Al(OH)_3$	alumosilicates with bridging hydroxyl groups	[141]
Ion pair potentials (formal charges)	H_4SiO_4 H_8SiO_6	SiO_2 polymorphs, stishovite	[180]
Shell model potential	numerous silicic acids and SBU's (also Al containing)	alumosilicates with bridging hydroxyl groups	[181]
	numerous silicic acids and SBU's (also Al containing)	SiO_2 polymorphs and alumosilicates	[182]
Molecular mechanics potentials	$(HO)_3SiOSi(OH)_3$ $H_6Si_2O_6$	SiO_2 polymorphs, stishovite	[180]
	numerous silicic acids and SBU's (also Al containing)	SiO_2 polymorphs and alumosilicates	[143, 144]

functions for bond stretching and angle bending and usually some kind of Fourier expansion for torsions. A Coulomb and a Van der Waals are normally also added to provide terms for intermolecular interactions. A number of molecular mechanics force fields makes also extensive use of coupling constants.

Tsuneyuki et al. tried to derive an ion pair potential from ab initio data for the SiO_4^{4-} ion embedded in point charges [176, 177]. They obtain a potential which can predict lattice constants and compressibility of α-quartz, α-cristobalite, coesite, and stishovite quite well, but Si–O distances and Si–O–Si angles are reproduced rather badly. However, the potential can simulate phase transitions at high pressures where the coordination of the silicon atoms increases from four to six. Della Valle et al. further examined Tsuneyuki's potential [183]. They find that calculated vibrational spectra show systematic derivations. The Si–O force constant is too small, the O$\cdots$O and Si$\cdots$Si interactions too strong. Nevertheless, this simple potential reproduces the spectra of different SiO_2 polymorphs quite well.

Van Beest et al. attempted to derive an ion pair potential from ab initio data for the H_4TO_4 models (T=Al, Si, P). However, the potential obtained does not yield reasonable results in lattice energy minimizations. The result of the minimization of α-quartz, e. g., has the symmetry of β-quartz. Only after the inclusion of experimental data of α-quartz Van Beest et al. are able to reproduce lattice and elasticity constants for different SiO_2 polymorphs very well [179, 184]. By applying an analogous procedure Van Beest et al. were also able to derive potential parameters for aluminum phosphates [179]. The calculated IR and Raman spectra also show rather large deviations from the observed data here [185]. An improvement is obtained by de Man et al. only by calculating the vibrational frequencies using a different, generalized valence force field at the geometry obtained with their potential. However, an agreement with the observed spectrum can only be obtained for parts of the spectrum [186].

Tse et al. [187] have compared the potentials of Tsuneyuki and Van Beest. They conclude that Van Beest's potential is slightly better than Tsuneyuki's. It reproduces the stability of the structure and phase transitions, it is excellent in the prediction of elasticity constants and in the equation of state for α-quartz and it results in vibrational spectra which are closer to experiment. Both potentials are in general less satisfactory for the computation of vibrational spectra.

Schröder and Sauer [181] and Sierka and Sauer [182] derived an ion pair potential following the systematic procedure explained above. Schröder's potential is based on Hartree–Fock calculations for a rather large number of silicic acids and structural units of zeolites. The same ab initio data have also been used to fit a molecular mechanics potential (vide infra). Sierka's potential uses the same structural units as Schröder's, but employs density functional calculations as the quantum chemical method. Both potentials reproduce very accurately the structure of a large number of silica polymorphs and also predict the vibrational spectra very well.

While density functional geometry optimizations for zeolites are now routinely being performed direct geometry optimizations of zeolites with Hartree–Fock calculations are still the exception. It is therefore not possible to compare the performance of both methods directly. However, the availability of potentials derived in the same way from Hartree–Fock and density functional calculations allows for the comparison of the structural predictions of both methods since the force field based on a certain method will carry the same information as the method itself. The Hartree–Fock based potential generally has difficulties to predict the symmetry of different SiO_2 polymorphs correctly. α-quartz, e. g., is predicted to have the symmetry of β-quartz. Silica sodalite is predicted by the

Hartree–Fock based potential to have space group symmetry $I\overline{M3M}$ with the four oxygen atoms of the four-membered rings all being in one plane. In experiment only template filled sodalite possesses this space group. If the template is removed the sodalite structure "collapses" to lower symmetry. Silicalite (all-silica ZSM–5) possess two modifications, a low-temperature monoclinic ($P2_1/n11$) and a high-temperature orthorhombic one (Pnma). The Hartree–Fock based potential predicts only the orthorhombic modification to be stable. The density functional based potential, on the other hand, predicts sodalite to have space group $I\overline{43M}$ and "collapsed" four-membered rings. It also predicts the monoclinic modification of silicalite to be more stable than the orthorhombic one.

Hartree–Fock calculations are therefore probably not suited to predict the structure of silica polymorphs correctly. Electron correlation effects are important. As mentioned above Van Beest et al. were not able to correctly obtain the structure of α-quartz with their Hartree–Fock based potential. Only after scaling with experimental data the potential is able to reproduce the structure of α-quartz. The experimental data contain, of course, electron correlation effects. Therefore it seems more straightforward to use density functional calculations in the derivation of force fields instead of Hartree–Fock calculations with subsequent scaling if a force field which reproduces experimental data is the target.

The simulation of aluminum containing zeolites can be quite difficult using ion pair potentials since the protons, which are necessary for charge equilibration, form bridging hydroxyl groups, which are mainly covalent. Kramer et al. therefore tried – to use their potential – to replace the oxygen atom of the bridging hydroxyl group with an effective atom whose parameters were determined from ab initio calculations on $(OH)_3Si(OH)Al(OH)_3$. However, to describe the Si(OH)Al angle correctly they had to use a three-body term [141].

A few studies deal with the development of a molecular mechanics potential. Mabilia et al. use a molecular mechanics potential with parameters [188] determined by Grigoras and Lane through ab initio calculations on small molecules [189] in their study of the stability of the sodalite cage as a function of the silicon/aluminum ratio. But these parameters were derived for silicon-containing organic compounds. Correspondingly, Mabilia et al. obtain results contradictory to experiment (increasing stability of the sodalite cage with increasing aluminum content).

Lasaga and Gibbs compare the performance of a molecular mechanics potential with an ion pair potential by molecular mechanics calculations for quartz and cristobalite. They find that the molecular mechanics potential describes the structure better, but cannot be used for high pressures [180].

Nicholas et al. achieve an excellent modeling of structure and vibrational spectra of silicates with their potential. They use harmonic terms for bond stretching and angle deformations as well as a quartic term for the Si–O–Si angle deformation. Furthermore, their potential has a coupling term for the coupling of the

Si–O bond with the Si–O–Si angle. Some of the parameters in their potential have been transferred from other potentials and have been derived from experimental results, while the remaining parameters are based on ab initio calculations. They use full Mulliken charges, but differing dielectric constants. The agreement with experimental results becomes better with increasing dielectric constants; the use of the Ewald summation has, according to their opinion, only a small effect on the structure [190].

Hill and Sauer [143, 144] have derived a molecular mechanics potential for both silica polymorphs and aluminum containing zeolites based on the quantum chemical approach outlined previously. This potential uses the functional form of the Consistent Force Field (CFF) [191, 192] (cf. p. 220), which includes not only terms for bond stretching, angle bending, and torsions, but also a large number of coupling terms. It has been shown that coupling terms are important for the reproduction of vibrational spectra in organic molecules [175, 193]. However, the CFF for zeolites does not provide reasonable agreement of vibrational spectra with experiment [181]. This is most likely not a problem of the functional form of the potential, but the problem of an insufficient number of force constants included in the parameter fit. Unit cells and internal coordinates of a large number of zeolites are, however, reproduced well. This force field has also been used to study processes during zeolite synthesis [194] and the role of organic templates in the synthesis. These studies could be carried out mainly due to the functional form of this force field and the availability of parameters for organic molecules, which serve as templates. Hartree–Fock calculations have also been used to derive force field parameters for the interaction of water with a zeolite [195].

Advantages and disadvantages of ion pair and molecular mechanics potentials as found in the literature are compared in Chapter 7 on page 224. In addition to what is shown there ion pair and shell model potentials are inconsistent in describing bridging hydroxyl groups while molecular mechanics potentials are not.

3.2.4   Embedding

Mechanical embedding as described in Chapter 7 has been used to predict the structure of disilicic acid as a function of different crystallographic environments. Table 3.3 shows results for disilicic acid embedded in four different positions in faujasite [196].

These data clearly show that embedding is able to predict differences in the geometry depending on the position at which an identical cluster is embedded in the framework. The embedded two-tetrahedra cluster would result in only one SiO bond length and one SiOSi bond angle if it would be calculated in the gas phase. Despite the simplicity of the mechanical embedding used to obtain the results shown, the agreement with observed bond lengths and angles is remark-

Table 3.3: Faujasite structures obtained for disilicic acid embedded using a shell model potential; r is the SiO bond length in pm and θ is the SiOSi bond angle in degrees

	O1		O2		O3		O4	
	r	θ	r	θ	r	θ	r	θ
Obsd.[a]	162.9	140.1	159.7	149.3	160.4	145.8	161.4	141.4
Embd.[b]	162.2	137.2	160.8	155.3	162.2	149.4	161.2	142.8
Force Field[c]	162.5	140.5	160.9	154.2	161.8	151.7	161.5	144.7

[a] Ref. [197], [b] Ref. [196], [c] Ref. [181]

able. Further applications of this embedding scheme have looked at absolute and site specific acidities in zeolites [198], have studied the faujasite/NH_3 [199] and Cu^+/NO_2/zeolite interactions [200], investigated the influence of the framework structure on acidity differences of zeolites [201], studied the location of copper ions [202, 203] and proton mobility [204] in zeolites, looked at the adsorption of n-heptane in H–ZSM-5 [205], and investigated the role of carbenium ions in hydrocarbon conversion reactions [206].

Other embedding schemes, in particular such that include electronic effects, have also been used in zeolite simulations. The reactivity of Ti(IV) centers in zeolite ZSM-5 has been studied employing the ONIOM method [207]. Large differences in the binding energies for water and ammonia compared to cluster calculations are observed indicating the importance of the inclusion of the framework in such simulations.

An embedding scheme which accounts for electronic effects of the quantum part on the molecular mechanical part was used to predict the isosteric heats of adsorption for N_2 and O_2 [208].

3.2.5 Reaction Energetics

The ability to calculate structures is one important issue, but chemical reactions are not only structure dependent. To study reactivity knowledge of the energies to break or form bonds is also required. Since reactions in nearly all cases involve changes to the electronic structure of a solid or molecule, calculation of reaction energies is generally only possible using quantum mechanical methods.

The reactivity of zeolitic catalysts is determined by two factors. First, the ze-

Table 3.4: Geometries of methanol and silanol obtained with standard and highly accurate methods compared to observed data

	SCF/6-31G*	SCF/DZP	CPF/TZ2d1f	Obsd.
	Methanol			
r(OH)/[pm]	94.6	94.6	95.9	96.3
r(CO)/[pm]	140.0	139.8	141.9	142.1
∠COH/[deg]	109.4	109.6	108.1	108.0
	Silanol			
r(OH)/[pm]	94.6	94.6	95.8	
r(SiO)/[pm]	165.4	164.6	165.0	
∠SiOH/[deg]	118.1	120.4	117.7	

olite framework must allow molecules to reach the catalytically active site. This factor can be determined by structural considerations of, e. g., the size of the molecules and the diameter of the zeolite channels. The second factor is the intrinsic reactivity of the active site itself. In a Brønsted acid zeolite this can, e. g., be the deprotonation energy or intrinsic acidity [209] of a bridging hydroxyl group. The intrinsic reactivity of an active site is itself dependent on the crystal environment and therefore the crystal structure. We have to determine the crystal structure first and then study how the active sites will react. This problem can become rather complex quickly. The adsorption of molecules will, of course, have an effect on the zeolite framework itself. Fortunately, in cases of zeolites this effect is usually small since the cages and channels in a zeolite are large enough to accommodate at least small molecules rather easily.

Highly Accurate Calculations

To establish a baseline for the calculation of deprotonation energies in zeolites highly accurate calculations have been carried out [210]. Ab initio calculations can be systematically improved by increasing the size of the basis set (the Hartree–Fock limit is reached with an infinite basis) and by including electron correlation. Electron correlation can also be systematically improved by including more and more excitations until at the end a full configuration interaction (CI) calculation is performed. Unfortunately, full CI calculations require so much computer power, that they are currently impossible for all but the smallest molecules.

The highly accurate calculations performed to study deprotonation energies

Table 3.5: Deprotonation energies of methanol and silanol calculated with different theoretical methods; δ is the mean error of a method defined as $\delta = E_{CPF} - E_{Method}$

Method	Methanol	Silanol	δ
CPF/TZ2d1f	1644	1531	0
SCF/DZP	1691	1576	46±2
SCF/6-31(+)G*(*)	1678	1561	32±2
SCF/3-21G	1768	1637	115±10

used methanol and silanol as the smallest possible models. Silanol can serve as a model for a hydroxyl group in a zeolite while methanol is the silanol analogue which can be studied experimentally. Thus performing the same calculations on both molecules can provide trust in the method (by comparison with experiment) and provide the data desired as well as an estimate of their accuracy. Sauer and Ahlrichs [210] used the Coupled Pair Functional and a triple-ζ basis set with 2 d and 1 f function. They did a full geometry optimization on methanol, silanol, and the corresponding anions. Table 3.4 compares the geometries obtained from these accurate calculations with those from methods more commonly applied.

From these data it can be concluded that SCF calculations using a DZP basis set yield bond lengths and angles which deviate typically by less than 2 pm and 3°, respectively, from the accurate result. Bond lengths are usually predicted too short and bond angles too wide.

Table 3.5 shows deprotonation energies calculated for methanol and silanol using different theoretical methods [211]. The error in the deprotonation energy caused by a certain theoretical approximation is fairly constant. Therefore it is possible to use a less accurate method and correct for the method's intrinsic errors by applying a correction derived from highly accurate calculations. Using this method energies calculated with different methods can be made comparable [211].

Deprotonation Energies of Zeolites

The first studies of zeolite reactivity consisted of the calculation of deprotonation energies of terminal and bridging hydroxyl groups. While these deprotonation energies very likely depend on the crystallographic environment of the bridging hydroxyl group this could not be accounted for in early calculations simply because of the size of the system. Rather, cluster models as used for the study of the structure of bridging hydroxyl groups were applied. The calculation of the depro-

Table 3.6: Structures of different models before and after deprotonation

Model/Basis set	Ref.	r(SiO)	r(AlO)	∠(SiOAl)
$H_3Si(OH)AlH_3/3$-21G	[212]	173.4	192.7	130.7
$H_3SiO^-AlH_3/3$-21G	[212]	161.7	175.6	180.0
H_3SiOH/extended	[210]	163.2		
$H3SiO^-$/extended	[210]	153.9		
$(HO)_3Si(OH)Al(OH)_3$/STO-3G	[213]	167.0	179.5	133
$(HO)_3SiO^-Al(OH)_3$/STO-3G	[213]	159.0	170.0	130
double four-membered ring with 1 Al/6-31G*	[214]	171.9	194.6	131.1
double four-membered ring anion with 1 Al/6-31G*	[214]	160.0	178.5	138.4

tonation energy is then pretty straightforward. The geometry of the model cluster with the proton attached is optimized, the proton is removed, and the geometry of the resulting anion is optimized again. The deprotonation energy is then just the difference of the energy of the anion and of the parent system. While this seems to be simple, there are a few pitfalls to avoid along the way. First, it is essential to optimize the geometries of both the anion and the parent system. The rearrangement of the atoms after the removal of the proton is significant (cf. Table 3.6) and no reliable energies can be obtained unless the geometry is optimized.

The second problem lies in the description of the electron distribution for both systems. As shown in Chapter 7 most quantum mechanical calculations build the molecular orbitals by linear combination of basis functions. Comparable energies can only be obtained if the basis sets are the same for all molecules in the calculation. However, if a proton is removed this is no longer the case. Basis functions are usually atom-centered; thus removal of the proton also removes the basis functions belonging to it. To calculate the deprotonation energy from the two molecules as described above will therefore introduce an error, which is called the basis set superposition error (BSSE).

The BSSE occurs in all calculations where interaction energies between different molecules are calculated. In our case we want to study the interaction energy of an anion and a proton, but to do the calculation for the combined system the basis set of the proton is superimposed on the basis set of the anion (and vice versa). The electrons from the anion now have more basis functions available and the

Table 3.7: Deprotonation energies of bridging hydroxyl groups in dependence of cluster size for clusters in vacuum and embedded in a faujasite lattice, kJ/mol. The energies have been corrected for basis set effects.

Model	SV basis	DZP basis
Cluster in vacuum, [211]		
$H_3Si(OH)AlH_3$	1299	1297
$H_3Si(OH)Al(OH)_3$		1321
$H_3Si(OH)Al(OH)_2OSiH_3$	1275	1273
$H_3Si(OH)Al(OSiH_3)_2OH$		1254
$(HO)_3Si(OH)Al(OH)_3$	1323	1316
$(HO)_3Si(OH)Al(OH)_2OSiH_3$		1231
$(HO)_3Si(OH)Al(OH)_2OSi(OH)_3$		1274
double four-membered ring with 1 Al	1260	
Cluster embedded in faujasite, [198]		
$(HO)_3Si(OH)Al(OH)_3$, O(1)H		1296
$(HO)_3Si(OH)Al(OH)_2OSi(OH)_3$, O(1)H		1298
three-shell model		1299
four-shell model		1299
five-shell model		1297

energy will be lowered artificially (as a consequence of the variational principle used in most quantum mechanical calculations). The amount of the lowering of the energy depends much on the size of the basis set used for the calculations. A sufficiently large basis set will have enough flexibility to accommodate the electrons and the lowering of the energy (the BSSE) will therefore be small. A small basis set will not have described the electron distribution well and the energy lowering will therefore be large. The BSSE can therefore also serve as a measure of the quality of a basis set. Calculations trying to obtain reliable deprotonation energies should use at least basis sets of double-ζ (DZ) quality. As Table 3.5 shows the use of polarization functions can reduce the error in the deprotonation energy by a factor of three. Polarization functions are therefore essential in such calculations.

It is possible to calculate a BSSE correction. In our example of the calculation of the deprotonation energy of a bridging hydroxyl group one would just remove the proton nucleus, but keep the basis functions in place for the calculation of the energy of the anion (this is often referred to as a *ghost orbital calculation*).

The difference of the energies of the anion calculated in the "normal" basis and with the "ghost functions" is then the BSSE. The true deprotonation energy is the energy calculated in the "normal" basis plus the BSSE. This scheme is called the *counterpoise correction* [215] and can be used to correct for the BSSE in any calculation of interaction energies (a more detailed explanation of the counterpoise correction for the BSSE can be found in Chapter 7 on page 198).

Another problem which might arise with the more wide spread use of density functional theory is that the density functionals commonly used have a fundamental problem with anions. There is a long list of publications which tries to access the reliability of density functionals for anions, e. g., [216–223]. The problem consists in the wrong long range behavior of the potential of an anion when calculated with these density functionals. It goes to $(N - Z)/r$ instead of to the electrostatically correct $(N - Z - 1)/r$ (Z number of protons, N number of electrons). This is caused by self-interactions of the electrons, which are not canceled exactly as in Hartree–Fock calculations.

The deprotonation energy for a bridging hydroxyl group depends strongly on the size of the cluster used in the calculation. Table 3.7 lists results obtained for clusters of increasing size [211]. These deprotonation energies have been corrected for the difference in the method used to calculate them according to the scheme shown in Table 3.5.

The deprotonation energy of the bridging hydroxyl group is obviously not yet converged even for the largest model studied so far. Therefore deprotonation energies have also been calculated using an embedding approach [198]. Eichler et al. embedded protonated and deprotonated di-tetrahedra models in faujasite and studied the effect of the lattice on the deprotonation energies for different sites. One problem encountered in such a calculation is that a periodic charged system has to be dealt with – a charge is created in the unit cell and this unit cell is repeated infinitely. This results in a clearly unphysical solid with an infinite charge where the Coulomb energy would be divergent. To eliminate this problem a uniform background charge is applied to the system and the interaction between the charged defects in the different unit cells is accounted for by a macroscopic approximation [118, 198]. The deprotonation energy for the O(1)H site is given in Table 3.7.[2] For the other sites the deprotonation energy is slightly lower (1272 ... 1293 kJ/mol after correcting for basis set effects). The correction introduced to account for the interaction of the charged defects in the different unit cells is 54 ... 55 kJ/mol (or 5 %) and therefore is considerable. Increasing the size of the cluster embedded in faujasite has very little effect on the deprotonation energy. It is interesting to note that the deprotonation energy of a free cluster is not much different from the embedded result. However, the structure constraints introduced by the embedding of, e. g., the $(HO)_3Si(OH)Al(OH)_2OSi(OH)_3$

---

[2]Note that the energies in Table 3.7 have been corrected for basis set effects.

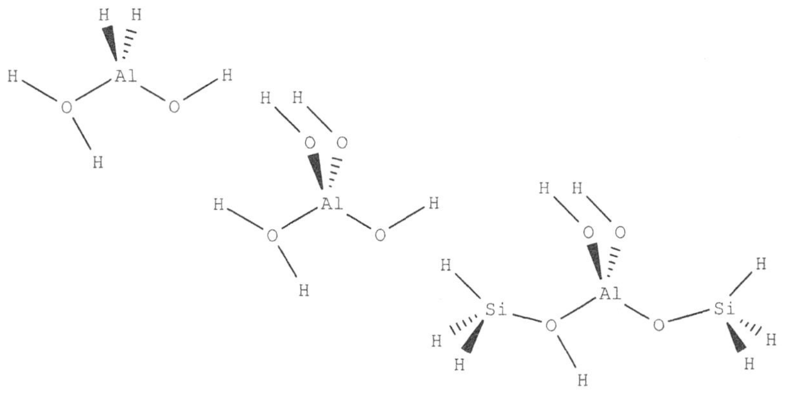

Figure 3.5: Models used to study intrazeolite proton jumps.

model, cause its quantum mechanically calculated deprotonation energy to increase to 1354 kJ/mol. The long-range corrections lower it to 1252 kJ/mol which after correcting for basis set effects becomes 1298 kJ/mol. That means that the free relaxation of the cluster in the vacuum case can make up for the long-range effects and may explain why calculations on free clusters have been successful in predicting deprotonation energies, but this might have been a fortunate cancellation of errors one should not rely on.

Proton Transfer Reactions

The initial step of a catalytic reaction in a zeolite is a proton transfer from the Brønsted acid site to the substrate. A proton transfer can also occur within the zeolite framework by moving a proton from one oxygen atom to a neighboring one. The proton has to overcome an energy barrier for this move, which can provide information about the strength of the bond the proton has with the zeolite framework. Ab initio studies have been performed for intrazeolite proton jumps as well as for the protonation of water [224], ammonia [225], methane [224, 226], ethane [224], propane [224], methanol [224, 227], ethanol [224], and benzene [224, 228]. Methanol is the most important system here since it is converted to gasoline in the industrial important methanol-to-gasoline process (for more detailed studies cf. p. 98).

Most of these calculations have been performed using a cluster approach. A cluster is selected which represents the active site and its immediate environment. The sorbate is placed at the appropriate position and the interaction between these two systems is studied. The simplest cluster model for an active site which has

Table 3.8: The energy barriers for a proton jump between neighboring oxygen atoms in zeolites calculated for different model sizes

Model	Symmetry of		Energy barrier [kJ/mol]
	initial structure	transition structure	
SCF/DZP			
$H_3Si(OH)Al(OH)_2OSiH_3$	C_s	C_{2v}	67.9
$H(OH)Al(OH)_3$	C_s	C_{2v}	73.6
$H(OH)AlH_2OH$	C_s	C_{2v}	68.9
CPF/DZP			
$H(OH)Al(OH)_3$	C_s	C_{2v}	49.3
$H(OH)AlH_2OH$	C_s	C_{2v}	43.7

been used to study intrazeolite proton jumps is a single tetrahedrally coordinated aluminum atom. This single tetrahedron model can be enhanced by adding additional tetrahedra containing silicon (cf. Figure 3.5). This way the model can be made larger step by step and the effects on the barrier for the proton jump can be studied.

Table 3.8 contains the energy barriers calculated for different model sizes. The smallest model gives an energy barrier which is pretty close to the largest model studied. The medium sized model results in a slightly higher barrier. This is true regardless of the method used to calculate the barrier height. The use of a correlated method results in a significant decrease of the barrier height.

Significantly higher barriers were obtained by Fermann et al. using density functional calculations with B3LYP for the structures and MP2 with a large basis set augmented with MP4 with a smaller basis set for the barrier height [229]. The calculated barriers are higher than experimental ones. The reason for the difference is that proton tunneling is not considered in the interpretation of experimental results [230]. The barrier height and also the barrier curvature can be correlated with the OAlO angle the proton crosses. The barrier height increases when this angle gets larger. At the same time the tunneling probability decreases with this angle [231].

Systematic density functional studies of the protonation of various compounds have shown a correlation between the deprotonation energy of a species and the activation energy of proton jumps [224]. Figure 3.6 shows such a correlation for H–ZSM-5 and various hydrocarbons. Compounds which can donate a proton more easily lower the activation barrier for proton jumps, which also results in a higher rate constant for proton movements.

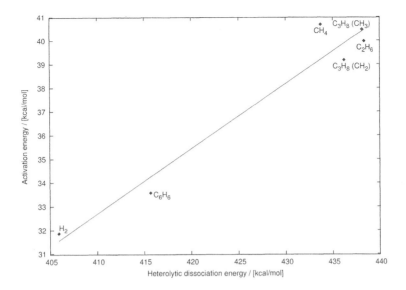

Figure 3.6: Deprotonation energy vs. activation energy of hydrogen exchange between H–ZSM-5 and RH.

3.2.6 Adsorption in Zeolites

Increasing computational power and methodological advances in the last few years, in particular the widespread adoption of density functional theory, have made it possible to study more realistic systems. Reactions of hydrocarbons in zeolites are of the greatest interest here since these are the reactions catalyzed by zeolites in oil refinement.

Zygmunt and co-workers have employed a cluster model to investigate the reaction mechanism of ethane cracking in H–ZSM-5 [232]. They have located stationary points along the reaction coordinate using a five tetrahedra model for the zeolite and the Hartree–Fock, B3LYP, and MP2 methods. The reaction occurs via protonation of the ethane molecule, which subsequently breaks into methane and a methoxy group bound to the zeolite framework. The barrier height can only be predicted in agreement with experiment if long-range electrostatic interactions are included in the model through embedding.

A more detailed study of the adsorption of linear hydrocarbons from C_1 to C_6 has been performed by Benco et al. [233]. They have used periodic calculations and found that the local density approximation overestimates the binding energy. The mechanisms of some zeolite catalyzed reactions of hydrocarbons involving the formation and rearrangement of carbonium ion intermediates were studied by

Boronat and co-workers using MP2 and density functional calculations [234,235]. The same authors have studied the protonation of ethene, propene, 1-butene, and iso-butene by the zeolite theta-1 using cluster models as well as a single point periodic calculation [236]. Both the Hartree–Fock and B3PW91 density functional approximations have been used. The stability of the alkoxide complexes formed is very sensitive to the local geometry of the active site and the nature of the carbon atom forming the bond with the zeolite framework. By comparison between cluster and periodic calculations it could be concluded that the cluster models should not be fully relaxed if realistic energies are to be calculated. Long-range interactions are less important than inclusion of electron correlation and the degree of geometry relaxation. Similar studies have been performed by Correa and Mota [237] and Limtrakul et al. [238].

The dehydrogenation of iso-butane has been studied by Milas and Nascimento using a cluster approach and the B3LYP hybrid density functional [239]. They conclude that there is no intermediate alkoxide as seen in a number of other studies, but a direct collapse of the carbocation into iso-butene and a proton restoring the zeolite's hydroxyl group.

Zaragoza et al. have studied the protonation of benzene in H–ZSM-5 as the first step in cracking [228]. They have used the BLYP density functional and can predict the formation of a carbocation. Benzene adsorption in mordenite has been investigated by Demuth and co-workers [240]. They have used both the local density approximation and a gradient-corrected functional and found that the adsorption strength correlates with the degree of structural distortion of the acid site of the zeolite. The methylation of methylbenzenes in zeolites was studied using a cluster approach and the B3LYP density functional [241]. The activation energy is found to decrease with an increasing number of methyl groups on the benzene.

The adsorption of methanol in chabazite was studied using a cluster model and gradient-corrected density functionals [242]. With the B3LYP and PW91 functionals used in this study the chemisorbed complex of methanol and the zeolite is found to be a minimum on the potential energy surface in accordance with periodic calculations.

The catalytic activity of zeolites with respect to the decomposition of nitrous oxide has been investigated using a cluster model. The activation energy and the preexponential factor for the decomposition of N_2O to N_2 and O_2 on Fe and Co containing ZSM-5 were calculated using the B3LYP density functional. Good agreement with experiment was obtained for the activation energy [243].

Density functional calculations have also been used to calculate Henry constants, separation constants, and isosteric heats of adsorption for N_2, O_2, argon, and CO in zeolite NaY [244]. Such calculations are still demanding due to the required sampling of the configuration space, but are becoming more and more feasible. However, as shown in ref. [245], isosteric heats of adsorption were found to be systematically too small.

Stability of Secondary Building Units

Zeolites can be thought of being composed of secondary building units (SBUs). Secondary building units, such as sodalite cages or double six-membered rings, are also natural choices as cluster models. A number of Hartree–Fock calculations have therefore been performed on such secondary building units as well as on single rings and the energies obtained can be used to predict the stability of the corresponding units [143, 144, 246]. Higher stability of certain secondary building units can be related to a higher probability of finding them in actual zeolite frameworks. Single rings in an all-silica framework can formally be formed by condensing orthosilicic acid

$$m \; H_4SiO_4 \rightarrow [SiO(OH)_2]_m + m \; H_2O \qquad (3.1)$$

while secondary building units can be formed by

$$m \; H_4SiO_4 \rightarrow [SiO_{3/2}(OH)]_m + 3m/2 \; H_2O \qquad (3.2)$$

For aluminum containing zeolites equivalent reactions can be written for single rings

$$m \; H_4SiO_4 + n \; HAl(OH)_4 \rightarrow [SiO(OH)_2]_m[Al(OH)_3]_n + (m + n) \; H_2O \qquad (3.3)$$

as well as for secondary building units

$$m \; H_4SiO_4 \;\; + \;\; n \; HAl(OH)_4 \rightarrow \qquad\qquad\qquad (3.4)$$
$$[SiO_{3/2}(OH)]_m[AlO_{3/2}(OH)H]_n + 3(m + n)/2 \; H_2O$$

In these reactions $HAl(OH)_4$ is a complex of water and $Al(OH)_3$. All these reactions belong to the class of isodesmic reactions (reactions in which the character of electron pairs is not changed) for which the Hartree–Fock approximation was shown to yield reliable results [247].

Table 3.9 shows the relative stabilities of a number of single rings and secondary building units. It is obvious that small rings are energetically less favorable than five- and six-membered rings. Ring size analyses of framework topologies confirm that these rings can indeed be found more often in zeolites [248]. For all-silica zeolites sodalite cages and double six-membered rings should be the preferred building units. For aluminum containing systems smaller rings with the maximum number of aluminum atoms should be favored. Large rings with the maximum number of aluminum atoms become less preferable since a higher number of rather long AlO bonds (cf. Table 3.1) has to be fitted into the ring.

Table 3.9: Relative stability of different single rings and secondary building units (SBUs) of zeolites per mole TO bond (T=Al, Si) formed, in kJ/mol

SBU	Ref.	m	n	energy
all-silica single rings				
three-membered ring	[144]	3		−4.6
four-membered ring	[144]	4		−15.1
five-membered ring	[144]	5		−17.0
six-membered ring	[144]	6		−17.1
all-silica SBUs				
Sodalite cage	[144]	24		−22.9
double six-membered ring	[144]	12		−22.9
double four-membered ring	[144]	8		−20.4
Al containing single rings				
four-membered ring	[143]	3	1	−16.2
four-membered ring	[143]	2	2	−24.2
six-membered ring	[143]	5	1	−23.0
six-membered ring	[143]	3	3	−10.8
Al containing SBUs				
double four-membered ring	[143]	6	2	−24.6
double four-membered ring	[143]	4	4	−32.3
double six-membered ring	[143]	10	2	−26.3

In contrast, for the double rings the double four-membered ring with maximum aluminum content is clearly the most preferred one. The stability of aluminum in zeolites can also be judged by looking at a formal substitution reaction of the form

$$(HO)_2Si - O - Si(OH)_2 \quad + \quad (HO)_2Si - \overset{H}{\underset{\;}{O}} - Al(OH)_2 \quad \rightarrow 2 \quad (HO)_2Si - O - Si(OH)_2$$

for which the reaction energy is 27.2 kJ/mol [143]. The formation of the single aluminum substituted four-membered ring is not favored. This is contrary to the so-called Dempsey rule empirically derived based on electrostatic arguments which states that aluminum atoms will take positions as far as possible from each other. Other examples which prove that the Dempsey rule is not valid have been provided by Schröder and Sauer [246].

3.2.7 Nuclear Magnetic Resonance Spectra

Nuclear magnetic resonance (NMR) is an analytical technique without which modern chemistry would be unthinkable. In case of microporous materials mainly proton and ^{29}Si resonance are of importance, but ^{27}Al resonance has also been used. With the technical improvements made in solid state NMR in the past few years, NMR experiments on microporous materials and the compounds involved in their synthesis provide more and more insight into the structure of this class of compounds. Two-dimensional NMR allows, e. g., to determine the connectivity pattern of the SiO$_4$ units. Consequently, the calculation of NMR chemical shifts has seen a substantial development to aid in the assignment of the signals and the determination of structures. These calculations span the range from empirical correlations [249–251] to ab initio computation of shielding constants [149, 182, 198, 252–256].

The shielding a nucleus experiences is an electronic phenomenon. Unless empirical structure/shift correlations are used only electronic structure calculations will be useful for the determination of chemical shifts. However, only recently a computer code has been developed which allows to compute chemical shifts for periodic systems [257]. Thus, an alternate approach has been used so far. Two methods have been employed. The chemical shifts are either calculated for a cluster alone or for a cluster whose geometry has been determined by embedding it into a periodic lattice, which was described with a force field (cf. p. 237). The idea behind both of these methods is that the most important effects responsible for a certain chemical shift are localized and can be adequately accounted for with a finite cluster large enough. The correct structure of the cluster is eventually obtained by including its interaction with the periodic lattice.

Sauer and Hill [149] performed calculations on the proton shift for a number of isolated clusters. Eichler et al. have performed proton shift calculations on cluster models embedded in faujasite and ZSM-5 [198]. The calculated shifts agree well with observed shifts in both cases. Moravetski et al. [252] studied the ^{29}Si chemical shifts of species which occur in silica gels. They find that the ^{29}Si chemical shifts are very sensitive to the environment; e. g., the chemical shift of orthosilicic acid changes from -72.1 to -70.8 ppm if four water molecules are added. Highly accurate ^{29}Si chemical shifts are obtained by using orthosilicic acid as an internal standard. Moravetski et al. were also able to question the assignment of certain observed signals and to provide better suggestions for them.

Bussemer et al. [254] have systematically studied the ^{29}Si chemical shifts in all-silica zeolites using an embedding approach. They used an ab initio based shell model potential to determine the structure of the framework. Then they cut clusters of increasing size out of the framework and calculated chemical shifts for them. This way they were able to study the convergence behavior of the chemical shift depending on the cluster size. They found that a three-shell model already provides a reasonable accurate chemical shift with a reasonable amount of computer resources. Calculated ^{29}Si chemical shifts are within less than 5 ppm of the observed values.

Bull et al. [255] have calculated ^{17}O and ^{29}Si NMR parameters for siliceous ferrierite and find excellent agreement with experiment for the ^{29}Si chemical shifts. The ^{29}Si chemical shifts are extremely sensitive to the accuracy of the structure used in the calculation and are therefore well suited to assess the quality of the crystal structure. The theoretical predictions of the chemical shifts, quadrupolar coupling constants, and asymmetry parameters for ^{17}O show only qualitative agreement with the experimental NMR spectra.

Ehresmann and co-workers have modeled the adsorption complexes of several bases in H–ZSM-5 and calculated the quadrupole coupling constants and asymmetry parameters for aluminum [256]. They were able to deduce a relation between the quadrupole coupling constant and the degree of deprotonation of the zeolite framework.

3.2.8   Ab Initio Molecular Dynamics

Increasing computer power and improvements in algorithms have made it possible to perform ab initio molecular dynamics calculations for zeolites in the last few years. In these calculations the evolution in time of the system is studied by electronic structure methods. Since electronic structure calculations are significantly more expensive than force field calculations used traditionally in molecular dynamics the length of the simulations is limited to a few picoseconds. Nevertheless, interesting results can be obtained even from these rather short simulations.

Both structural aspects and the behavior of adsorbed molecules in zeolites have been investigated. The way methanol adsorbs in chabazite [258, 259] and ferrierite and silicalite [259] has been studied. More details about these calculations can be found in Section 3.4. Acetonitrile hydrogen bonding in chabazite was investigated by Trout et al. [260]. The surface reactivity of β-cristobalite has been studied [261].

The mobility of water adsorbed in AlPO$_4$-34 has been investigated by Poulet and co-workers [262] using density functional based molecular dynamics calculations. A dynamic interpretation of the experimental temperature factors could be obtained. The stability of the hydrated structure is not created by the individual interaction of water molecules with the AlPO$_4$ channel, but is ensured by the formation of a collective hydrogen-bond network. This is the reason why the system shows an abrupt transition between an empty and a full phase. Molecular dynamics calculations can provide information about the location of extra-framework water molecules which is not available from x-ray diffraction.

3.3 Force Field Calculations

3.3.1 Structures

Previously, we have looked at the use of quantum mechanical methods to derive potentials for the simulation of microporous structures, but we have not shown any results of their application. Both the potentials based on quantum mechanical calculations and the potentials derived empirically have been used in lattice energy minimizations of dense silica and microporous structures. The emphasis in most of the studies was on a correct description of the lattice parameters and of some measurable properties, e. g., elastic constants. Only a few studies considered the arrangement of the atoms in the unit cell, e. g., by looking at bond lengths and bond angles. Since the structural diversity of microporous materials is caused by the ease of changing certain bond angles (Si–O–Si), a correct description of these angles by a potential is, in our opinion, necessary. Potentials which are based on quantum mechanical calculations are able to predict bond lengths and bond angles quite well. This is not surprising, considering that they were fitted to models which contain these bonds and angles. In contrast, empirically derived potentials are fitted to macroscopic observables and are therefore only indirectly related to bond lengths and angles.

A problem which often occurs while comparing lattice energy minimized structures and experimental data is that the experiment usually cannot distinguish between aluminum and silicon atoms and that the positions of the oxygen atoms are calculated based on the assumption of certain models. There are also often no experimental data available for calculated structures since, e. g., for zeolites, completely dealuminated compounds are not always obtainable. In such a case extrapolations from a number of structures with different Al/Si ratios can be made to a structure where this ratio is 0 [263].

Tables 3.10 to 3.13 contain the results of lattice energy minimizations for dense silica and different microporous solids. These tables are divided into two parts, one for ion pair potentials (including shell model potentials) and one for molecular mechanics potentials. Some authors have used different types of potentials, but the same methodology to derive the parameters. Therefore a comparison can be made between the different functional forms of the potentials. In addition to the potentials which have been derived by the same authors it should be noted that also the shell model potential of Schröder [181] and the molecular mechanics potential of Hill [143] are based on the same quantum mechanical calculations.

The best agreement between the observed and calculated unit cells of α-quartz is obtained for the molecular mechanics potential of de Vos Burchardt et al. fol-

Table 3.10: The results of lattice energy minimizations for α-quartz

α-quartz	Ref.	a=b [pm]	c [pm]	$\langle r(SiO) \rangle$ [pm]	$\langle \angle(SiOSi) \rangle$ [deg]
Exp.	[264]	491	540	161	144
Schröder	[265]	484	535	161	139
Schröder	[181]	499	551		148
Van Beest	[179, 184]	494	545	160	148
Van Beest	[187]	498	548		148
Tsuneyuki	[176]	502	555	164	147
Catlow (shell model)	[266]	484	535	161	147
Catlow (rigid ion)	[266]	531	590		166
Lasaga (ionic)	[180]	515	571	162	164
Sierka	[182]	489	545		
Lasaga (covalent)	[180]	492	532	160	144
de Vos Burchardt	[267, 268]	494	537		147
Hill	[144]	473	563	160	144
Sefcik	[269]	497	543		

lowed by the density functional based shell model potential of Sierka and Sauer. As far as information on bond lengths and bond angles is available some of the rigid ion potentials have problems to predict the SiOSi angle correctly. In general, the agreement for the SiO bond length and the SiOSi bond angle is quite good for most of the potentials.

Table 3.11 shows results obtained for some of the less common dense silica modifications α-cristobalite, coesite, and stishovite. Stishovite is a silica modification which occurs only under high pressure. It is much more dense than the other silica modifications and possesses six-fold coordinated silicon atoms. Most of the molecular mechanics potentials are therefore not applicable here since they were fitted to four-fold coordinated silicon only. The only exception is Lasaga's potential. This potential was fitted to ab initio data of H_8SiO_6, but cannot be used for the study of the change in coordination from four to six. Due to the functional form of a molecular mechanics potential, which assumes a certain connectivity, the parameters derived depend on this connectivity (the force constant for SiO bonds in Lasaga's potential is, e. g., 1041.6 kcal mol^{-1} Å^{-2} for four-fold coordinated silicon, but 879.15 kcal mol^{-1} Å^{-2} for six-fold coordinated silicon).

The deviations between calculated and observed unit cell parameters is some-

Table 3.11: The results of lattice energy minimizations for different dense silica modifications

α-cristobalite	Ref.	a [pm]	b [pm]	c [pm]	⟨r(SiO)⟩ [pm]	⟨∠(SiOSi)⟩ [deg]
Exp.	[270]	498		695	160	147
Schröder	[181]	513		727		
Van Beest	[184]	492		660	160	144
Van Beest	[187]	498		676		145
Tsuneyuki	[176]	499		666	163	142
Catlow	[266]	497		701	158	154
Lasaga (ionic)	[180]	526		744	161	180
Sierka	[182]	498		717		
Lasaga (covalent)	[180]	497		666	160	144
Sefcik	[269]	496		669		

coesite	Ref.	a [pm]	b [pm]	c [pm]	⟨r(SiO)⟩ [pm]	⟨∠(SiOSi)⟩ [deg]
Exp.	[271]	714	1237	717		
Schröder	[181]	719	1254	739		
Van Beest	[184]	714	1249	717	160	180
						151
Tsuneyuki	[176]	723	1274	743		
Catlow	[266]	703	1229	731	158	180/147
Sefcik	[269]	715	1231	706		

stishovite	Ref.	a [pm]	b [pm]	c [pm]	⟨r(SiO)⟩ [pm]	⟨∠(SiOSi)⟩ [deg]
Exp.	[272]	418		267	178	
Van Beest	[184]	415		266	177	
Van Beest	[187]	416		269		131
Tsuneyuki	[176]	427		275	183	
Catlow	[184]	397		284	176	
Lasaga (ionic)	[180]	427		269	183	
Lasaga (covalent)	[180]	425		249	176	

Table 3.12: The results of lattice energy minimizations for different zeolites

sodalite	Ref.	a [pm]	b [pm]	c [pm]	$\langle r(SiO) \rangle$ [pm]	$\langle \angle(SiOSi) \rangle$ [deg]
Exp.	[273]		883		159	160
Schröder	[265]		882		159	156
Schröder	[181]		895			
Van Beest	[184]		899			
Tsuneyuki	[184]		920			
Catlow	[184]		877			
Akporiaye	[178]		891			
Sierka	[182]		885			
Hill	[144]		889		162	152

faujasite	Ref.	a [pm]	b [pm]	c [pm]	$\langle r(SiO) \rangle$ [pm]	$\langle \angle(SiOSi) \rangle$ [deg]
Exp.	[263]		2412		162	139
Schröder	[265]		2428		161	143
Schröder	[181]		2463			
Sierka	[182]		2466			
de Vos Burchardt	[267]		2430			
Hill	[144]		2410		161	146
Sefcik	[269]		2402			

mordenite	Ref.	a [pm]	b [pm]	c [pm]	$\langle r(SiO) \rangle$ [pm]	$\langle \angle(SiOSi) \rangle$ [deg]
Exp.	[274]	1809	2052	752	162	153
Catlow	[266]	1802	2004	743		
Schröder	[181]	1830	2049	758		
Akporiaye	[178]	1840	2031	757		
Sierka	[182]	1812	2069	762		
de Vos Burchardt	[267]	1805	2015	741		
Hill	[144]	1935	2058	764	165	157

Table 3.12: continued

silicalite	Ref.	a [pm]	b [pm]	c [pm]	$\langle r(SiO)\rangle$ [pm]	$\langle \angle(SiOSi)\rangle$ [deg]
Exp.	[275]	2007	1992	1342	159	155
Schröder	[181]	2043	2021	1363		
Van Beest	[184]	2037	2033	1368	161	162
Tsuneyuki	[184]	2088	2081	1400	164	163
Bell	[276]	1999	1975	1332	160	149
Akporiaye	[178]	2032	2025	1363		
Sierka	[182]	2020	1996	1348		
de Vos Burchardt	[267]	2005	1978	1325		
Hill	[144]	2053	2041	1372		
Sefcik	[269]	2002	1985	1333		

what larger for the dense silica modifications considered here compared to α-quartz. The potential of van Beest [184] performs best for all dense silica modifications. It is the best for coesite and stishovite and second best for α-cristobalite. Surprisingly, some potentials which work well for one system do not work so well for another system. This indicates a lack of transferability.

Table 3.12 shows results for some commonly studied zeolites. Faujasite and silicalite are industrial important catalysts if they contain aluminum. Sodalite is predicted best by Schröder's empirical potential [265]. For faujasite all potentials are very close to the observed unit cell dimensions while for mordenite Akporiaye's potential [178] performs best. In the case of silicalite Sierka's potential outperforms all others. In general, molecular mechanics potentials perform slightly better for zeolites than ion pair and shell model potentials.

It is interesting to compare the performance of empirically derived potentials with those potentials derived from quantum mechanical calculations. Sierka and Sauer have provided a comparison of Jackson and Catlow's empirical shell model potential with Schröder and Sauer's Hartree–Fock based potential, as well as their own density functional based shell model potential. The mean deviation from the observed unit cell parameters was found to be 0.7 %, 1.9 %, and 1.4 %, respectively. That means that the empirical shell model potential is twice as accurate as the best quantum chemically derived one. However, the calculated vibrational spectra of silicalite are in good agreement with experiment for both quantum chemically derived potentials [182] while they are not for the empirical potential [181].

Table 3.13: The results of lattice energy minimization of different AlPO's

AlPO-8	Ref.	a [pm]	b [pm]	c [pm]
Exp.	[277]	3329	1476	825.7
de Man	[278]	3428	1466	836
de Vos Burchardt	[279]	3435	1432	829
VPI-5	Ref.	a [pm]	b [pm]	c [pm]
Exp.	[280]	1852		833
de Man	[278]	1901		848
Van Beest	[184]	1905		853
de Vos Burchardt	[279]	1864		833
berlinite	Ref.	a [pm]	b [pm]	c [pm]
Exp.	[281]	494		1097
Van Beest	[184]	505		1117
de Vos Burchardt	[279]	500		1083

Another criterion for the quality of a potential is often its ability to describe vibrational spectra. Table 3.14 contains a comparison of observed and calculated vibrational spectra of α-quartz. This table shows that the molecular mechanics potential of de Vos Burchardt is clearly superior to Tsuneyuki's ion pair potential.

In recent years simulations of zeolite structures have seen a larger emphasis on understanding the effects of structure directing agents used in zeolite synthesis. It has been possible to explain, e. g., the aluminum distribution by force field calculations [283].

Table 3.14: Observed and calculated vibrational frequencies for α-quartz

Symmetry	Exp. ref. [282]	Tsuneyuki ref. [183]	de Vos Burchardt ref. [268]
A_1	207	243	200
	356	403	329
	466	541	541
	1082	980	1123
A_2	364	437	371
	495	622	495
	778	696	798
	1080	1007	1105
E	128	163	140
	265	301	294
	394	426	423
	450	580	465
	697	663	671
	795	706	708
	1072	990	1097
	1162	1110	1196
root mean square deviation		19	9

3.3.2 Lattice Dynamics

In the previous section we have seen how force field calculations can be used to predict the structure of zeolites. These calculations determine the minimum of the potential energy surface for a given structure. For the minimum force constants can be obtained, which are useful for predicting vibrational spectra in the harmonic approximation. All these calculations are, however, limited to properties at 0 K and cannot predict the behavior of the lattice at higher temperatures. In its industrial use zeolites are usually employed at temperatures of the order of a few hundred degrees Celsius. Therefore, e. g., phase changes which can occur between absolute zero and the operating temperature cannot be predicted. Lattice dynamics calculations offer a method to study zeolites at temperatures different from zero. These calculations are almost exclusively force field based molecular dynamics simulations.

Molecular dynamics simulations have certain requirements for a force field, which make not all force fields used for structure predictions suitable to dynamics calculations. In particular the most successful shell model potentials are very difficult to use in molecular dynamics calculations since the shells do not have a mass. Therefore, the Newtonian equations cannot be solved for the shells and a rather time-consuming minimization procedure has to be performed in each time step to locate the optimum position of each shell. It would also be possible to assign part of the mass of the atom to the shell, but then the problem arises that energy can be redistributed into the non-physical core-shell vibration. To our knowledge no molecular dynamics simulations on zeolites have been performed using shell model potentials.

Another problem usually encountered with force field parameters in molecular dynamics simulations is the implicit temperature dependence of parameters derived from experimental data. Since experimental data are usually collected at room temperature, parameters fitted to reproduce these experimental data will do so also at room temperature. In molecular dynamics simulations the temperature is, however, a variable and it is assumed that the force field reproduces, e. g., a structure at 0 K. The temperature effects are then added on top so that parameters derived from data measured at a different temperature will necessarily introduce a small error. Parameters derived from ab initio calculations are free of these errors since they are fitted to data for 0 K.

Demontis et al. started molecular dynamics simulations of zeolite frameworks by employing a simple force field where the nearest neighbor atoms were connected by both harmonic and anharmonic springs [284, 285]. However, such a simple force field is only applicable to systems where electrostatic effects do not play a role. The presence of cations or sorbate molecules prevents the application of such a force field. Nevertheless, Demontis et al. are able to reach a reasonable agreement between calculated and observed IR spectra for the zeolites Linde 4A and silicalite. In constant pressure molecular dynamics simulations the framework collapses, however, due to the lack of any repulsive term in the spring model.

Tsuneyuki et al. have studied pressure induced structural transitions at room temperature in a number of dense silica polymorphs [176, 177]. They were able to observe changes in the coordination of the silicon atoms from four to six. Song et al. were able to study the melting of zeolite A using molecular dynamics [286]. Nicholas et al. obtained an excellent agreement between the IR spectrum calculated from a molecular dynamics simulation and the observed one [190]. They used a molecular mechanics potential, which also reproduces the structure of sodalite and silicalite.

Tse and Klug used van Beest's force field to perform molecular dynamics studies on α-quartz, α-cristobalite, and stishovite [187]. They can model the pressure-induced crystalline $\rightarrow$ amorphous transition of α-quartz, obtaining satisfactory vibrational spectra and very good elastic constants. Ermoshin et al. have been able to get good IR spectra for faujasites with and without protons using molecular dynamics simulations and a molecular mechanics potential [287].

Bornhauser and Bougeard have used molecular dynamics simulations together with a model based on electro-optical parameters derived from ab initio calculations on cluster models to predict the intensities in IR spectra of zeolites [288]. They obtain good agreement between experiment and simulation. Charge transfer has been studied in zeolite NaY [289] and silicalite [290] by Martínez Morales et al. using charge transfer molecular dynamics [291]. They obtain excellent agreement with charge distributions deduced from ^{29}Si MAS NMR experiments. The same method was applied to the study of dynamic properties of cristobalite resulting in excellent agreement of infrared spectra and densities of states between theory and experiment [292]. Inelastic neutron spectra could be reproduced in molecular dynamics simulations for zeolites Na-ZSM-5, X, and Y [293].

In general the area of molecular dynamics calculations on zeolite frameworks has seen comparatively little work (in contrast to the study of diffusion of sorbate molecule in zeolites, vide infra). Molecular dynamics calculations are able to predict vibrational spectra for zeolites very well. Phase changes in dense silica polymorphs and zeolites can also be modeled to some degree.

3.3.3 Active Sites

Previously, we discussed quantum mechanical studies that have been performed on bridging hydroxyl groups in clusters, which served as models for Brønsted active sites in zeolites. In structure determination experiments on zeolites it is rarely possible to locate such bridging hydroxyl groups. The concentration of these active sites is very small and they are not distributed regularly throughout the solid. A few experimental results on the existence of such groups stem from IR spectroscopy, which shows OH stretch vibrations [294, 295]. The deformation vibrations caused by such a bridging hydroxyl group are already covered by lattice vibrations and therefore not visible. Other results can be obtained from NMR spectroscopy where coupling constants can be used to provide structural proposals for bridging hydroxyl groups [296–298].

Bridging hydroxyl groups have been modeled using a number of different force fields. A particular well studied example is the bridging hydroxyl group in faujasite. Faujasite contains only one crystallographic T atom site and four crystallographic different oxygen atom sites. Therefore there are only four possibilities for a bridging hydroxyl group. It is experimentally known that not all of the possible sites are occupied in nature [296, 299–305]. Czjzek et al., e. g., find an occupation of 3:1:1.6:0 for the sites O1 to O4 [305].

Schröder et al. performed lattice energy minimizations for faujasites with a bridging hydroxyl group on all four different positions using an empirical shell model potential [306]. They find significant differences in the lattice energies depending on the position of the bridging hydroxyl group (cf. Table 3.15). The positions O1 and O3 have a relatively low energy, while O2 and O4 have energies

Table 3.15: Relative energies and OH harmonic stretching frequencies of the bridging hydroxyl group in faujasite obtained using various shell model potentials, in kJ/mol and cm^{-1}

	ΔE_{rel}			ν_{OH}				
	empir-ical[a]	HF[b]	DFT[c]	empir-ical	HF[d]	DFT	obsd. harm.[e]	anharm.[f]
O1H	5.3	8.8	11.1	3772	3752	3723	3787	3623
O2H	19.8	18.0	17.1	3702	3628	3602	–	–
O3H	0.0	0.0	0.0	3736	3694	3644	3707	3550
O4H	23.7	27.9	24.7	3751	3697	3673	–	–

[a] Ref. [306], [b] Ref. [181], [c] Ref. [182], [d] frequencies scaled by factor 0.9, [e] Ref. [307], [f] Ref. [295].

which are significantly higher. The calculated OH stretch frequencies of 3772 and 3736 for O1 and O3, respectively, agree well with the two observed IR bands at 3787 and 3707 cm^{-1}.

Very similar results were found for these hydroxyl groups using shell model potentials derived based on ab initio and density functional calculations. Table 3.15 includes results for Schröder and Sauer's Hartree–Fock based potential as well as for Sierka and Sauer's density functional based potential.

Hill and Sauer also performed a study of the geometry of these bridging hydroxyl groups using a molecular mechanics potential [143]. Such studies have later been extended to zeolites with larger unit cells such as H–ZSM-5 [308, 309].

3.3.4   Adsorption of Molecules in Zeolites

The structure and properties of zeolites is an area in itself which has found widespread attention in molecular modeling. But zeolites are technically so valuable because they can adsorb other molecules and these molecules can undergo reactions within the zeolite framework. Therefore adsorption properties and the behavior of other molecules in zeolites have also been widely studied. In this section we will see what molecular modeling methodologies have been applied in this area.

The two basic methods used are Monte Carlo calculations, which allow to obtain, e. g., adsorption energies and preferred locations for the adsorbed species and molecular dynamics simulations, which result in dynamical properties such as diffusion coefficients. Both methods require to calculate the energy of the system many times and are therefore still an area where force fields dominate. A few studies have been performed using density functional calculations, e. g., [258], but these calculations still require super-computer power.

One of the problems faced while trying to simulate the adsorption of molecules in zeolites is the availability of a suitable force field. We have seen previously that a lot of effort has gone into the development of force fields for the zeolite framework. A much larger effort has gone into the development of force fields for organic molecules (a small selection of organic force fields is, e. g., [191, 192, 310–317]). One would think that it is simply a matter of combining two of these force fields to be able to study the adsorption of organic molecules in zeolites. Unfortunately, this is not the case. First, most organic force fields are molecular mechanics force fields, which describe molecules by bonds, bond angle, torsions, and cross-terms between these in addition to a Van der Waals and a Coulomb term. The most successful zeolite force fields are, however, shell model potentials, which basically consists only of a Van der Waals and a Coulomb term plus some core-shell interaction. They usually use formal charges while the molecular mechanics force fields use much smaller charges. A combination of both, even if mathematically possible, will most likely result in an unbalanced description of the system. Even force fields for zeolites which employ the same functional form as the molecular mechanics potentials (e. g., the zeolite extension of CFF [143, 144]) have problems in predicting adsorption energies correctly [194]. This is mainly due to the fact that both the organic and the zeolite part of the force field have been developed independently and no data about the interaction of organic groups with zeolites have gone into the derivation of the parameters. On the other hand the functional form of a shell model potential is not well suited for organic molecules. Most studies of adsorption of molecules in zeolites have been performed using some empirical extension of the zeolite force field by requiring the sorbate molecules to be rigid. In such a case the molecular mechanics force fields are also reduced to a Van der Waals and a Coulomb term since the other terms are only used for intramolecular interactions. However, the approach of rigid sorbate molecules is necessarily limited to small molecules. Larger molecules will adapt to the shape of the channels of the zeolite and results obtained using a rigid molecule approach will most likely be wrong.

Molecular dynamics simulations have been focused in the beginning on studying spherically shaped particles in zeolites. These have been either noble gases, like argon [318, 319], krypton [319], and xenon or small molecules, like methane. For these simulations the sorbates have been treated as soft spheres interacting with the zeolite lattice via a Lennard-Jones potential. Most often the aluminum and silicon atoms in the framework are considered shielded by the oxygen atoms and no interactions with the sorbates are included. Most of the studies concentrated on commercially important zeolites such as zeolites A and Y and silicalite (all-silica ZSM-5) where also a wealth of experimental information is available.

The first attempts to predict macroscopic properties such as Henry constants, initial isosteric heats of adsorption or changes in the standard differential entropies for noble gases, $H_2, N_2, O_2, CO, H_2O, NH_3, CO_2$, n-alkanes up to C_6, ethylene, acetylene, and benzene in silicalite were performed by Kiselev and coworkers at the beginning of the 80's [320–323]. They used potential parameters derived using the Kirkwood-Müller formula and a simple equilibrium condition. Since there was not much experimental information available at this time, Kiselev and co-workers were not able to perform detailed comparisons. Later, the same methodology has been used to study preferred adsorption sites and potential maps of small molecules such as methane, water, and methanol adsorbed in zeolites [324–327]. The sorption isotherms for CO_2 and N_2 in silicalite have been determined by grand canonical Monte Carlo calculations in very good agreement with experiment [328]. Manos et al. are able to predict sorption isotherms for ethane, hexane, and heptane in silicalite using commensurate transitions in a lattice model in very good agreement with experiment and Monte Carlo calculations [329]. The unusual shapes of the isotherms for these molecules can be correctly reproduced.

Monte Carlo techniques have also been used by Maurin and co-workers to study the locations of monovalent cations in zeolites with various Si/Al ratios [330]. They have used data from ^{29}Si NMR to build mordenite models with realistic Si/Al populations and then used these models to predict the locations of sodium cations in agreement with experimental x-ray diffraction data. The distribution of sodium cations in zeolites Y and X was studied by Buttefey et al. [331] employing Monte Carlo calculations. The effect of cation exchange on the adsorption properties of faujasite with respect to xylene mixtures was investigated by Lachet and co-workers using Monte Carlo techniques [332].

Monte Carlo techniques have been used for the prediction of the average positions of methane in zeolite Y [333], ZSM-5, and mordenite [334]. These simulations were able to predict the heats of adsorption in agreement with experiment. Monte Carlo calculations have also been used by Derouane et al. to study water in ferrierite [335]. They found reasonable agreement with experiment for the heats of adsorption. The same method has been used by Hou, Zhu, and Xu to determine the adsorption energies and preferred adsorption sites of benzene in the zeolite ITQ-1 [336]. Paschek and Krishna have used kinetic and configurational-biased Monte Carlo calculations to study the adsorption of 2-methylhexane [337], methane, ethane, propane, n- and isobutane [338, 339] in silicalite. In the latter case they find evidence that the repulsive interactions between adsorbed molecules play a role in enhancing the mobility of the sorbates. The adsorption of binary mixtures of various hydrocarbons in silicalite [338, 340] and of CO_2 and N_2 in silicalite, ITQ-3, and ITQ-7 [341] were also investigated. The importance of using a flexible lattice in Monte Carlo simulations has been pointed out by Vlugt and Schenk in their simulation of adsorption properties of various hydrocarbons in silicalite [342]. At low loadings the effect of the flexible lattice is small, but with higher loadings it becomes important. The uptake of branched hydrocarbons might be higher by a factor of three using a flexible lattice. An improved method for including lattice vibrations in Monte Carlo simulations has been applied to benzene in silicalite [343] resulting in significant time savings to converge the calculation without loss of accuracy.

A large number of studies have been performed on the interaction of small molecules with various zeolites, especially the diffusion of methane [285, 344–366] and water [195, 335, 367–375] (which is usually present in zeolites and important as solvent for ion exchange), but also CO_2 and N_2 [328], rare gases [366], CF_4 and SF_6 [366], benzene and toluene [336, 376–382], acetonitrile [383], nitroaniline [378], hydrofluorocarbons [384, 385], and higher hydrocarbons up to C_8 [327, 350, 386–395] have been simulated in zeolite frameworks. The diffusion mechanism can be understood by plotting the logarithm of the mean square displacement as a function of the logarithm of the simulation time. Such a plot will result in linear regions, which denote distinct diffusive regimes. A straight line of slope 2 represents Newtonian dynamics (nearly free motion) while a straight line of slope 1 is characteristic for Einstein diffusion. Such an analysis has been applied by El Amrani and Kolb for the first time for molecules diffusing in a zeolite [396].

It is interesting to compare the results of the simulations obtained using spherical and non-spherical (structured) representations of the sorbates. At the same time the effect of a rigid and a flexible lattice should be considered. June et al. can reproduce self-diffusivities and isosteric heats of adsorption of methane and xenon in silicalite in good agreement with experimental results employing a rigid but structured methane molecule despite the fact that they do not use a Coulomb term in their potential [349]. Even when treating methane as a sphere it is still possible to predict experimentally observed diffusion coefficients well [353]. Demontis et al. [285, 348, 354, 355, 357, 397] have used a flexible framework, but represented methane as a sphere. They find that the flexible framework has a significant effect on the behavior of methane. The framework is able to exchange energy with the methane molecules, practically acting as a heat bath. The flexible walls of the zeolite channels are also softer, which leads to a decreased repulsion of methane molecules. This results in a reduced number of high-energy collisions between different methane molecules and collisions which involve more than two methane molecules are nearly inhibited. On the other hand, Thomson et al. do not observe a significant difference in methane diffusivity in $AlPO_4$-5 if they use a rigid and a flexible framework [362]. A detailed study of whether lattice vibration drives diffusion in zeolites was carried out by Kopelevich and Chang [398]. They found that no general conclusions could be drawn since for some zeolite–sorbate combinations lattice vibration is important while it is not for others.

Another interesting result obtained using simulations of spherical sorbates is that a sorbate with a diameter close to the diameter of a zeolite channel experiences a "levitation" or super-diffusivity effect, which is caused by the lack of any radial force on the sorbate in such a setting [399, 400].

Rigid lattices and rigid, but structured sorbates can also cause problems if the sorbate is rather large. Nowak et al. studied the diffusion of methane in silicalite, mordenite, and EU-1 and of ethane and propane in silicalite. They found that the diffusion coefficient of propane does not agree with experiment (it is much larger in experiment than in the molecular dynamics calculation) [350]. However, the calculated enthalpies of adsorption agree well with experimental results. Their simulations of propane are probably too short to reach a conclusive decision about the need for flexible sorbates, but later work [344, 359, 361] using flexible propane molecules predicts both the diffusion coefficient and the heat of adsorption in excellent agreement with experiment. Kawano et al. found that using a rigid framework, but flexible methane molecules can result in an increase of the diffusion coefficient by 70 % [356]. In contrast, framework flexibility has nearly no effect on the diffusion coefficient if methane is described as a sphere.

Titiloye et al. have used a partially flexible framework as well as flexible sorbate molecules and found remarkable differences compared to calculations with rigid systems. Lattice relaxation is especially important for aluminum containing systems; the heats of adsorption agree with experiment only when the framework is allowed to relax and the preferred adsorption sites are completely different [327]. The influence of a flexible framework has also been made clear by Schrimpf et al. [401] and Leroy et al. [402]. Sastre et al. have studied the diffusion of *o*- and *p*-xylene in purely siliceous zeolite CIT-1 at different loadings and compared their findings with experimental results [388]. They used both a flexible lattice and flexible sorbates. However, the diffusion coefficients they obtain from the simulation are three orders of magnitude larger than experimentally determined diffusion coefficients. Possible reasons for this difference might be that purely siliceous zeolite CIT-1 could not be obtained in experiment, that extraframework material might have been present, and that the applied experimental technique (FTIR) does not measure exactly the same property as calculated in the simulation. However, Demontis et al. can show that using a flexible lattice while studying ethane diffusion in $AlPO_4$-5 is needed to correctly reproduce the energy exchange mechanism [393].

Vibrations of water in natrolite have been studied by Demontis et al. [403]. A poor reproduction of the vibrational spectrum was observed and attributed to the effect the electric field of the zeolite has on the water molecules. The authors developed later an electric field dependent force field for water [374, 404] to improve the simulation capabilities for hydrated zeolites. Water in ferrierite was studied in great detail by Leherte and co-workers [335, 367–372] and water complexes in cation-exchanged zeolites were investigated by Dil'mukhambetov et al. [405].

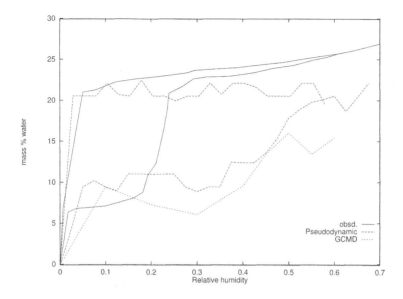

Figure 3.7: The water adsorption isotherm in zeolite Na-MAP as observed and simulated using a combination of Monte Carlo calculations and lattice relaxation (pseudodynamic) and grand canonical molecular dynamics (GCMD) [373] (reproduced by permission of the PCCP Owner Societies).

The effect of water adsorption on the framework of zeolite Na-MAP was studied by Hill and co-workers [373]. The water adsorption isotherm of this zeolite could be reproduced qualitatively by grand canonical molecular dynamics calculations and a combination of Monte Carlo calculations and lattice relaxation (Fig. 3.7). It was essential to include the flexibility of the zeolite framework in the calculations to be able to simulate the observed isotherm. Grand canonical molecular dynamics simulations are rather demanding and using Monte Carlo calculations to determine the loading of the zeolite framework with water followed by a lattice relaxation under the presence of the water molecules determined in the Monte Carlo step proved to be a more efficient approach.

Zeolites can contain charge compensating cations. These cations can be quite mobile and can be exchanged, usually by using an aqueous solution. Molecular dynamics simulations of the cation mobility have been performed by Shin et al. [406] in anhydrous zeolite A. The most complete studies of a real zeolite are so far molecular dynamics simulations [407–409] of zeolite A containing both sodium cations and water and energy minimizations of zeolite A with water and Na^+, Cs^+, Ca^{2+}, Ba^{2+}, Cd^{2+}, and Sr^{2+} [410].

Diffusion of sorbates in zeolites simulated using molecular dynamics tech-

niques usually results in a good agreement between observed and calculated diffusion coefficients. These simulations have also provided valuable insight into the
behavior of the sorbate molecules within the zeolite channels and cages, which
could not be obtained by experiment. With respect to the simulation methodology previous calculations have shown that rigid frameworks should only be used
when the guest molecules are smaller than the critical diameters of the channels,
are well represented by spheres, highly diluted, and collisions with the framework are not important for the required results. The state-of-the-art in performing
molecular dynamics calculations of zeolites has been reviewed by Demontis and
Suffritti [411].

Force fields available for the study of dense silica polymorphs and zeolites
can in general be used to obtain reliable structures and vibrational spectra for
the frameworks. While at the beginning of the development of these force fields
it seemed necessary to use molecular mechanics type force fields to be able to
predict vibrational spectra correctly, it has recently been shown [181, 182] that
shell model potentials are also capable of that. Shell model potentials, therefore,
seem to be the best suited force fields for silica polymorphs. They model the more
ionic character of the SiO bond, include polarization, allow coordination changes,
and contain only a few parameters. Molecular mechanics potentials might still
have their value in molecular dynamics calculations where shell model potentials
are cumbersome to use. They also allow for an easier extension with respect to
studying the adsorption behavior of zeolites.

3.4   A Case Study – Methanol Adsorption on Bridg-ing Hydroxyl Groups

One of the most important areas of application of zeolites in industry is their
use as catalysts in the production of gasoline. The Brønsted acidity of certain
zeolites is, e. g., used in the methanol-to-gasoline (MTG) process [412]. In this
process methanol is converted to gasoline using the zeolite ZSM–5 as a catalyst.
The initial step of the conversion is thought to be the adsorption of methanol on
a bridging hydroxyl group in the zeolite. Very little is known about that step
experimentally; in particular it is not clear what the adsorbed methanol looks
like. It could be simply physically adsorbed by forming a hydrogen bond with the
zeolite framework (Figure 3.8a). It could also be protonated by the zeolite, thus
forming a methoxonium ion and becoming chemisorbed (Figure 3.8b).

To help in the interpretation of experimental data numerous theoretical studies
have been performed on this system. We will follow the progress of these studies
along the time line they have been done. This will show us how increases in the
available computational power have been applied to gain better understanding of
the underlying physics of this industrially important process.

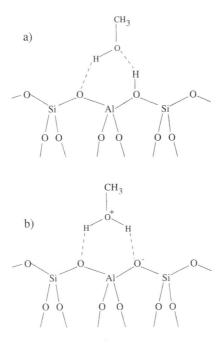

Figure 3.8: Adsorption of methanol on a bridging hydroxyl group, physisorption (a) and chemisorption (b).

The first ab initio calculations to be performed on methanol adsorbed on a bridging hydroxyl group used cluster models for the zeolite [211,413,414]. Sauer et al. [211] and Haase and Sauer [414] used clusters consisting of an AlO_4 tetrahedron saturated with either hydrogen atoms, two SiH_3 groups and two hydrogen atoms, or four SiH_3 groups. Haase and Sauer also used a larger four-condensed-ring model cut out of the zeolite faujasite [414]. Both Hartree–Fock and MP2 calculations were performed. Only the physisorbed complex was found to be a minimum. The methoxonium ion was found to be a transition state for the proton transfer from one framework oxygen atom to a neighboring one, but the energy difference between both structures is only a few kJ/mol. NMR chemical shifts calculated for both complexes show that the shift for the physisorbed complex agrees much better with observed chemical shifts than the shift for the chemisorbed complex. The calculated vibrational spectrum could not provide a consistent explanation for the observed spectrum for the chemisorbed complex. For the physisorbed complex an explanation could be found assuming Fermi resonance. Therefore, it was concluded that the physisorbed complex is the species formed on methanol adsorption in zeolites.

With the availability of more powerful computers periodic quantum mechanical calculations became possible for small zeolites. Schwarz et al. performed molecular dynamics simulations of methanol in sodalite using a density functional plane wave approach [415–417]. While the cages in sodalite are too small to allow methanol to reach a bridging hydroxyl group this system is a better model for methanol adsorption in zeolites than a simple cluster. Schwarz et al. observed proton exchange between methanol and the sodalite framework. Their simulation showed that within 1 ps the system would stay for 0.3 ps in one form, proton exchange would then occur within less than 0.1 ps and after another 0.3 ps the proton would move back to the original site.

Shah et al. used a density functional plane wave calculation to study the adsorption of methanol on a bridging hydroxyl group in chabazite [418], which shows catalytic activity in the MTG process. They found that the proton from the bridging hydroxyl group is readily transferred to methanol, thus forming a chemisorbed complex. There seems to be little or no barrier for proton transfer. Shah et al. were not able to locate a physisorbed complex, but do not rule out its existence in the form of a local minimum.

Haase and Sauer studied the adsorption of methanol in chabazite using first principles molecular dynamics simulations based on density functional theory with a plane wave basis set and pseudopotentials [258]. They come to the conclusion that methanol is not chemisorbed as Shah et al. thought, but that there is another deeper minimum where the Brønsted acid proton is significantly delocalized in the region between the methanol and the framework oxygen atom of the zeolite. The motion of the proton shows two superimposed patterns. One is the OH stretch motion, the other is a much slower change in the OH distance, which is anti-correlated with the change in the OO distance. This superposition is the reason for the band broadening of the OH stretch band observed in IR spectroscopy on hydrogen bond formation.

Further first principles molecular dynamics studies for methanol in chabazite, ferrierite, and silicalite have been performed by Stich et al. [259]. Reaction pathways have been investigated by Andzelm and co-workers [419]. Both studies find that the nature of the adsorption complex depends on the type of zeolite framework and on the loading. The chemisorbed methoxonium ion is more stable at higher loadings and in ferrierite's eight-ring channels at any loading studied. At higher loadings a significant weakening of the CO bond can be observed, which is thought to increase the likelihood of a nucleophilic attack on this bond.

Thus, while the first studies using static cluster calculations basically predicted the right adsorption complex the real system is much more complex. The use of the most advanced modeling techniques is necessary to obtain a description of the system which explains all observed facts.

Chapter 4

Glass

4.1 Introduction

Amorphous materials, such as glass, present a great challenge to the application of computer simulation methods. Experimental methods do not provide accurate atomic positions but give average radial distribution functions (from X-ray diffraction methods) or local arrangements (from extended X-ray absorption fine structure and nuclear magnetic resonance experiments). Structural information can be obtained from molecular modeling studies. In order to ensure that the atomic scale information from these calculations is reliable, it is important to validate the modeling results by calculating some macroscopic properties (e. g., densities, IR spectra) and comparing these with experimental results.

4.2 Simulation of Silica Glass

The procedure employed to construct model glass structures is similar to laboratory glass preparation. Computer modeling of glass structures involves high temperature molecular dynamics of component oxides followed by quenching to room temperature, thereby freezing the structure into a disordered glassy phase. The system size employed in these calculations is by necessity smaller than the roughly 10^{26} atoms of a typical glass sample (simulation cells typically contain several hundred atoms). Similarly, the heating and quenching rates exceed those relevant in the laboratory.

The importance of vitreous silica is not only in the traditional ceramic and glass industry but also in its application in optical fibers, microelectronics, and catalysis industries. A detailed understanding of the bulk and the surface structure is needed to improve optical fiber coatings, microelectronic devices, and catalytic supports. First, a reliable bulk structure needs to be simulated and validated with experimental data before surface reactions on glass can be modeled.

The structure of vitreous silica is a continuous random network of corner-sharing $SiO_{4/2}$ tetrahedra. In order to simulate vitreous silica, a crystalline structure is first melted at 6000 K, which is followed by a constant energy simulation to achieve internal equilibrium. The system is then cooled stepwise to 4000 K, 2000 K, 1000 K, and finally to 300 K. After each step the system is allowed to equilibrate. Several independent systems are simulated and the results averaged. Such vitreous silica models have been constructed by several ion pair potentials [420–424]. The structural results in terms of radial distribution functions from these molecular dynamics simulations of vitreous silica compare well with X-ray diffraction data and neutron scattering data. However, this procedure results in a broad distribution and an overestimation of the $O - Si - O$ and $Si - O - Si$ angles. This shortcoming can be traced to the absence of any direction-dependent term in the two-body potentials. In order to circumvent this problem, many researchers have used a three-body potential which includes a direction-dependent term to model partial covalence, e. g., Feuston and Garofalini's molecular dynamics simulation of the bulk [425] and surface structures of vitreous silica [426]. These authors as well as Rosenthal et al. [427] added a three-body interaction term to a modified form of the Born-Mayer-Huggins (BMH) ion potential. The three-body potential results in narrower $O - Si - O$ angle distribution with a peak at the tetrahedral angle and a half-width of the distribution less than $7°$.

Using a three-body potential Vessal [428] used a constant-volume molecular dynamics method to simulate vitreous silica. A Buckingham potential is used to model the short-range interaction between different ions and the three-body potential consists of an angle bending spring constant term, and two power terms.

$$E_{ijk} = k_{ijk}[\theta^n (\theta - \theta_0)^2 (\theta + \theta_0 - 2\pi)^2 - n/2\pi^{n-1}(\theta - \theta_0)^2 (\pi - \theta_0)^3] \quad (4.1)$$

where k_{ijk} is the three-body spring constant, θ_0 is the equilibrium bond angle, and θ is the calculated bond angle. A cubic simulation box containing 216 Si^{4+} and 432 O^{2-} ions is used. The experimental density of vitreous silica is used. Similar to the procedure already described, the system is melted at high temperature and cooled to 300 K in a step-wise manner with equilibration between the cooling steps with a total time of 36 ps for the molecular dynamics annealing. The simulated structure of vitreous silica is seen to compare well with high-quality neutron scattering data. Table 4.1 lists a few of the simulated results for the equilibrium distances of $Si - O$, $O - O$, and $Si - Si$.

In a paper by Litton and Garofalini [432], the direction-dependent bonding is again described by modified Born-Mayer-Huggins ion potential functions along with three-body interaction terms. The bulk vitreous model is built similar to the procedure mentioned above using the molecular dynamics melt-quench method.

Table 4.1: Equilibrium distances obtained from MD simulation of vitreous silica

	r_{Si-O} [Å]	r_{O-O} [Å]	r_{Si-Si} [Å]
Soules [429]	1.61	2.55	3.18
Zirl [430]	1.62	2.63	3.20
Feuston [425]	1.62	2.64	3.14
Vessal [428]	1.62	2.62	3.10
Expt [431]	1.60-1.62	2.62-2.65	3.11-3.13

The number of atoms used in their calculation is 648 in an 21.4 Å cubic cell. The kinetics of self-diffusion of Si and O in the bulk of vitreous silica is then studied for the final 90 ps simulation. Diffusion is activated by breaking of a single $Si - O$ bridging bond. The slope of the mean squared displacement r^2 as a function of time is given by the equation

$$D = \frac{r^2}{2\alpha t} \qquad (4.2)$$

where α is the dimensionality (=3 for bulk) that is used to calculate the diffusion coefficient. From regression analysis of D as a function of (1/temperature) the activation energies for diffusion, Q, and the diffusivity pre-exponential factor, D_o, are calculated:

$$D = D_o \exp(-Q/RT) \qquad (4.3)$$

The activation energies for self-diffusion of Si and O in the bulk vitreous silica is 115 kcal/mol and this is in agreement with experimental data obtained from diffusion-couple tracer studies.

The work by Benoit et al. [433] reports the simulation of vitreous silica using a combination of first principles Car–Parrinello [434] and classical molecular dynamics methods. This combination allows both the explicit treatment of the electronic structure of disordered silica as well as accessible dynamics of vitreous silica. Using classical molecular dynamics the glass structure is first simulated using the melt-quench procedure. The glass equilibrated with the classical potential is then used as the initial configuration for the short-time quantum dynamics runs based on the Car–Parrinello method.

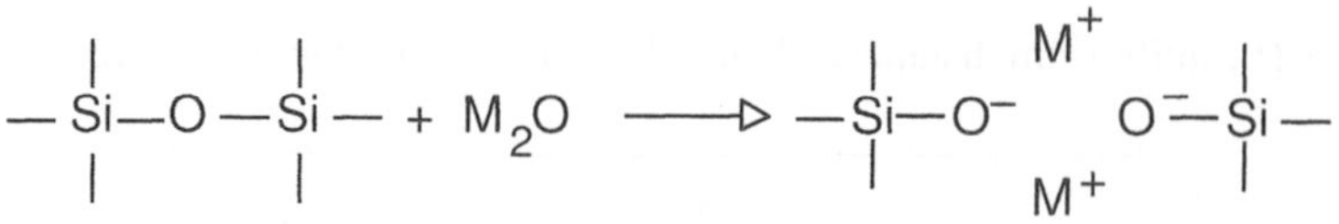

$$-\overset{|}{\underset{|}{Si}}-O-\overset{|}{\underset{|}{Si}}- \ + \ M_2O \ \longrightarrow \ -\overset{|}{\underset{|}{Si}}-O^- \quad \overset{M^+}{\underset{M^+}{O=\overset{|}{\underset{|}{Si}}-}}$$

Figure 4.1: The addition of one unit of M_2O (M = alkali metal) to corner-sharing SiO_2 tetrahedra causes a bridging oxygen atom to be replaced by two non-bridging oxygen atoms.

4.3 Alkali Silicate Glasses

Extending the simple vitreous silica model to alkali-silicate glasses requires that the alkali cations be regarded as network modifiers as shown in Figure 4.1. Each alkali–oxide unit added results in one bridging oxygen replaced by two non-bridging oxygen atoms.

Smith, Greaves and Gillan [435] studied the structure and dynamics of sodium disilicate glass by molecular dynamics simulations. The simulations were performed on a 1080 atom system containing 240 sodium ions, 240 silicon ions, and 600 oxygen ions. Their study showed that the glass structure is inhomogeneous with the sodium ions segregated into pores within the structure. A strong association is present between the non-bridging oxygen atoms and the sodium ions. The simulated radial distribution functions compared well with those from experiments and showed sharp peaks at 1.7 Å corresponding to $Si - O$, 2.6 Å corresponding to $O - O$, and 2.4 Å corresponding to $Na - O$.

Vessal and co-workers [436, 437] have simulated a series of alkali silicates to help interpret highly accurate neutron diffraction data.

Several compositions of sodium silicate, potassium silicate, and mixtures of sodium potassium silicate glasses were simulated using a short range Buckingham potential and three-body potentials to model the $O - Si - O$ and the $Si - O - Si$ angles [428]. The work by Huang and Cormack [438] on sodium silicate glasses showed that the sodium ions are associated with non-bridging oxygen atoms, and that there are silica rich and alkali rich regions. The work by Smith et al. [439] on sodium disilicate glass revealed alkali clustering effects, as well as the mobility of sodium ions via a hopping mechanism.

Simulation results on mixed sodium potassium silicate glasses showed a non-uniform distribution of the alkali ions. A similar study on a variety of sodium and rubidium containing silicate glasses [440] indicated that although the alkali oxides are minor components in the glass, they do form clusters within the simulated networks. Within these clusters the two alkalis are intimately mixed. A similar conclusion was obtained on the $(K, Cs)_2Si_2O_5$ glasses from EXAFS experimental studies [441].

Vessal and co-workers [440], using molecular dynamics simulation techniques, investigated an interesting phenomenon known as the mixed alkali effect, where the ionic conductivity of alkali glasses decreases dramatically in the presence of more than one type of alkali ions. Due to the segregation of alkali ions into clusters, though within the cluster, the ionic mobility is high, the mixing of alkalis impedes the hopping process of a given alkali ion. This increased activation energy for the hopping alkali results in lowering of the total ionic conductivity.

Cormack [442] studied the effects of glass structure on alkali-ion diffusion behavior in mixed alkali silicate glasses using atomistic simulations. As the amount of alkali increases, the separation of alkali-rich regions and silica-rich regions becomes less clear. The structural features underlying the mixed alkali effect lead to a deepening of the potential felt by the smaller alkali ion and by a separation of alkalis into distinct sites.

Experimental studies such as ion-scattering spectroscopy (ISS) indicate that fracture surfaces of potassium silicate glass and sodium silicate glass have higher concentrations of alkali ions compared to the bulk either due to crack propagation along alkali-rich regions or due to surface rearrangements. A molecular dynamics simulation study [443] of the fracture surface of $K_2O \cdot 3SiO_2$ glass shows that potassium ions can build up at the fracture site within 6 ps after surface formation. A similar behavior of sodium ions moving to the fracture surface is seen from molecular dynamics simulations of $Na_2O \cdot 3SiO_2$ [424]. However, in the case of lithium silicate glasses, no excess of Li ions is seen at the surface both from simulation [424] and ISS experiments.

Experimental X-ray diffraction and X-ray adsorption fine structure (EXAFS) studies on sodium silicate glasses have demonstrated that about five oxygen atoms are at a distance of 2.3-2.4 Å from one sodium atom. Such multiple coordination of sodium is possible only with bridging and non-bridging oxygen atoms. In order to gain insight into the actual coordination environment of alkali cations in glass, Uchino and Yoko [444] have carried out ab initio molecular orbital calculations on a cluster of atoms containing five $SiO_{3/2}O^-Na^+$ units connected at the corners by siloxane bonds $(Si - O - Si)$. The cluster geometry was optimized using a Hartree–Fock 6-31G(d) split-valence basis set. Using the optimized geometry, a normal-mode analysis was performed to obtain the harmonic frequencies. This work revealed that the bond between sodium and the non-bridging oxygen atoms is shorter and is characterized by strong covalent interaction and Coulombic interaction. This indicates that when such a glass is formed the structure will be so constructed as to maximize the number of non-bridging oxygen atoms around sodium ions to compensate for the positive charge most effectively. This explains the reason for the formation of alkali rich regions in silicate networks.

The size, shape, and distribution of metallic inclusions, e. g., Ag particles, affect the physical properties of glasses such as the color and polarization. Timpel and co-workers [445] have generated a model based on molecular dynamics simulation for a Na-Ag ion-exchanged sodium trisilicate glass using a three-body Born-Mayer-Huggins potential for the glass and embedded atom potentials for Ag [446]. The Ag-glass interaction potential is empirically fitted to experimental structural modifications of ion-exchanged glass. The sodium atoms in the glass were found to form channels while silver particles were found to form clusters. Their subsequent work [447] which is a detailed study of the diffusion process revealed that both sodium atoms and silver atoms exhibit similar diffusion mechanisms involving discrete hopping between stable sites.

The understanding of the structural effects in iron containing glasses is of primary importance in the control of technological products since it is a main chromophore. Spectroscopic studies of these glasses are complicated by the multivalent chemistry of iron. In a study by Rossano, Ramos and Delaye [448] on the environment of ferrous iron in $CaFeSi_2O_6$ glass both EXAFS and molecular dynamics simulations were performed. They used the Born-Mayer-Huggins interatomic pair potentials to reproduce the ionic bond, with additional three-body terms to describe the covalent Si-O bond. The initial value used for the Fe-O repulsive term was that given by Belashchenko [449]. It was optimized to give the best match between the calculated and the experimental radial distribution function. The 4-fold coordination, associated with distorted tetrahedra, and the 5-fold coordination, associated with trigonal bipyramids, are observed. Although the medium-range order is characterized by the presence of Fe^{2+} and Ca^{2+} at short distances, as indicative of edge-sharing polyhedra, no cation segregation was noticed.

4.4   Aluminosilicate, Borosilicate, and Other Glasses

One of the interesting properties of sodium aluminosilicate glasses is that as the ratio of sodium to aluminum changes, the viscosity and electrical conductivity changes drastically. This is attributed to structural changes as the ratio is changed. Zirl and Garofalini [430] showed using molecular dynamics simulations that the structural changes could occur without changing the coordination of Al from tetrahedral to octahedral as previously considered. Their simulation studies on a series of sodium aluminosilicate glasses with different Na:Al ratio matched experimental results. The elimination of non-bridging oxygen, introduction of oxygen triclusters, and the change in the ring structure distribution were found to be the cause for changes in viscosity with different Na:Al ratio.

Due to the ambiguity in the interpretation of NMR, IR, and Raman spectra of hydrous aluminosilicate glasses, Sykes, Kubicki and Farrar [253] performed ab initio molecular orbital calculations on large aluminosilicate clusters using 3-

21G**, 6-31G*, and 6-311+G** basis sets. The calculated isotropic chemical shifts relative to TMS was able to predict the experimental trends seen. This work also reports the basis set dependency of the calculated Al and Si chemical shifts.

Modeling of vitreous borate or borosilicate glasses poses another problem due to the fact that the B-O bond is partially covalent and boron can change its coordination number from 3 to 4 depending on the environment. Takada et al. [450,451] have derived interatomic potentials using a partial ionic model and included a B-O-B bending three-body term to reproduce the structure of vitreous B_2O_3. Molecular dynamics simulation of B_2O_3 glass using different potentials [452] revealed the necessity of three-body interactions for a satisfactory reproduction of vibrational modes.

Soules and Varshneya [453] used a Born-Mayer-Huggins potential of the form

$$V_{ij}(r) = A_{ij} \exp(-r/\rho) + (z_i z_j e^2/r)\mathrm{erfc}(r/\eta L) \tag{4.4}$$

where

$$A_{ij} = (1 + z_i/n_i + z_j/n_j)b \exp((r_i + r_j)/\rho) \tag{4.5}$$

z is the electronic charge, n is the number of valence electrons, L is the length of the simulation cube, b and ρ are compressibility parameters. The borosilicate glass model generated contained a near-random network of SiO_4, BO_3, and BO_4 polyhedra. Soules and Varshneya were able to reproduce the trigonal to tetrahedral conversion of boron with increasing amount of Na_2O in the glass which agrees with NMR results.

Delaye and Galeb [454] used a Born-Mayer-Huggins potential with three-body terms to model the nuclear waste glass $SiO_2 + B_2O_3 + Na_2O + ZrO_2$. The model was able to reproduce the experimental density, thermal expansion coefficient, sodium diffusion, and viscosity. Takahashi and co-workers [455] used the pair potential functions

$$V_{ij} = \frac{Z_i Z_j e^2}{r} + f_0(b_i + b_j) \exp[(a_i + a_j - r_{ij})/(b_i + b_j)] \tag{4.6}$$

where Z_i is the ionic charge, a and b are parameters related to the radius and compressibility, f_0 is a constant. They studied Li^+ diffusion in $Li_4SiO_4 - Li_3BO_3$ for different compositions.

Due to the high refractive index lead-containing glasses are used as nuclear scintillators and in up-conversion laser devices. Glasses containing over 30 mol% lead form PbO_3 and PbO_4 units. Using molecular dynamics simulations, Cormier and co-workers [456] were able to simulate the bulk structural features of lead silicate glass. There are two networks present in the system, the silicate tetrahedra and the Pb-O polyhedra. In their study on Yb^{3+} doped lead silicate glass, the Yb^{3+} ions are primarily found in the Pb-O network.

Gruenhut and co-workers [457] have studied fluoride glasses by molecular dynamics simulations. These fluoride glasses are superior to silica glasses since they have potential to be low loss optical fibers. They used a Buckingham pair potential of the form

$$V_{ij} = \frac{q_i q_j}{4\pi\epsilon_0 r_{ij}} + A_{ij} e^{[-r_{ij}/\rho_{ij}]} \tag{4.7}$$

where q_i and q_j are the charges of the individual ions, r_{ij} is the distance between the ions, A_{ij} is the depth of the potential well, and ρ_{ij} is the slope of the short range exponential repulsion [458]. The potential was initially tuned using the X-ray structure of $BaNaZr_2F_{11}$. The A_{ij} value for each pair potential was determined by comparing the radial distribution function generated from the simulated structure with those from X-ray diffraction experiments. The simulation program that was used is the FUNGUS code of BIOSYM/MSI [459]. Different simulation procedures were employed all at constant volume. The use of the NVT ensemble did not lead to a minimum energy configuration. However, a combination of NVT and NVE ensembles did provide the lowest potential energy configuration. In order to determine the point of homogeneity or equilibrium in the simulations, Gruenhut and co-workers developed tools such as spherically averaging of all ions and spherically averaging of each ion type within a sphere of radius half the simulation box length. This involves integrals of the form

$$n_{ij}(r) = \int_0^r \rho_j g_{ij}(r) 4\pi r^2 dr \tag{4.8}$$

where ρ_j is the density of the ions being examined and $g_{ij}(r)$ is the radial distribution function of atom type j around atom type i within a radius r. They successfully demonstrated that the effects of insufficient diffusion during the quenching process could lead to insufficient structural relaxation which could then result in clustering. From their study, they concluded that the simulation method is as important as the potential used to obtain glass structures. Horbach and co-workers [460] found that the time-dependent properties such as diffusion are affected by system simulation size.

4.5   Simulation of Glass Surface and Diffusion

In order to simulate a glass surface one needs to first simulate the bulk glass structure using the melt-quench procedure. Surface simulation begins by fracturing the bulk along the z direction. This is accomplished by increasing the cell dimension in the z direction to more than twice the original dimension. The model now has a "surface" in contact with vacuum. Atoms in a small volume element at the bottom of the cell are not allowed to move during simulations. These atoms are considered to model the bulk. The mobile layers are annealed to about 1000 K to allow structural relaxation of the new surface followed by stepwise cooling to room temperature. In order to remove the stress at the mobile-immobile interface, the immobile layer is incrementally increased during the annealing process. Vitreous silica surfaces simulated by Feuston and Garofalini [426] exhibit many structural features seen by experimental methods. The outermost species on the glass surface is found to be oxygen. The experimental concentration of surface hydroxyls matches the number of defect sites (such as non-bridging oxygen atoms, undercoordinated silicon, three-member rings) in the glass surface model.

Litton and Garofalini [432] have studied the surface self-diffusion of vitreous silica. Removing the periodic boundary conditions in the z-direction and freezing the bottom half of the cell with respect to the z direction leads to the formation of the surface after the molecular dynamics run. In their 648 atom simulation in a $(21.4\text{Å})^3$ box the reflecting boundary was maintained at 10 Å above the surface so that any atom that desorbed from the surface during heating could get well beyond the interaction cut-off of 5.5 Å before being reflected back towards the surface. Only the atoms on the surface are considered for calculating the diffusion at the surface. This is achieved by counting only those atoms with z-coordinates greater than a specified value, thus creating a volume of a certain thickness parallel to the surface. In the work by Litton and Garofalini, it is seen that overcoordinated defects were not important precursors to diffusive jumps for surface self-diffusion since they are annealed out of the low density surface regions unlike that for a bulk diffusion. Rotations of undercoordinated species result in larger jump distances on the surface compared to bulk. In general, they found that surface diffusion coefficients are larger than those of bulk diffusion coefficients by a factor of 2.

In general, many transport properties such as diffusion coefficients can be calculated from molecular dynamics simulations. The atom positions are stored in regular intervals during the dynamics run. The time evolution of the mean square displacement can be obtained from the stored positions and using the Einstein relation, the diffusion coefficient is calculated from a least squares fit of the plot of mean square displacement versus time,

$$\langle |r(t) - r(0)|^2 \rangle = B + 6Dt \tag{4.9}$$

The left hand side of the equation represents the mean square displacement, i. e., displacement of an atom from its initial position averaged over all atoms and over time. B is the thermal vibration contribution, D is the diffusion coefficient, and t is the time. The molecular dynamics simulation should be carried out for a fairly long time, typically 200-300ps. This will ensure that the calculated D value is not affected by the starting configuration. Also the errors in D can depend on the nature of the system. All these factors need to be considered when calculated diffusion coefficients are compared with experimentally measured values.

Simulation of adsorption of alkali ions on glass surfaces and their diffusion into subsurface layers is important to understand the effects of contaminants on the properties of optical fibers. Garofalini and Zirl [461] studied the adsorption and diffusion of K and Li on vitreous silica. From their simulation studies it was seen that potassium ions diffuse over the surface and preferentially adsorb on the non-bridging oxygen while lithium ions initially adsorb on both the non-bridging and bridging oxygen atoms on the surface. On saturation of the non-bridging oxygen sites, the potassium ions attack the siloxane bonds while the lithium ions diffuse into the subsurface regions through the channels. Even though the channels are large enough, the potassium ions do not enter the subsurface regions. These simulation results were verified using X-ray photoelectron spectroscopy and ion-scattering spectroscopy techniques by Garofalini's group.

Interfaces between metal and glass is another area of importance since such features are present in glass fiber optic coatings and electronic packaging. Webb and Garofalini [462] have studied the effect of adsorption of Pt atom on the surface of a sodium aluminosilicate (NAS) glass using molecular dynamics simulations. The substrate-substrate interaction involves a modified Born-Mayer-Huggins pair potential and a three-body term. The Pt-Pt interaction is described using the embedded atom potential method (EAM) [25]. The EAM potential approach is a semi-empirical method developed from density functional theory and subsequent effective medium theory [463]. No electronic transfer between the adsorbate atom and the glass substrate is assumed. The adsorbate-substrate interaction is described by a two-body Lennard-Jones potential. The significant result from this study is the fact that sodium atoms move away from the region where the deposited Pt atoms reside on the glass surface.

4.6   Calculation of Glass Properties

In principle, molecular dynamics simulations provide complete information about the system under study, including the position, velocity, and potential energy of each particle. With such data, correlation functions and thermodynamic parameters can be calculated with relative ease. Also the frequency-dependent thermal conductivity and the constant volume specific heat are accessible through such simulations. These are generally difficult to measure in the laboratory.

While simulation studies offer us a powerful tool, statistical mechanics concerns itself with large numbers of particles. Present day computers allow meaningful simulations of the order of 1000 particles, far fewer than the 10^{23} particles that would be ideally required. Even the longest computer simulations, which may take a few days or weeks of clock time, are limited to the equivalent of nanoseconds of system time.

It is generally found [428,436,437] that the techniques employed to computationally construct glass models result in structures with high fictive temperatures.

The main problem with the technique is in the time scale of the quench which is several orders of magnitude greater than the rates achieved experimentally. Therefore, glass transition temperatures are typically higher by a factor of 2 compared to the experimental Tg. In general, the glass transition temperature is defined in terms of a kinetic process involving volume contraction [464]. Thus a plot of specific volume versus temperature is expected to yield the Tg of glass.

The protocol generally used is to perform a constant pressure molecular dynamics simulation while heating and cooling the system in a stepwise manner. The volume is measured during the heating and the cooling cycles. For a crystalline system, a plot of volume versus temperature will show a sharp break, indicative of melting. However, in the case of a glass, there is no sharp break in the curve. The glass transition temperature can be roughly estimated from tangents drawn to the different portions of the heating or cooling curve.

Thermoelastic properties such as the thermal expansion coefficient and isothermal compressibility can be calculated using long molecular dynamics simulations at constant pressure. The following equations may be used to calculate the isothermal compressibility, β_T, and the thermal expansion coefficient, α_P:

$$\beta_T = \frac{1}{KTV}\langle\delta V^2\rangle \qquad (4.10)$$

$$\alpha_P = \frac{1}{KT^2V}\langle\delta V \delta(KE + PE + PV)\rangle \qquad (4.11)$$

where KE and PE are the kinetic energy and potential energy, respectively, P, V, and T are the thermodynamic state variables, and the quantity within the angular brackets denotes the equilibrium ensemble average.

Thermal diffusion in bulk glasses is simulated by first subjecting the ensemble to molecular dynamics at constant temperature and volume (NVT). Then the simulation box is extended in one direction to create an extended bulk model of the melt. The two ends of the simulation box are then subjected to different instantaneous temperatures and another molecular dynamics simulation is run at constant energy. This allows the profile to spread in response to thermal diffusion. Finally, uniform equilibrium conditions can be established within the material by an NVT run. The energy at the two ends can then be measured during this simulation and from the energy versus time plot, the thermal diffusivity, κ, can be estimated.

The work by Takase, Akiyama and Ohtori [465] reports evaluating the thermal conductivity in SiO_2 glass. They performed molecular dynamics calculations on a 300 atom cell of silicon atoms and oxygen atoms with a density of 2.2×10^3 kgm^{-3} at 300 K using the NVE ensemble. The interatomic potentials were of the Born-Mayer-Huggins form with additional dispersion terms. The Coulombic interactions were treated using the conventional Ewald method with periodic boundary conditions [466]. The initial structure was heated to 20000 K and kept in the molten state for 100000 time steps with a time step of 1 fs in order to obtain a disordered structure. It was cooled to 300 K in a step-wise fashion keeping the density a constant. The coordinates and velocities were obtained for various temperatures. A similar procedure was repeated for different number of atoms, 600 and 900.

The thermal conductivity for an insulating binary ionic solid is given by the Green-Kubo formula [467, 468].

$$\lambda = \frac{L_{ee}}{k_B T^2} \tag{4.12}$$

$$L_{ee} = \frac{1}{V} \int_0^\infty C_{ee}(t)dt \tag{4.13}$$

$$C_{ee}(t) = \frac{1}{3} \langle J^e(t) J^e(0) \rangle \tag{4.14}$$

where λ is the thermal conductivity, L_{ee} is the time integral of the autocorrelation function of the energy current C_{ee}, T is the temperature, V is the volume, k_B is the Boltzmann constant, J^e is the energy current. At lower temperatures, the evaluated values of L_{ee} were comparable to the statistical fluctuations leading to uncertainty in the calculation of the thermal conductivity at lower temperatures. The thermal conductivity did not depend on the number of atoms used in the simulation. The estimated uncertainty for the thermal conductivity is 25 % at 981 K and 8 % at 1978 K. The calculated value was in agreement with experimental results within 20-35 % in the 981-1981 K temperature range.

The shear viscosity of glassy melts can be calculated using non-equilibrium molecular dynamics methods based on the original work reported by Ashurst and Hoover [469]. Further adaptation of this technique to simple hydrocarbons [470–473] leads to the calculation of their shear viscosity. It may be possible to use the same protocol for glassy melts in order to calculate their shear viscosity. There are no publications yet on glassy melts; however, the protocol that could be followed is described below.

The glass system for which the shear viscosity is to be calculated consists of three regions, namely, the lower wall, the fluid, and the upper wall. The upper and lower walls are then translated in opposite directions to each other at a constant velocity, thus effectively shearing the glass melt. The atoms in the wall regions are maintained at a constant temperature by velocity scaling. The wall atoms are permitted to vibrate about their equilibrium positions. The atoms in the fluid region are free to move. This leads to the development of a temperature profile between the two walls as the liquid increases in temperature due to frictional heat from the shear. The resulting temperature profile is parabolic, as heat is conducted out through the walls.

Before any shear dynamics can be performed on the system it must be equilibrated in order to allow the atom positions to be optimized for that particular temperature. This is achieved by a simple NVT dynamics for 20 ps at the desired temperature. An extended production run, typically 500 ps, is then performed, during which a number of properties such as velocity and temperature profiles as well as shear stress are calculated. These properties are averaged over successive "blocks" of the simulation, with the block length and intervals specified as input parameters to the protocol. The shear rate, γ, is the gradient of the block-averaged velocity. The shear viscosity, η, can then be calculated by dividing the modulus of shear stress in the zx direction by the shear rate, as shown in the equation below.

$$\eta = \frac{-\langle P_{(zx)}\rangle}{\gamma} \tag{4.15}$$

$$\gamma = \frac{\delta v_x}{\delta z} \tag{4.16}$$

Further research is being carried out in many commercial and academic groups to improve force-field parameters for various atom types as well as in describing interfaces better. With increasing power of compute engines, larger, more realistic glass models can be built to provide better simulation results.

Chapter 5

Semiconductors and Superconductors

5.1   Semiconductors

A semiconductor is a crystal with a narrow energy gap between a filled valence band and an empty conduction band. Common examples of semiconductors include both elemental (Si, Ge) and compound (GaAs, GaP, AsAs, CdTe, CdSe, InAs, InP, ZnSe, etc.) materials. At T = 0 a semiconductor is a perfect insulator. At room temperature some electrons from the valence band get thermally excited into the conduction band, resulting in a small electrical conductivity. The conductivity arises from the mobility of excited electrons in the conduction band, as well as from the mobility of the "holes" that the excited electrons leave behind in the valence band. In the pure material (i. e., no impurities) the number of electrons in the conduction band is equal to the number of holes in the valence band, and this number decreases exponentially as a function of the energy gap of the semiconductor. However, the number of charged carriers (electrons or holes) can be dramatically altered by the addition of impurity atoms with a valence different from that of the host atom through a process called "doping". For instance, consider Si, the most commonly used semiconductor in commercial microelectronics. Si crystal has a diamond structure, with each atom covalently bonded to four Si-neighbors. When a pentavalent impurity (e. g., As) replaces a Si-atom, four of its five "valence" electrons form covalent bonds with the four Si-neighbors, while the fifth electron is only weakly bound to the As-center. This electron can easily get excited into the conduction band, much more easily so than a valence electron localized on a Si-Si bond. Under typical As concentrations, the number of electrons donated by the As-centers to the conduction band is much larger than the thermally excited electrons of the pure Si. This results in a majority of carriers with a negative charge (i. e., electrons), and such a system is called an n-type or an n-doped semiconductor, while the pentavalent As impurities are called n-dopants. Similarly, a three-valent impurity (e. g., B) is a p-dopant, and gives rise to a majority of positive-charged carriers (i. e., holes) by accepting electrons from the valence band in order to satisfy the covalent bonds between the B-center and its four Si neighbors.

5.1.1 Electronic Band Structure in Solids

In order to understand the structure-property relationships in solid-state materials, an essential concept is that of electronic "bands". A good starting point is an isolated atom with electronic orbitals (i. e., energy eigenstates) corresponding to discrete energy levels. As many atoms are brought progressively closer to each other in order to form a crystal, the "outer" or "valence" orbitals of the neighboring atoms begin to overlap. A more appropriate description of the electrons in a crystal is then given by extended "electronic states" or "crystal orbitals", which are a linear combination of the atomic orbitals, a concept similar to the construction of molecular orbitals in molecules or atomic clusters (see section 7.1.5). However, the crucial differences between a molecule and a crystal are that: (1) number of atoms (and therefore electrons) in a macroscopic crystal is essentially infinite; and (2) a crystal has translational periodicity.

Point (2) above implies that a defect-free crystal can be completely described just by the atoms in its unit cell, which, acted upon by the lattice translation vectors, generates the whole crystal. This, in turn, implies that one effectively needs to solve the quantum mechanical problem only for the atoms inside the unit cell, with the wave function outside the cell related to that inside the cell through the celebrated Bloch's theorem. As discussed in section 7.1.11, the Bloch's theorem introduces the concept of wave vectors $\vec{k}$, which can be used as a natural representation (i. e., a good quantum number) of electronic states in a crystal. Essentially, the Bloch's theorem recasts the problem of an infinite number of electrons in a macroscopic crystal to that of a finite number of electrons (inside the unit cell) and a continuum of $\vec{k}$-points. From the translational symmetry of the crystal it turns out that only a finite range of $\vec{k}$ values is needed to generate all possible electronic states in the crystal. This region of unique $\vec{k}$-points in the wave vector space is known as the First Brillouin Zone, often called just the Brillouin Zone for simplicity, and is defined as the Wigner-Seitz primitive cell of the reciprocal lattice of the crystal [474]. The shape of the Brillouin Zone depends on the space-group symmetry of the crystal. For more details on this, refer to [475, 476].

Solving the Schrödinger equation for isolated atoms yields electronic eigenstates (i. e., orbitals) corresponding to discrete energy levels. Similarly, solving the Schrödinger equation for a crystal results in multiple electronic states corresponding to discrete energy levels for each $\vec{k}$-point within the Brillouin Zone. Suppose that these levels are indexed by a quantum number n. One could then group together the electronic states at all $\vec{k}$-points within the Brillouin Zone corresponding to a given value of n. Such a grouping of states is called the n-th electronic "band" of the crystal. Each band is a continuous spectrum of states within a range of energies called the band width. One way to interpret the formation of electronic bands in solids is that the n-th electronic orbital of each constituent atom overlaps with the n-th orbital of the neighboring atoms in the solid, and the resulting interaction "broadens" the degenerate n-th energy level of the isolated atoms into the n-th band of the solid. Such an approach to band formation is called the "Tight binding theory" (see section 7.1.12). The above picture holds for elemental solids, where all atoms are of the same element type. In reality, however, the situation can be more complicated, because several closely spaced energy levels belonging to different element types can interact with each other, leading to bands of mixed or hybrid characters. Electronic bands in solids have some very important general properties: (1) in a given band, energies of electronic states at wave vectors $\vec{k}$ and $-\vec{k}$ are equal, as follows from time-reversal symmetry; (2) the mean velocity of an electron in state $\vec{k}$ is proportional to the derivative of the energy with respect to the wave vector $\vec{k}$; (3) the total electrical current carried by a band completely filled with electrons is zero; (4) wider the band, the more delocalized the electrons are; (5) the $\vec{k}$-points corresponding to the minimum and maximum of energy in a band are either a high-symmetry point in the Brillouin Zone or lie along a line joining two high-symmetry points.

Since electrons are Fermions, it follows from Pauli's exclusion principle that a molecule in the ground state (i. e., T=0) has only the lowest few orbitals occupied by electrons. The Highest Occupied Molecular Orbital (HOMO) can be fully or partly occupied. All levels below the HOMO are completely filled, while all levels above the HOMO are completely empty. The first empty level above the HOMO is called the Lowest Unoccupied Molecular Orbital (LUMO). Similarly, in solids at T=0, all energy levels up to a certain energy, called the Fermi energy E_f, are completely filled with electrons, and all levels above E_f are empty. If the material is such that E_f lies within the energy spectrum (i. e., band width) of one or more bands, then these bands are partially occupied with electrons. Such bands can support a net charge transport even under small electric field, and the system is a metal. On the other hand, if the material is such that the value of E_f does not fall within the spectrum of energies defined by any band, then a few bands are completely filled, while the higher-energy bands are completely empty. Since filled bands cannot support any net electrical current, such systems are non-metallic, i. e., either semiconductors or insulators. For metals, the partially filled band(s) containing the E_f, and for non-metals (i. e., semiconductors and insulators), the empty band(s) just above E_f, are called the conduction band(s). Filled bands below the conduction bands are called valence bands. For non-metals, the con-

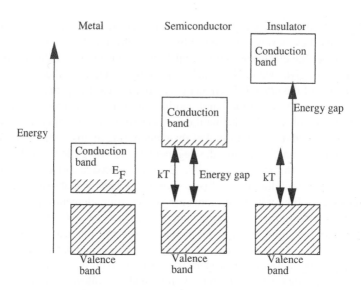

Figure 5.1: Schematic representation of conduction and valence bands in metals, semiconductors, and insulators.

duction bands are strictly empty (and the valence bands strictly full) only at T=0. At a finite T, some electrons are thermally excited from the valence to the conduction bands, with the equilibrium number of electrons governed by Fermi-Dirac statistics. In a non-metallic system, the difference between the energy-minimum of the conduction band and the energy maximum of the valence band is called the "band gap" of the material. The above two energy extrema can occur either at the same $\vec{k}$-point ("direct" band gap, as in GaAs, InSb, etc.) or at different $\vec{k}$-points (indirect band gap, as in Si, Ge, etc.). Insulators are materials with large band gaps (at least a few eV), while semiconductors are materials with small to medium-sized band gaps (typically less than 1.5 eV), although there are examples of wide-band-gap semiconductors with band gaps of 2 eV or more. Figure 5.1 is a schematic energy diagram for metals, semiconductors, and insulators. At T=0, the Fermi energy for metals is at the highest energy level occupied by the electrons, while for semiconductors and insulators, it is in the middle of the band gap. In general, all bands close to the Fermi energy are important in determining the electronic and other properties.

A similar picture can be drawn for a doped semiconductor. Figure 5.2 shows schematically the effect of a dopant added to the semiconductor. In a n-type semiconductor, the excess electrons loosely bound to the dopant atoms form a narrow band of "donor" levels just below the conduction band. At room temperature

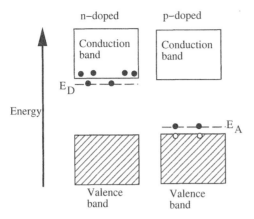

Figure 5.2: Schematic representation of band structures in pure, *n*-doped, and *p*-doped semiconductors.

many of these electrons get thermally excited to the conduction band, leading to electrical conductivity. Similarly, in a *p*-type semiconductor a band of "acceptor" levels form just above the valence band. Each acceptor level corresponds to an extra bound electron that a *p*-dopant center can accommodate. At T=0 the acceptor levels are unoccupied. However, at room temperature, many electrons from the valence band get excited into the acceptor levels, leaving mobile "holes" in the valence band to conduct electricity. For a finite T analysis of the electron and hole population in various bands of a doped semiconductor we refer the reader to Wolfe at al. [477] and other standard textbooks in semiconductor physics.

5.1.2 Modern Computational Methods of Electronic Structure

Thanks to increasing processor speeds at falling hardware prices, and concomitant developments in electron structure theory and numerical algorithms, it is now possible to routinely compute the electronic band-structure of complex systems, with hundreds of atoms in an unit cell. Accurate electronic band structure, Density of States (DOS), and total energy allows one to compute structural, mechanical, optical, and magnetic properties, including lattice parameters, elastic constants, magnetic moments, and the frequency-dependent dielectric function. In addition to electronic band structure, modern-day codes allow one to accurately compute forces on atoms and the stresses on the unit cell. Therefore, geometry optimization, molecular dynamics, transition state calculations, and cell deformation under external pressure have all become standard calculations among the present-day researchers. Some of the recent theoretical developments that have transformed electronic structure methods into a powerful multifaceted research tool include:

1. Description of a many-electron system through the Density Functional Theory (DFT), with validated expressions for the exchange-correlation energy and gradient corrections (see section 7.1.10).

2. Definition of a periodic "supercell", repeated in three dimensions using the Born–von Kármán periodic boundary condition. This allows one to treat periodic solids, surfaces, molecules, clusters, and liquids for all types of materials [478].

3. Approximating the $\vec{k}$-space integration over the whole Brillouin zone to a simple weighted summation over a set of special $\vec{k}$-points chosen according to the symmetry of the periodic supercell [479]. Metals typically require a higher density of these special $\vec{k}$-points than non-metals. For large supercells (of lengths 20 Å or larger in each dimension) a single $\vec{k}$-point (usually chosen at the origin of the reciprocal lattice, known as the Γ-point) is sufficient. This is known as the Γ-point approximation.

4. Replacing the inert "core" electrons with non-local pseudopotentials (section 7.1.13). This not only limits the calculations to the chemically important valence electrons, but also replaces large oscillations of valence wave function near the atomic core with smooth functions representing "pseudo" wave functions, which simplifies expansion in a plane wave basis set.

5. Development of "iterative" diagonalization techniques of the Hamiltonian matrix in order to ensure fast relaxation of electrons to the ground state (also known as the Born–Oppenheimer surface). Commonly used iterative diagonalization techniques include the molecular-dynamics method of Car–Parrinello [434] and the conjugate-gradients minimization method of Payne et al. [478]. This is particularly important for codes involving the Plane wave basis set, as conventional matrix diagonalization is extremely memory- and CPU-intensive.

6. Development of a linear response theory based on the Density Functional Perturbation Theory (DFPT) [480]. This makes possible the accurate computation of phonon frequencies, from which one could calculate thermodynamic and derived quantities (entropy, Free Energy, heat capacity, thermal expansion coefficient, T-dependence of band gap, and so on). One could also use this formalism to compute other types of properties, e. g., NMR response, elastic constants, dielectric function, and so on.

Electronic Band structure in solids is typically displayed along lines joining high-symmetry $\vec{k}$-points, as shown in Figure 5.3(a) for Si. This band structure was

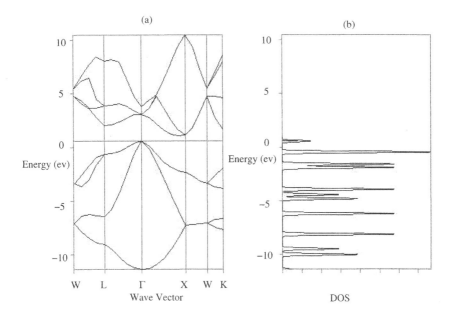

Figure 5.3: (a) Calculated band structure and (b) density of states for Si.

computed using the code CASTEP [478], [481], [482], which is based on a Plane-Wave Pseudopotential approach. Si is a diamond lattice, with a Bravais lattice that is face-centered-cubic. The Brillouin zone is therefore a truncated octahedron, with eight hexagonal and six square facets. The various high-symmetry $\vec{k}$-points for such lattice are as follows: Γ at the center of the Brillouin zone; X is at the center of a square facet; L at the center of a hexagonal facet; W at a vertex (shared by two hexagons and a square); K at the middle of an edge shared by two hexagons. As Figure 5.3(a) shows, Si has a doubly degenerate valence band maximum at the Γ-point, while the conduction band has a minimum at about 80 % on the way from Γ to the Brillouin zone boundary at the X-points. Since the valence band maximum and the conduction band minimum occur at different $\vec{k}$-points, the band gap in Si is indirect.

A quantity useful to many calculations is the derivative of the number of electronic states with respect to energy, commonly known as the density of States (DOS). Figure 5.3(b) displays the Si DOS corresponding to the occupied (i. e., valence) bands in Figure 5.3(a). Computing DOS requires a more accurate $\vec{k}$-space integration than that required for the total energy calculation. A review article by Leckey and Riley [483] compares calculated band structures with experimental data.

Although Si is the most commonly used semiconductor in microelectronics, its band structure is quite distinct from other semiconductors. Thus, for instance, most III-V semiconductors (GaAs, InP, InSb, etc.) are direct-band-gap materials with both valence-band maximum and conduction-band minimum occurring at the Γ-point.

A typical plane wave pseudopotential calculation using the Density Functional Theory (DFT) proceeds as follows.

1. First, one needs to specify all the input parameters either through a Graphics User Interface (GUI) or through one or more input files. The input parameters include the supercell information (lattice parameters, element types and atomic coordinates, space-group symmetry, total charge and spin), plane-wave cutoff, density of special k-points (for Brillouin zone sampling), type of pseudopotential, e. g., "norm conserving" [484] or "ultrasoft" [485], the Exchange-correlation functional (parameterization, gradient correction), tolerances defining various convergence criteria, and so on.

2. Next one needs to select what type of job is to be run, e. g., relaxing just the electrons to the ground state, relaxing the atoms/ions, performing molecular dynamics, optimizing to a saddle point (transition state), etc.

3. Next one needs to specify the properties to be calculated, e. g., elastic constants, band structure, DOS, optical properties, phonon spectrum, etc.

As the job starts executing, first it attempts to relax the electrons to the ground state (i. e., Born–Oppenheimer surface) corresponding to the initial position of the atoms, which is mathematically equivalent to minimizing the Kohn–Sham functional. This relaxation procedure involves multiple steps in which the total electron density, the DFT Hamiltonian, and the electronic wave functions are iterated "self-consistently" until all of them converge, and the electron wave functions form an orthonormal set within a specified tolerance. Such a procedure of attaining self-consistency is known as the self-consistent field (SCF) procedure, as has been alluded to several times within the book. Once the electrons are relaxed to the ground state, the program computes the total energy of the system, as well as the forces on each atom and the stress tensor on the supercell, if necessary. Such information would then be used for geometry optimization of the atoms and the cell, or carrying out molecular dynamics simulations, depending on the type of calculation specified by the user. At the end, the program uses the final electron density to compute other useful quantities like the band structure, DOS, optical absorption spectra, and other properties.

Before getting into specific application examples of molecular modeling in the field of semiconductors, we would like to point out a known drawback of DFT. The original theory proposed by Hohenberg, Kohn and Sham (section 7.1.10) is a ground state theory of electrons. Therefore, it does not guarantee the accuracy of

Table 5.1: The energy gaps calculated for a few semiconductors using different methods

System	Band Gap at Γ [eV] [494]		
	Expt.	LDA	sX-LDA
Si	3.05, 3.4	2.54	3.37
Ge	0.89	−0.06	0.28
GaAs	1.52	0.44	1.11
InP	1.42	0.94	1.60
InSb	0.24	−0.32	0.21

computed excited-state properties. For electronic band structure, this means that the position of the conduction bands should not necessarily be correct. In fact, it is well known from numerous calculations that band gaps predicted from the Local Density Approximation (LDA) are often severely underestimated as compared to experimental values. For example, in the case of GaN, the experimental band gap of the wurtzite structure is 3.39 eV [486] while the plane-wave pseudopotential method using LDA results in a band gap of 2.75 eV using experimental lattice parameters [487]. Gradient corrections (see section 7.1.10) can sometimes improve the band gap, but still not enough to be comparable with experimental accuracy. Since in the density functional formalism the eigenvalues of the Kohn–Sham equations are not single particle energies, Koopman's theorem [488] for excitation energies is not valid unlike in Hartre–Fock theory [489], [490]. Therefore, in order to calculate band gaps and electronic excitations, quasiparticle-based many-body theory often becomes necessary [491–493]. We will not go into the details of these methods and the interested reader can refer to the literature. Table 5.1 lists the experimental band gaps for a few of the semiconductors. The calculated band gaps from simple LDA and screened-exchange LDA (sX-LDA) are also listed here. In the sX-LDA approach, the approximate LDA expression for the screened-exchange contribution is replaced by an exact non-local Hartree–Fock term [494]. While LDA underestimates the band gap to a large extent such that Ge and InSb are predicted to be metallic, the screened exchange LDA approach is able to open up the band gap.

Due to potential applications in optoelectronic devices and high-speed electronic circuits, III-V semiconductor alloys are being studied by various groups. An example of such an alloy is the InGaAs. It is not clear if there is clustering of like atoms or if the In and Ga atoms aggregate in these systems. A paper by Picozzi and co-workers [495] reports the study of $In_xGa_{1-x}X$ where x=0.25, 0.75 and X is As or Sb. Full-potential linearized augmented plane-wave (FLAPW) calculations [496, 497] with screened-exchange were performed on these systems. The authors studied the effects of volume and composition changes on the direct and indirect band-gaps. They found that either by changing the lattice parameter or by varying the ratio of Ga to In, the band gaps could be tuned.

5.1.3   Chemical Vapor Deposition

Chemical vapor deposition (CVD) is a process by which thin films of a large variety of materials can be deposited. CVD has been applied to many different manufacturing processes ranging from the fabrication of microelectronic devices to the deposition of protective coatings to the recent synthesis of carbon nanotubes. In a typical CVD process reactant gases at room temperature enter the reaction chamber. The gas mixture is then heated as it approaches the deposition surface. The reactant gases undergo homogeneous chemical reactions in the vapor phase before striking the surface. The reactive intermediate species interact with the surface and result in surface deposition of the material. Gaseous reaction by-products are then transported out of the reaction chamber.

Figure 5.4 shows a schematic of a CVD system for growth of GaAs semiconductors. Volatile precursors, in this case $Ga(CH_3)_3$ and AsH_3, are introduced into the reactor. The heated reactants undergo a series of gas-phase radical reactions. The reactive intermediates such as $Ga(CH_3)$ and AsH_2 react with the surface As and Ga atoms, respectively, resulting in the growth of a layer or islands of GaAs. Methane and other by-products from the gaseous reaction are removed from the reaction chamber. Details of CVD are explained by Hitchman and Jensen [498].

In order to model CVD, a complete picture requires both a microscopic and a macroscopic view [499]. Data for the microscopic processes, such as thermochemistry and reaction rates, can be obtained using semi-empirical, density functional, or ab initio methods. The macroscopic transport and reaction processes can be simulated using finite-element methods for fluid flow, heat transfer, and mass transfer models. Reaction Design's CHEMKIN package [500], e. g., allows macro-level simulations of the transport properties using the results of atomistic calculations as input.

Ho et al. [501] calculated the heats of formation, $\Delta H_f(0K)$ and $\Delta H_f(298K)$, for a variety of $H - Si - Cl$ compounds using ab initio methods. The geometry of the various structures were optimized and the harmonic vibrational frequencies were calculated using the HF/6-31G* (Hartree–Fock with a split valence plus polarization 6-31G* basis set) level of theory. This level of theory has been shown [502] to provide accurate geometries but appears to fail for total energy calculations, which involve covalent bond breaking. Therefore, fourth order Møller–Plesset (MP4) perturbation theory with the 6-31G** basis set was

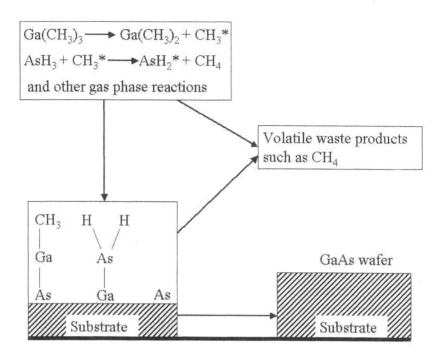

Figure 5.4: Schematic diagram of chemical vapor deposition.

used to compute the total energies. An empirical correction factor to correct for the basis set truncation errors was used in this work. The heat of formation at different temperatures is a combination of the correction factor and the total atomization enthalpies (enthalpy required to entirely dissociate the molecule into its component atoms) at these temperatures. The application of this method to larger systems becomes impractical due to the computationally demanding MP4 calculations that are required for electron correlation.

A more recent study on similar systems was performed by Hay [503] using density functional calculations. The effects of electron correlation are included in the exchange-correlation Becke–Lee–Yang–Parr (BLYP) functional [504, 505] and, therefore, no extra computational effort is required. The magnitude of the overall computational expense in this case is roughly the same as that for a Hartree–Fock calculation. The atomization energy calculated using the gradient corrected BLYP functional and the hybrid B3LYP functional is within 10 kcal/mol compared to experimental results for the first-row compounds. This is better than the MP4 results.

Outlined below is the general procedure to calculate thermochemical properties of polyatomic molecules needed to understand CVD processes at the atomistic
level.

Step 1. Choose level of theory, one that allows reasonably accurate estima-
tion of thermodynamic properties, e. g., density functional method with gradient
corrections. Electron correlation effects are important to obtain reliable bond energies.

Step 2. Optimize the geometry of each molecule using a suitable basis set,
e. g., 6-31G** or 6-31+G**.

Step 3. Calculate the total energy, E_{Total}, and the zero-point vibrational energy, E_{ZP}, of each species.

Step 4. Calculate the atomization energy, D_0, of, say, AB_2C_3

$$D_0 = E_{Total}(AB_2C_3) - E_{Total}(A) - 2E_{Total}(B) - 3E_{Total}(C) + E_{ZP}(AB_2C_3)$$

This is the atomization energy at 0 K. At higher temperatures, additional terms
resulting from rotational and translational degrees of freedom have to be added.
Using the ideal gas approximation, standard statistical mechanics can be used to
deduce this term. An entropy term due to change in volume, i. e., difference
between the number of moles of products and reactants is also required.

Step 5. Calculate the reaction thermochemistry, for example, for the reaction
$SiH_4 \rightarrow SiH_2 + H_2$

$$\Delta H_{reaction} = (D_0)(SiH_2) + (D_0)(H_2) - (D_0)(SiH_4)$$

Ho et al. [506] have calculated transition state energies for possible reaction
pathways for thermal decomposition of disilane. They also report the calculation
of reaction energies for various possible dissociation products of Si_2H_6. The
lowest energy pathway leads to SiH_4 and SiH_2 as the reaction products. A 1,2-H_2
elimination followed by rearrangement to give H_2SiSiH_2 and H_2 can also occur.

The work by Sato and co-workers [507] discusses in detail the use of ab initio
molecular orbital calculations to the hydrogen elimination reactions on methane
plasma in comparison to silane plasma CVD process in the formation of amorphous silicon (*a*-Si:H) films. They found that the activation energy of the hydrogen elimination step in the carbon system is much higher than that in the silicon system indicating that the hydrogen displacement mechanism just below the
surface of methane plasma is different from that in silane plasma. This is in accordance with experimental observation that at low substrate temperatures (below
600 °C) the carbon network contain more sp^2 structure than sp^3 structure while
the silicon network contain more sp^3 structure.

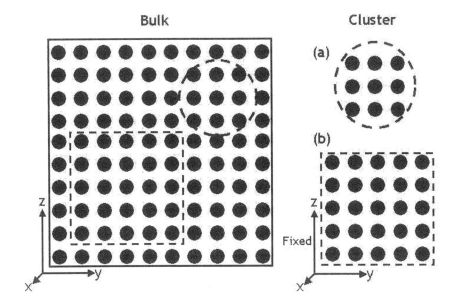

Figure 5.5: Schematic diagram of cluster models.

5.1.4 Adsorption Energetics and Dynamics

Behavior of adsorbed atoms on semiconductor surfaces and their surface diffusion is crucial to understanding crystal growth and surface structural transitions. Molecular beam epitaxy (MBE), a leading method to grow semiconductor devices, is a technique for controlled deposition of atomic layers on a substrate material.

In order to simulate a surface, either a cluster model or a slab model is generally used. A cluster model can be built by carving out a piece from the bulk as shown in Figure 5.5.

Cluster model of, say, Si, can be built by carving out a spherical model from the bulk as shown in Figure 5.5a. About 30-40 atoms in the cluster model allows the construction of a network containing a core of fourfold coordinated atoms similar to bulk Si surrounded by surface atoms which are threefold coordinated. The dangling bonds can be saturated with hydrogen atoms maintaining the original bond angles in order to model bulk Si with a cluster. Kaxiras [508] have modeled reconstructions on Si(111) surface using such cluster models. In the cluster model (b) shown in figure, all the dangling bonds are terminated with hydrogen atoms except on the top. The top layer is the surface, the second layer is the subsurface layer and the atoms in these layers are allowed to move during simulation. The

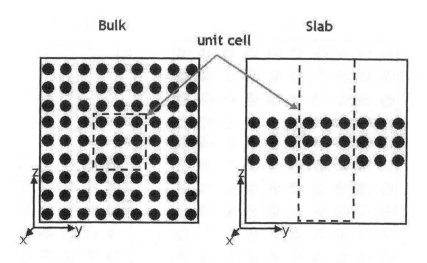

Figure 5.6: Schematic diagram of bulk and slab model.

atoms in the third and fourth layer are fixed and these represent the bulk. In the xy direction in model (b) the outside atoms are fixed and terminated with hydrogens and only the central atoms are allowed to move. The advantage of this model is that it is a molecular model and any code that works on discrete molecular systems can be used. The disadvantage of this model is the loss of accuracy in representing an infinite solid surface by a discrete cluster. Moreover, the use of ab-initio codes to handle too many atoms is too time consuming. Therefore, it is very important that suitable cluster size is used. The property that is to be studied should be calculated for different cluster sizes until convergence is achieved as a function of the cluster-size.

A slab model is built from the crystal structure of the solid surface. In a slab model the periodicity is maintained in the xy direction similar to an actual surface. In the z direction, the unit cell size is increased by twice or three times the original cell to accommodate vacuum space as shown in Figure 5.6. The top layer and the bottom layers represent the surface. The central layers represent the bulk. The advantage of this model is that periodicity is maintained in the x and y directions to mimic the actual extended surface. This model is much preferred to represent a surface rather than the cluster model.

Shiraishi et al. [509, 510] have investigated the adsorption behavior of GaAs during molecular beam epitaxy. They used a pseudopotential method and the local density approximation. A slab model with periodic boundary conditions was employed to simulate the GaAs (001) surface. It is important to determine the optimum thickness of the slab used for such simulations. In order to eliminate interaction of the surfaces, the surface bonds are usually terminated with hydrogen atoms. To determine the adsorption behavior during the growth process, Shiraishi et al. computed the migration potential and the adsorption energies of adatoms on the reconstructed As-rich surface. The migration potential of a Ga adatom depends on the adsorption site and the number of surface Ga atoms. The most stable sites were determined by relaxing the z-coordinate of the adatom as well as the substrate atoms, for each (x, y) position of the adatom. The adsorption energy on the stable sites can be determined as the difference in total energy between the structures with Ga adatom on the stable site and Ga adatom at infinite separation. Shiraishi et al. report that the most stable site for Ga adsorption changes with the number of Ga adatoms. The adsorption energy oscillates with the adsorption of every other atom. When there are a few Ga adatoms, the surface distortions determine the migration potential. However, with an increasing number of Ga adatoms, the stabilization of band energies becomes more important than the surface distortions.

A similar study of Si adatoms on Si(001) was carried out by Brocks [511] et al. using a density functional method. By mapping out the total energy as a function of position on the surface, they show several saddle points for the migration of the adatom. They report 0.6 eV activation energy for diffusion parallel to the dimer rows on the surface and 1.0 eV for diffusion perpendicular to the dimer rows on the surface. Other studies of the same system were performed using empirical Stillinger-Weber potentials [512, 513]. They report that the barriers to diffusion are highly anisotropic and strongly favor migration along the rows.

Using an eight-layer slab model containing over a 1000 atoms, Zhang and co-workers [514] simulated Si dimers adsorbing on the Si (001) surface. They used empirical Stillinger-Weber potentials to estimate the adsorption energies. The usual modeling strategy in these cases is to fix the bottom layers of the slab and let the inner region of the top few layers and the adatoms relax.

5.1.5 Grain Boundaries and Dislocations in Semiconductors

Grain boundaries are internal interfaces between perfect crystalline regions (grains) at different orientations. Dislocations are line-defects along which the atoms are under high stress resulting from unfavorable coordination or strained bonds. Grain boundaries are inherently present in all polycrystalline materials. Dislocations can result from plastic deformation, or under strains arising from a lattice mismatch, e. g., in a heterostructure interface, or at a grain boundary. In fact, grain boundaries, in particular small-angle grain boundaries, can be considered to be an array of parallel dislocations separated by perfect crystalline regions. Unlike point-defects, grain boundaries and dislocations extend over many lattice constants and are therefore, respectively 2-D and 1-D examples of extended defects. Grain boundaries and dislocations occur in all types of crystalline materials: metals, semiconductors, insulators, and ceramics, and control a number of important macroscopic properties: mechanical, electrical, doping, microstructural, atomic, and electronic transport, just to name a few.

From a microelectronics industry point of view it is important to study the role of grain boundaries and dislocations in polycrystalline Si (commonly called Polysilicon), an important component of many modern-day devices, especially memory, opto-electronic, and MEMS devices. Grain boundaries and dislocations may provide preferential sites for the segregation of dopants and point defects, trap sites for charge carriers (electrons and holes), and low-energy pathways for atomic diffusion.

Transmission electron microscopy has become a powerful tool for imaging with atomic resolution, and has been extremely useful in determining atomistic structure of well-defined grain boundaries in different types of materials. Such imaging has led to a number of important results. For instance, in Si grain boundaries all atoms are known to be fourfold-coordinated, just as in the bulk crystal. However, experimental methods alone are often not sufficient to identify individual impurities segregated at grain boundaries, or to determine the structure of an isolated dislocation core. Molecular modeling methods, in conjunction with atomic resolution images can often provide a much more complete picture. There have been a large number of computational investigations on various aspects of grain boundaries: structure, formation energies, dynamics, impurity segregation, and so on. Below we mention a few First-Principles calculations relevant to segregation in polycrystalline elemental semiconductors.

It has been known for some time that n-type dopants like As or P segregate at Si or Ge grain boundaries in electrically inactive configurations. The fraction of dopants in grain boundaries (as compared to bulk crystal) is governed by the segregation energy, i. e., the energy difference between a dopant atom in the grain boundary and a dopant atom in the bulk crystal. For the specific case of As segregation in Si or Ge gain boundaries, experimental values of segregation energies range from 0.4-0.65 eV.

Arias and Joannopoulos [515] used a plane-wave density functional theory approach [516] to calculate the segregation energies of isolated As (n-type dopant) or Ga (p-type dopant) atoms placed at different substitutional sites in a $\Sigma = 5$ symmetric tilt boundary (tilt angle = 36.9 degrees) of Ge with the tilt axis parallel to the (001) direction, and the interface parallel to the (310) plane of the original crystal (shown in Figure 5.7).

These calculations employed a local exchange-correlation functional as parameterized by Perdew and Zunger (see section 7.1.10). The atomic cores were represented by a local pseudopotential due to Starkloff and Joannopoulos [517] and a non-local psedopotential for Ga [518]. A periodic supercell of 68 atoms containing two grain-boundaries was used. First the all-Ge structure was relaxed atomically, and it was found that all sites at the grain boundary were fourfold-

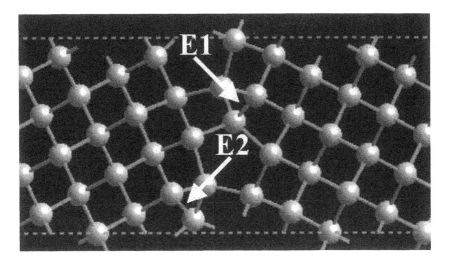

Figure 5.7: Ground-state structure for a $\Sigma = 5$ symmetric tilt boundary in Si, showing Burger's vectors of two constituent edge dislocation cores. The tilt axis is parallel to the (001) direction (normal to the plane of the paper), and the (310) planes of each grain face each other at the interface. The tilt angle is approximately 36.9 degrees.

coordinated, just as in the bulk crystal. Then, various Ge-sites in the grain boundary were substituted for by isolated As and Ga impurities. Interestingly, it was found that this grain boundary binds only n-type, but not p-type impurities. However, the computed segregation energy was only ≈ 0.1 eV, much smaller than the experimental value, with a small site-to-site variation. It was postulated that the much higher experimental segregation energy was due to the presence of steps or other defects to which the dopants have a larger binding.

Further investigation by Maiti and co-workers [519] on the same grain boundary in Si revealed that segregation is possible at defect-free boundaries through the cooperative incorporation of As in 3-fold coordinated configurations. This study revealed that cooperative phenomena involving chains of 3-fold coordinated As atoms or dimers can result in much larger segregation energies than isolated As atoms, and one does not require the presence of steps or other defects. Larger segregation energies can also lead to structural transformation of grain boundaries, as was nicely demonstrated in this work as well. Segregation of As-atoms in the form of dimers was further substantiated by Z-contrast imaging of As impurities in a $\Sigma = 13$ Si grain boundary studied by Chisholm and co-workers [520]. Through a combination of image intensity analysis, first-principles calculations, and statis-

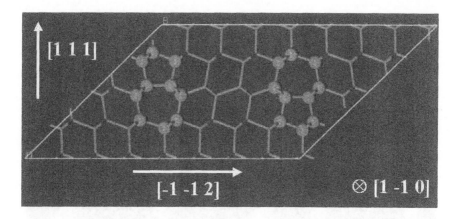

Figure 5.8: A periodic supercell showing oppositely-oriented 90-degree partial dislocations in Silicon. Directions of various axes are indicated.

tical mechanics, these authors were able to establish that segregation occurs in the form of dimers of threefold-coordinated As. Similar calculations performed on an isolated dislocation core (shown in Figure 5.8) revealed much larger segregation energies with a much more pronounced site-to-site variation [521].

In Si and Ge, most dislocations occur in the closely packed 111 planes. Jones and co-workers [522] report first-principles calculations of dislocations in Si semiconductors. Using a 72-atom cluster model with H atoms saturating dangling bonds, the density functional method was used to relax the system. Effects of impurities such as B, P, N, As, and C were also studied. In pure Si, the dislocation was found to be strongly reconstructed. However, the presence of P, N, B, and As impurities appeared to destroy the reconstruction.

5.2   Superconductors

Superconductors are materials which have the ability to conduct electricity without resistance below a critical temperature which is larger than absolute zero. The phenomenon of superconductivity was first seen in mercury at liquid helium temperatures. The discovery (in the late 1980s by Muller and Bednorz) that certain ceramic-like materials exhibit superconductivity at critical temperatures greater than 30 K has given rise to a tremendous interest and activity in this area. The excitement increased further when P. C. W. Chu and co-workers found yttrium barium copper oxide (YBCO) to be superconducting even above liquid nitrogen temperatures. There are various articles and books devoted to this subject [523–525] and we will discuss only some of the simulation aspects here.

After extensive studies over the years on copper oxide superconductors, the following is known [526]:

a) Superconductivity occurs within two-dimensional CuO_2 arrays, which are formed by square planar corner sharing oxygen atoms.

b) Either through addition of positive or negative charges (hole or electron) created by chemical doping, the intermediate spacer layers between the superconducting planes act as charge reservoirs.

c) Due to an energy match between the O 2p orbitals and Cu 3d orbitals the Fermi level is highly hybridized.

d) In all cases seen so far the original antiferromagnetic insulator becomes metallic upon doping to the correct level.

The actual electronic mechanism of superconductivity is still under debate and there is no conclusive evidence on whether the pairing occurs due to magnetic or electronic reasons. However, electronic structure and classical molecular mechanics calculations have provided important insight into complex atomic geometries, charge distributions, and how they change with varying dopant types and amounts. Diffusion of oxygen atoms in the superconductors can also be studied using appropriate model potentials. In the following subsections we will deal with a few examples that will highlight the use of atomistic simulations applied to understanding the static and dynamic properties of superconductors.

5.2.1 Structure

Allan and Mackrodt [527,528] as well as Tokura and co-workers [529] have studied doped lanthanum and neodymium copper oxide systems. As shown in Figure 5.9, both have planar CuO_2 units which is considered an essential feature for high temperature superconductors. The Cu atoms are six-fold coordinated in La_2CuO_4, while they are four-fold coordinated in Nd_2CuO_4.

For ionic oxides such as these, atomistic simulations can be performed using empirical ion pair potentials. Defect structures due to oxygen vacancies created by substitution of La^{3+} by divalent cations such as Ca^{2+}, Sr^{2+}, or Ba^{2+} can be simulated using these potentials. Evain et al. [530] used empirical atom-atom potentials to simulate the crystal structure of La_2CuO_4. They studied the tetragonal to orthorhombic distortion of this system as well as the high temperature superconductor $La_{2-x}M_xCuO_4$ (M = Ba, Sr). The crystal structure of La_2CuO_4 is tetragonal above 533 K and orthorhombic below this temperature. This distortion occurs by the buckling of the planar CuO_4 units due to ionic interaction involving La^{3+} cations. The ion pair potentials are described as consisting of a Coulomb

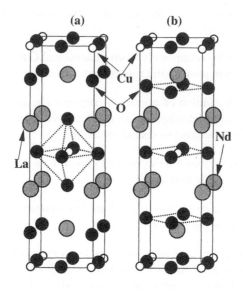

Figure 5.9: Structure of (a) La_2CuO_4 and (b) Nd_2CuO_4.

term (V_c), a non-bond repulsion term (V_{nr}), and a non-bond dispersion term (V_v) expressed as

$$V_c = q_iq_j/r_{ij} \tag{5.1}$$

$$V_{nr} = B_{ij}\exp(-r_{ij}/\rho_{ij}) \tag{5.2}$$

$$V_v = -C_{ij}/r_{ij}^6 \tag{5.3}$$

where q_i and q_j are the charges on the ions, r_{ij} is the distance of separation, B_{ij}, C_{ij}, and ρ_{ij} are adjustable parameters to be fitted using the experimental data.

$$1/\rho_{ij} = (1/\rho_{ii} + 1/\rho_{jj})/2 \tag{5.4}$$

and

$$C_{ii} = 3I_iP_i^2/4 \tag{5.5}$$

where I_i is the ionization potential and P_i is the polarizability of ion i. The strategy is to fit B_{ij} and ρ_{ij} of $Cu^{2+}\cdots Cu^{2+}$, $O^{2-}\cdots O^{2-}$, and the $La^{3+}\cdots La^{3+}$ pairs using the experimental unit cell of CuO and La_2O_3. These adjustable parameters are changed until the simulated lattice constant and the atom positions match the experimental data. These parameters can then be used in the simulations on the La_2CuO_4 structure. In order to simulate the $La_{2-x}M_xCuO_4$ (M = Ba, Sr) structure the strategy employed by Evain et al. was to increase the $La^{3+} \cdots La^{3+}$ B_{ij} value and decrease that for the $Cu^{2+} \cdots Cu^{2+}$. This was justified since both Ba^{2+} and Sr^{2+} are much larger than the La^{3+} ion and since in the doped system the copper atoms are usually in a higher oxidation state to compensate for excess charge. Based on these assumptions the tetragonal to orthorhombic structural distortions were simulated by Evain et al. for $La_{1.85}Ba_{0.15}CuO_4$ and for $La_{1.85}Sr_{0.15}CuO_4$ and these were found to be in agreement with experiments.

5.2.2   Oxygen Vacancies

In the case of Tl-Ba-Ca-Cu-O superconductors there are two phases; one in which the lattice is primarily tetragonal with the ideal chemical formula $TlBa_2Ca_{n-1}Cu_nO_{2n+2.5}$ (n=1,2,3,4) and a second phase, which belongs to a body-centered tetragonal lattice with an ideal chemical formula $Tl_2Ba_2Ca_{n-1}Cu_nO_{2n+4}$ (n=1,2,3,4). In the former phase each unit cell contains 4.5 oxygen atoms if n=1, i. e., there is half an oxygen vacancy per unit cell. In order to simulate this, for simplicity, randomness is omitted and the vacancy can be repeated periodically. By building two unit cells and omitting one oxygen in the super cell it is possible to construct a model of periodic defects in $TlBa_2CuO_{4.5}$ [531].

The work by Jia [531] on this system was performed using the extended Hückel tight binding method. No geometry optimization around the vacancy was performed. Different positions of the oxygen atom vacancy were considered and for each position the net charge on each atom and the electric field gradient were calculated. It is seen that one of the positions for the oxygen atom vacancy (as shown in the figure) results in +2.4 and +0.62 for the net charge of the two copper atoms. This is similar to $YBa_2Cu_3O_{7-x}$ where both Cu^{3+} and Cu^+ cations exist. Also the electric field gradient calculated for this oxygen atom vacancy is the largest indicating that the movement of the electron and the hole are most facilitated in this structure.

5.2.3   Oxygen Diffusion

Oxygen diffusion has been studied by various workers both experimentally and theoretically. However, until recently, the nature of oxygen diffusion paths in superconductors was a matter of controversy. Figure 5.10 shows the structure of a superconductor, $YBa_2Cu_3O_7$.

By calculating the defect energy of the mobile ion along the diffusion path, Islam [532] showed that the most favorable energetic path for oxygen diffusion was O1 to O4 to O1. On the other hand, Rothman and co-workers [533] suggested that oxygen diffusion occurs via an O1 oxygen jump to an O5 empty site and moves along empty O5 sites. Catlow and co-workers [534] have studied oxygen diffu-

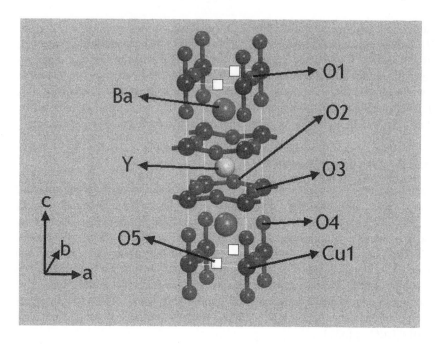

Figure 5.10: Structure of $YBa_2Cu_3O_7$.

sion in $YBa_2Cu_3O_{(7-x)}$ (x = 0.09 - 0.27) using molecular dynamics simulations. All the work was based on ion pair potential. Catlow and co-workers derived the potential parameters by fitting a rigid ion model to reproduce the calculated structure within 5 % of the experimental structure. Molecular dynamics calculations at constant volume were performed on the compositions $YBa_2Cu_3O_{6.91}$, $YBa_2Cu_3O_{6.82}$, and $YBa_2Cu_3O_{6.73}$ for 100 ps each at different temperatures ranging from 1350-1500 K [534]. These high temperatures are necessary in simulations to obtain accurate diffusion coefficients. Their molecular dynamics simulations revealed that no O5 to O5 jump was present but the paths O1 to O5, O1 to O4, and O4 to O5 were all possible. At lower temperatures they suggest that the O1-O4-O1 path is the main contributor, while at high temperatures the O1-O5-O1 path becomes important.

Chapter 6

Nanomaterials

6.1 Introduction

Nanotechnology [535–539] is a generic name given to the science and technology of systems/machines with at least one of the dimensions being tens of nanometers or less (1 nanometer = 10^{-9} m = 10 Å). Materials systems designed to be used in nanotechnology are called nanomaterials. Nanotechnology is a rapidly emerging field in which new ideas are being born and discoveries being made at a breathtaking pace. New materials systems with novel properties are being designed or invented at an incredible rate in laboratories throughout the world. The tremendous interest and activity in nanotechnology stems from the facts that: (1) properties at the nanoscale can be very different as compared to bulk materials, including structural, mechanical, electronic, optical, magnetic, thermal, chemical, and catalytic properties; (2) many of these properties could be tuned in a controlled way simply upon changing the size of the nanomaterial or by attaching chemical functional groups to it; (3) present-day experimentalists have the tools to create/synthesize/manipulate nanomaterials of diverse chemical compositions, shapes, and sizes; and (4) nanomaterials possess tremendous potential for technological applications.

6.1.1 Different Types of Nanomaterials

Nanomaterials come in all shapes and sizes, span a large part of the periodic table, and cover a diverse range of materials types including organics, metals, polymers, semiconductors, ceramics, glasses, minerals, and composites. Table 6.1 lists a fascinatingly diverse array of nanomaterials that have been synthesized so far.

Table 6.1: Examples of technologically important nanomaterials

Classes of Nanomaterials	Specific Examples
Quasi-one-dimensional	Carbon nanotubes (CNTs) [540, 541], other (BN, MoS_2, metal-oxide) nanotubes [542], metal-oxide (ZnO, SnO_2) nanoribbons [543], Semiconducting and metallic nanowires [543, 544]
Particles	C_{60} and the Fullerenes [545], nanoparticles [546], nanoshells [547], Quantum dots [544, 548–551], Organic molecules, Molecular wires (rotaxanes, catenanes, & variants thereof) [552, 553], phosphors, dyes, light-emitting polymers [554], drugs [555], lubricants [556]
Macromolecules	Polymers, micelles, dendrimers [557], DNA, RNA, Proteins, organic nanotubes [558], biochips [559–561], bio-mimetic systems [562]
Nanocomposites	Polymer-CNT [563, 564], Polymer-layered-silicate, Polymer-ceramic, nanoparticle-implants, clays [565]

6.1.2   Synthesis Methods and Potential Uses

Nanotechnology usually implies an intelligently directed association of molecular/chemical/biological components. Broadly speaking, the synthesis of nanomaterials takes two different design routes, i. e., top-down design, and the bottom-up design. The top-down design involves etching, chiseling, or sculpting nanoscale features into an existing structure, typically accomplished by microscopy (Atomic Force microscope (AFM), Scanning Tunneling microscope (STM)) or electron-beam lithography. Unfortunately, this method is not parallel or scalable, and therefore time-consuming, and very expensive. Such a method could perhaps be useful in precision building of nano-templates, from which thousands to millions of cheaper copies could be created. However, large-scale synthesis of systems and devices is typically accomplished by a bottom-up design, which involves building up nanostructures from atomic or molecular building blocks through some sort of physical or chemical self-assembly. This method is massively parallel (and therefore fast), cheap, and makes clever use of electrostatic, van der Waals, H-bonding, hydrophobic/hydrophilic, or electronic/chemical interactions. Most practically useful nanomaterials are created this way. Some examples of self-assembly methods include Dip-Pen-Nanolithography [566], molecular beam epitaxy (MBE) [567], vapor deposition techniques (physical vapor deposition (PVD) [568], chemical vapor deposition (CVD) [569], laser-assisted catalytic growth [570,571], nanofluidics-aided self-assembly, and DNA- and Protein-directed self-assembly [572–576].

The availability of nanomaterials with a wide range of intriguing properties has given rise to the possibility of a number of technological applications both in the Materials as well as in the Life sciences, including:

- Next-generation computer chips (molecule-based electronic switches, memory and logic devices, CNT-based transistors and interconnects, nanowire-based electronic and optoelectronic devices, spintronics-based Quantum computers)

- Light- and electron-emitting devices (CNT-based displays, organic light-emitting-devices, Quantum dots)

- Structural materials (ductile ceramics, bio-implants with stronger adhesion, nanocomposites with high strength-to-weight ratio)

- Energy conversion and storage (fuel cells, long-life Li-ion batteries, H-storage materials, solar cells)

- Catalysts (nanopowder with very high surface-to-volume ratio, supported metallic nanoclusters with high selectivity, nanoporous materials as molecular sieves)

- Medical implants (nanoparticle-enhanced replacement materials for bones, teeth and other orthopedic implants, heart-valves)

- Drugs, medical imaging (metallofullerene-based MRI contrast agents, buckyball/fullerene-based drugs)

- Drug delivery (liposome-, dendrimer-, polymer- and nanoshell-based drug delivery devices, controlled release with external stimulus, site-specific targeting, in-cell delivery of macro-molecules)

- Sensors (NEMS, photodetectors, chemical sensors, bio-sensors)

Several products enhanced with nanoparticles are either available commercially, or are close to being introduced into the market. Some of these include coatings, paints, adhesives, cement, sunscreen lotion, and textiles/ fabric.

6.1.3   Role of Computational Modeling

As with any new technology, Nanotechnology has many challenges to overcome, typically associated with our control and precision at the nanoscale. Some of the challenges include: device integration (interconnect failure, addressability issues), growth and synthesis (difficulty in size-dispersion control, requirement of novel assembly techniques, presence of structural defects), contact resistance (necessity of atomic-level structural precision at junctions), functionalization (challenges with chemical inertness), and doping (non-uniformity of dopant levels). Molecular modeling is a great approach to surmounting some of the above obstacles because it often provides deeper insight into the system properties as a function of size/shape, structural defects, added functional groups, and system surroundings. Often the whole nanosystem/device can be modeled in full atomistic detail on the computer, and its properties studied with a level of precision not possible experimentally. Growing success of Density Functional Theory (DFT) [577], availability of accurate interatomic potentials (Force-Fields) [578], development of Green's function-based electronic transport codes [579], and deployment of sophisticated graphics user interface (GUI) have all led to the emergence of molecular modeling as a powerful tool in Nanotech research. Molecular modeling applications to nanomaterials are too numerous to mention here, and we can only point to the literature for further reference [580–586]. Instead, the rest of this chapter focuses on two commercially important nanosystems, i. e., carbon nanotubes (CNTs) and metal-oxide nanoribbons. The well-characterized atomic structures of these two systems, as well as their high degree of structural purity, allow accurate computer modeling and in silico property prediction. In the following we illustrate some of the modern techniques of molecular modeling to study technologically important applications like displays, electromechanical sensing, and chemical sensing.

6.2   Carbon Nanotubes (CNTs)

6.2.1   Atomic and Electronic Structure

A CNT can be geometrically thought of as a graphite sheet rolled into a seamless cylinder. A necessary condition for the cylinder to be seamless is that upon rolling, a graphite lattice point (n1, n2) coincides with the origin (0, 0). Thus, if $\vec{a}_1$ and $\vec{a}_2$ are the two lattice vectors of graphite, the CNT circumference is equal to the length of the vector $(n_1\vec{a}_1 + n_2\vec{a}_2)$, while the CNT chiral angle θ is defined as the angle between vectors $(n_1\vec{a}_1 + n_2\vec{a}_2)$ and $\vec{a}_1$. With the choice of lattice vectors as in Fig. 6.1 (a), the chiral angle and diameter of a CNT are given respectively by the formulas:

$$\theta = \tan^{-1}[\sqrt{3}n_2/(2n_1 + n_2)], \text{and} \qquad (6.1)$$

(a) Graphite Sheet   (b) Armchair CNT   (c) Zigzag CNT

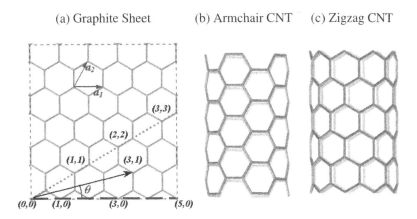

Figure 6.1: Carbon nanotube (CNT) basics. (a) A graphite sheet with lattice vectors a_1, a_2. A few lattice points are indicated, as is the chiral angle ($\theta \sim 13.9°$) for a (3, 1) CNT. Dotted and dashed lines are drawn along circumferences of armchair and zigzag tubes respectively; (b) a (5, 5) armchair tube; (c) a (9, 0) zigzag tube.

$$d = a\sqrt{(n_1^2 + n_1 n_2 + n_2^2)/\pi} \qquad (6.2)$$

where a = $|\vec{a_1}| = |\vec{a_2}| \sim 2.45$ Å is the lattice constant of graphite. The CNT diameter and chirality, and therefore its atomic geometry, is completely specified by the two integers (n_1, n_2), which are referred to as the chiral indices of the CNT. Due to the symmetry of the graphite lattice, a nanotube of any arbitrary chirality can be defined in the range $n_1 \geq n_2 \geq 0$ and $n_1 > 0$, which implies that the chiral angle θ for all CNTs lies between 0 and 30°. CNTs with the extreme chiral angles of 0 and 30° have special names: a CNT with $\theta = 0$ (i. e., $n_2 = 0$) is called zigzag, while a CNT with $\theta = 30°$ ($n_1 = n_2$) is called armchair. The names armchair and zigzag simply reflect the shape of the open edges of these CNTs (Figs. 6.1 (b, c)). CNTs with any other chiral angles (i. e., $0 < \theta < 30°$) are called chiral.

Armchair and zigzag CNTs possess small periodic repeat lengths along the nanotube axis, the repeat-length being only a (~ 2.45 Å) for armchair tubes and $\sqrt{3}a$ (~ 4.24 Å) for zigzag tubes. Chiral CNTs on the other hand can possess very long periodic repeat lengths depending on the ratio of its chiral indices. Thus, electronic structure calculations, especially those employing first-principles Quantum Mechanics (QM) with periodic boundary conditions, are rarely performed on chiral tubes. However, this does not turn out to be a serious limitation. Since a CNT is just a rolled-up graphite sheet, one can obtain a good approximation to the CNT electronic structure simply by applying an appropriate boundary condition to the electronic structure of a graphite sheet, with a small perturbation due to the finite cylindrical curvature of the CNT surface. The boundary condition for a CNT with chiral indices (n_1, n_2) corresponds to the coincidence of the (n_1, n_2) lattice point of graphite with the origin $(0, 0)$. It has been known for some time that a single sheet of graphite (also known as graphene) is neither a semiconductor nor a metal, but a semi-metal [587] (i. e., a zero-bandgap semiconductor). This peculiarity implies that the electronic states of graphene are very sensitive to additional boundary conditions that a CNT mandates. Taking into account small effects due to curvature, such boundary conditions lead to the important result [588–593] that: all armchair tubes are metallic; CNTs with $n_1 - n_2 = 3n$ (n = any positive integer), which includes the $(3n, 0)$ zigzag tubes as a special class, are quasi-metallic (small band-gap ~ 10 meV or less, arising from curvature effects); and CNTs with $n_1 - n_2 \neq 3n$ are semiconducting, with a band-gap decreasing as $1/d$ as a function of tube diameter d (thereby converging to the zero bandgap of graphite in the limit $d \to \infty$). Experimental measurements are often not able to make the distinction between metallic and quasi-metallic tubes because of the presence of contact resistance and thermal effects. Thus for simplicity, experimentalists often classify CNTs as either metallic or semiconducting, and we follow the same convention in the discussion below.

6.2.2 Synthesis Methods, Properties, Potential Applications

CNTs were originally discovered by Sumio Iijima in 1991 [594] in the soot produced by an arc discharge between graphite electrodes in a helium atmosphere. The original tubes [594–596] were multi-walled, i. e., that they consisted of several concentric cylinders with successive layers separated by a distance approximately equal to the inter-layer spacing of graphite. Subsequently single-walled CNTs were synthesized in the same arc-discharge apparatus with the addition of transition-metal catalysts [597, 598]. Presently, high-quality multi-walled and single-walled CNTs can be grown in well-defined directions using chemical vapor deposition (CVD) [599], while high-yield of single-walled CNTs can be obtained by several techniques, including: arc-discharge of Ni-Y catalyzed graphite electrodes [600], laser ablation of Ni-Co catalyzed graphite targets [601], and vapor-phase pyrolysis of CO and $Fe(CO)_5$ [602] (the so-called HiPCO process).

Over the last decade CNTs have become one of the hottest research areas in all of science and engineering because of a number of fascinating properties:

- Depending on their chiral indices, they can behave like metals or semiconductors, as discussed in section 6.2.1.

- Single- or multi-walled CNTs are exceptionally strong, could possess a Young's modulus as high as 1.2 Terapascal, six times the modulus of steel.

- CNTs are elastic to the highest degree, and do not display plasticity behavior even under large deformation including stretching, bending, or twisting.

- Metallic CNT are 1-dimensional quantum conductors where electrons travel ballistically: there is no heat dissipation along the length of the CNT. All dissipation occurs at the contacts.

- CNTs can have huge aspect ratio (i. e., length to diameter ratio), as large as 105. Field-emission of electrons can, therefore, be induced from the tip of long metallic CNTs in the presence of moderate electric fields.

- Depending on its chiral indices, metallic CNTs can undergo metal-to-semiconductor transition under small tensile or torsional strain. This effect is discussed in more detail in section 6.2.5.

- With a magnetic field parallel to its axis, a CNT can exhibit the Aharonov–Bohm effect.

- Atoms/molecules can be enclosed inside a CNT

- A CNT can be doped both p-type and (to a lesser degree) n-type.

All the above properties come with the promise of a host of commercial applications [603, 604], including: Field Emission-based Flat Panel displays [605–611], novel semiconducting devices and robust metallic interconnects in microelectronics [612, 613], hydrogen storage devices [614], structural reinforcement agents [615], chemical sensors [616–618], electromechanical sensors [619], nanoprobes [620], and nanotweezers [621].

6.2.3 Examples of Molecular Modeling

Ever since the discovery of CNTs, it has provided a fertile ground for theoretical simulations and analysis. In fact, the prediction of the dependence of CNT's electronic structure on its chirality [588–590] came within just a few months of the experimental discovery [594]. Since then there have been a huge number of theoretical investigations [582–584, 622–624] of growth mechanisms, structure and energetics of topological defects, mechanical and electrical response to various kinds physical perturbation, field-emission from tips of metallic CNTs, electronic effects of doping and gas adsorption, chemical reactivity, interaction with polymers, capillary effects, CNT-metal contacts, H- and Li-storage, thermal conductivity, and encapsulation of organic as well as inorganic material. Computational approaches used in the above work include solving diffusion equations, QM simulations (DFT, tight-binding, and semi-empirical methods), classical molecular dynamics, kinetic Monte Carlo, Genetic algorithms, and Green's-function-based electronic transport theory. The following two sections illustrate the use of some of the above techniques to explore two application areas of CNTs: (1) field-emission displays; and (2) nano-electromechanical sensors.

6.2.4   CNT-Based Displays: Effect of Adsorbates

Synthesis of CNTs with length-to-width ratio as large as 105, as well as advanced fabrication methods for generating self-aligned or patterned nanotube films on glass [625] or silicon [626] substrates have pushed CNT-based flat-panel displays to the brink of commercial reality [627–629]. One major challenge to overcome is to reduce the threshold field, defined as the electric field necessary to induce a current of 1 mA/cm^2. One way to achieve this is to introduce adsorbates that might effectively lower the ionization potential (IP) and facilitate the extraction of electrons. An important experimental work in this regard was done at Motorola [630], which studied changes in field emission behavior in the presence of various gases, i. e., CO, CO_2, H_2, and water vapor. Adsorbed water was found to significantly enhance the emission current, while the other residual gases did not have any noticeable impact. Water also appeared to be present at the CNT-tip up to very high temperatures ($\sim$ 900 K), and field-emission current was more enhanced at higher partial pressures of water.

In order to understand the difference between H_2 and water on the field-emission properties of CNTs, first-principles electronic structure calculations were carried out on a (5, 5) armchair tube, with a hemispherical tip represented by the half of a C_{60} molecule. A uniform external field E_{FE}, directed toward the tube from above, was chosen to represent the electric field close to the tube tip under field emission conditions. Since the actual field close to the tip is radial rather than uniform [631], the uniform field representation is appropriate only if the tube-stem in the simulation is confined to the same size as the tube diameter. For this reason, a short stem of three atomic layers (30 atoms) was used. A more realistic field around the tip could perhaps be simulated by means of a large number of point charges situated at some distance from the tip. Such an effort was not undertaken for two reasons: (1) in all flat panel display designs the field emission occurs not from isolated tubes, but from a film of closely-spaced nearly-aligned tubes, with the resulting E-field more closely resembling a uniform field above the tip assembly; (2) the essential chemistry that drives the effect of adsorbates on field emission behavior can be extracted from calculations using a uniform field, as illustrated below. For all calculations a magnitude of $E_{FE} = 1$ V/Å was used as a ballpark figure around which field emission is known to occur for these systems. Difference in electronegativities between H and C introduces an artificial dipole moment in the CNT if the dangling C-atoms at the stem-end are H-saturated. The C-atoms at the stem end were therefore left unsaturated. In addition, these C-atoms were fixed during the simulation in order to mimic the presence of a long stem in actual experiments. Structures of the CNT with or without adsor-

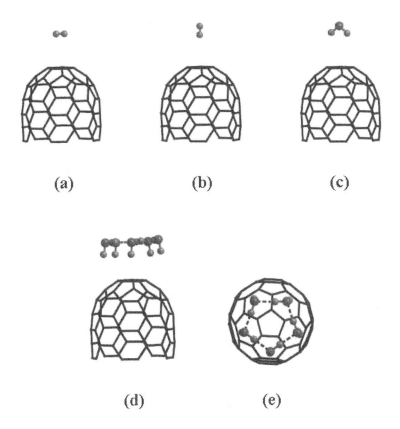

Figure 6.2: Optimized structures for adsorbates on a CNT tip in zero electric field: (a) H2 (flat); (b) H2 upright; (c) a single down water; (d) a cluster of 5 water molecules; (e) Fig. (d) viewed from top.

bates were optimized using the DFT code DMol^3 [454, 632–635], in which the electronic wavefunctions are expanded as a linear combination of atom-centered basis functions defined on a numerical grid. The present calculations explicitly considered all electrons in the system (i. e., core electrons were included), with wavefunctions expanded in the Double Numeric Polarized (DNP) basis set [632], and the exchange-correlation part of the electron-electron interactions represented by the gradient-corrected PBE functional. The main results from this study [636] can be summarized as follows:

1. A single H_2 molecule can either lie "flat" or "upright" on the CNT (Fig. 6.2 (a, b)). A single water molecule adsorbs in such a way that its plane

symmetrically cuts the CNT into two equal halves. The two H-atoms are symmetrically equidistant from the tip, and could be either closer or further from the surface as compared to the O-atom. We call these configurations "down" (Fig. 6.2 (c)) and "up" water respectively.

2. In zero electric field none of the adsorbates show any significant binding to the CNT tip. The H_2 has less than 0.1 kcal/mol binding in either configuration, while a single water has ~ 0.7 kcal/mol binding.

3. At a tip-field of 1 V/Å, the binding energies change significantly. Thus the H_2 has a binding energy of 1.1 and 3.5 eV respectively in the flat and the upright configurations, while the "down" water has a binding energy of ~ 20 kcal/mol, which is of the order of a chemical bond. The "up" water becomes unstable in the presence of a finite electric field at the CNT and spontaneously flips into a "down" water. The "up" water is therefore not considered further in the discussion below.

4. Because water molecules can form clusters through H-bonding, it was also important to consider the structure and adsorption energy of a cluster of water molecules on the CNT tip. Because of the five-fold symmetry of the (5, 5) CNT (with a C_{60}-half cap) about its axis, it was easy to determine the global minimum for a cluster of 5 water molecules. In the minimum energy structure the five water molecules form a "crown" on the top of the CNT (Fig. 6.2 (d, e)). Each water molecule in this crown is tilted with respect to the ideal "down" water configuration for a single water. However, the resulting loss in interaction energy with the CNT tip is almost exactly recovered from the gain in H-bonding to the nearest neighbor waters.

5. The adsorbate atoms in all cases are at a distance of 3 Å or more from the CNT surface. This strongly suggests that they are being physisorbed. This is supported by an analysis of charge density in the region between the adsorbate and the CNT tip, which is less than 0.004 el/$Å^3$. Even at a field as large as 1 V/Å, the changes to the adsorbate and CNT structures are small. In particular, the distance between the adsorbate and the CNT changes by less than 0.07 Å, and the bond-lengths change by less than 0.01 Å. The most significant geometrical change is in the H-O-H angle in water, which decreases from 103.7° in zero field to 98° on the tube tip at $E_{FE} = 1$ V/Å. Such a change in angle leads to only a 0.02 a.u. increase in the dipole moment of water.

6. An analysis of the various interaction energies and the dipole moments of the CNT as well as the adsorbates yield the following picture: in the absence of an electric field, the CNT has a very small dipole moment. This small

Figure 6.3: Highest Occupied Molecular Orbital (HOMO) for: isolated CNT; CNT + 1 adsorbed water; and CNT + 5 adsorbed water. Although the position and shape of the HOMO is not affected significantly by the adsorbates, the energy level and ionization potential change markedly.

dipole has only negligible interaction with any of the adsorbates. However, under field-emission conditions, large E-field at the CNT tip induces a significant dipole moment. This dipole interacts with the dipole of the adsorbates. Water molecules having a large intrinsic dipole moment interact strongly, while H_2 molecules have only a small field-induced dipole, and therefore a much weaker interaction.

7. Finally, the highest occupied molecular orbital (HOMO) for the isolated CNT, one water molecule on CNT, and five water molecules on the CNT were analyzed (Fig. 6.3). The HOMO essentially remains confined to the CNT tip, and its shape does not change significantly due to the presence of the adsorbates. However, when the Ionization Potential (IP) is computed, it is found to be 0.1 eV and 0.6 eV lower for the one water and five water cases respectively, as compared to the adsorbate-free CNT. This also correlates perfectly to the instability of the HOMO orbital. Thus, electrostatic interactions from the adsorbates make the HOMO level in the CNT more unstable, with a corresponding decrease in the IP, thereby lowering the operating voltage for field emission.

Table 6.2 displays computed IP of a (5, 5) CNT-tip (same as in Fig. 6.2) for a number of potential adsorbates [637], including water. It is clear that the lowering of IP is more pronounced, the larger the dipole moment of the adsorbate. This work calls for further experiments on polar adsorbates in order to optimize the performance of CNT-based field-emission displays. It would also be interesting to compute I-V characteristics due to field-emission from CNT-tips [638] in the presence of adsorbates.

Table 6.2: IP of a (5, 5) CNT for various molecules physisorbed on its tip

Adsorbate	Dipole Moment (Debye)	Ionization Potential (eV)
H_2	0.0	6.40
HCl	1.0	6.36
H_2O	2.0	6.30
HCN	3.0	6.20
LiH	5.9	6.12

6.2.5   CNT-Based Nano-Electromechanical Sensors

Interest in the application of carbon nanotubes as electromechanical sensors got a significant boost from the pioneering experiment of Tombler et al. [619], in which the middle part of the segment of a metallic CNT suspended over a trench was pushed with an AFM tip. Beyond a deformation angle of $\sim 13°$ the electrical conductance of the tube dropped by more than two orders of magnitude. The effect was found to be completely reversible, i. e., through repeated cycles of AFM-deformation and tip removal, the electrical conductance displayed a cyclical variation with constant amplitude. An interesting explanation was put forward by O(N) tight-binding calculations [639], which show that beyond a critical deformation several C-atoms close to the AFM tip become sp^3-coordinated. This leads to the tying up of π-electrons into localized σ-states, which would explain the large drop in electrical conductance.

Considering the significance of the above result, it was important to carry out an independent investigation using first-principles QM. Unfortunately, the smallest models required to simulate the AFM-deformation of a CNT typically involve a few thousand atoms, which makes first-principles QM simulations unfeasible. This necessitated a combination of first-principles DFT and classical molecular mechanics. Bond reconstruction, if any, is likely to occur only in the highly deformed, non-straight part of the tube close to the AFM-tip. For such atoms ($\sim$ 100-150 atoms including AFM-tip atoms), a DFT-based QM description was used, while the long and essentially straight part away from the middle was described accurately using the Universal Forcefield (UFF) [640] which had previously been used in CNT simulations [641].

Because of known differences in the electronic response of zigzag and armchair tubes to mechanical deformation, the simulations were performed on a (12, 0) zigzag and a (6, 6) armchair tube, each consisting of 2400 atoms. The AFM tip was modeled by a 6-layer deep 15-atom Li-needle normal to the (100) direction, terminating in an atomically sharp tip. To simulate AFM-tip-deformation, the Li-needle was initially aimed at the center of a hexagon on the bottom-side of the middle part of the tube. The Li-needle tip was then displaced by an amount δ toward the tube along the needle-axis, resulting in a deformation angle $\theta = \tan^{-1}(2\delta/l)$, l being the unstretched length of the tube. At each end of the tube, a contact region defined by a unit cell plus one atomic ring (a total of 36 and 60 atoms for the armchair and the zigzag tube respectively) was then fixed and the whole tube relaxed with the UFF, while constraining the needle atoms as well. The contact region atoms were fixed in order to simulate an ideal undeformed semi-infinite carbon nanotube lead, and to ensure that all possible contact modes are coupled to the deformed part of the tube. Following the UFF relaxation, a cluster of 132 C-atoms for the (6, 6) tube, and a cluster of 144 C-atoms for the (12, 0) tube were cut out from the middle of the tubes. These clusters, referred to below as the QM clusters (plus the 15 Li-tip atoms in tip-deformation simulations), were further relaxed with Accelrys' DFT-code DMol^3 [635], with the end atoms of the cluster plus the Li-tip atoms fixed at their respective classical positions. In order to cut down on CPU requirements, the DFT calculations were performed using the Harris functional [642, 643] and the local exchange-correlation potential due to Vosko, Wilk and Nusair [644].

Fig. 6.4 displays the tip-deformed QM-cluster for (6, 6) and (12, 0) tubes at the highest deformation angle of $25°$ considered in these simulations. Even under such large deformations, there is no indication of sp^3 bonding, and the structure was very similar to what was observed for a (5, 5) tube in a previous work [645]. The absence of sp^3 coordination is inferred based on an analysis of nearest-neighbor distances of the atoms with the highest displacements, i. e., the ones closest to the Li-tip. Although for each of these atoms the three nearest neighbor C-C bonds are stretched to between 1.45-1.75 Å, the distance of the fourth neighbor, required to induce sp^3 coordination, is greater than 2.2 Å for all tubes in our simulations.

Following the structural relaxation of the CNTs, the transmission and conductance were computed using the recursive Green's function formalism [579, 646]. A nearest-neighbor sp^3 tight-binding Hamiltonian in a non-orthogonal basis was chosen, and ideal semi-infinite contacts assumed at both ends. The parameterization scheme explicitly accounts for effects of strain in the system through a bond-length-dependence of the Hamiltonian and the overlap matrices H_{ij} and S_{ij}, [647]. First, the retarded Green's function G^R of the whole CNT was determined by solving the following equation:

$$(E \cdot S_{ij} - H_{ij} - \Sigma_{L,ij} - \Sigma_{R,ij})G^{R,jk} = \delta_i^k \qquad (6.3)$$

where $\Sigma_{L,R}$ are the retarded self-energies of the left and the right semi-infinite contacts. The transmission at each energy was then obtained [648] from the equa-

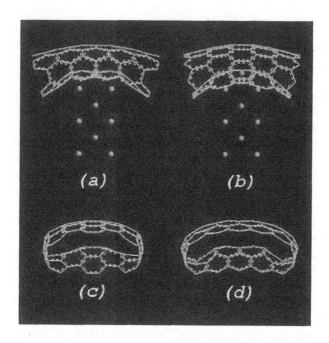

Figure 6.4: DMol^3-relaxed Li-tip-deformed QM clusters for: (a) the (6, 6) arm-chair (132 C-atoms); and (b) the (12, 0) zigzag (144 C-atoms), in side views. The deformation angle is 25° for both tubes. Figs. (c) and (d) are respective views along the tube length, with the Li-tip hidden for clarity.

tion:

$$T(E) = G^{R,ij}\Gamma_{L,jk}G^{A,kl}\Gamma_{R,li} \tag{6.4}$$

where $\Gamma_{L,R} = i(\Sigma_{L,R}^{R} - \Sigma_{L,R}^{A})$ are the couplings to the left and right leads. Finally, the total conductance of the tube was computed using Landauer–Büttiker formula:

$$G = \frac{2e^2}{h} \int_{-\infty}^{\infty} T(E) \left(-\frac{\partial f_o}{\partial E}\right) dE \tag{6.5}$$

where $f_o(E)$ is the Fermi–Dirac function. In Eqs. (6.3, 6.4), summation over the repeating Roman indices is implied. The lower and upper indices denote covariant and contravariant components of a tensor.

The simulations (Fig. 6.5) indicate that the conductance remains essentially constant for the (6, 6) armchair tube up to deformation as large as 25°. However, for the (12, 0) zigzag tube the conductance drops significantly, by two orders of magnitude at 20°, and 4 orders of magnitude at θ=25°. Since sp^3 coordination could be ruled out, what could be the cause for such a large conductance drop in

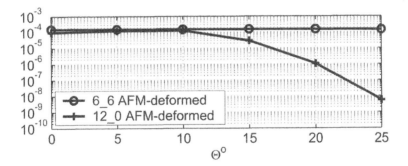

Figure 6.5: Computed conductance (in Siemens) of AFM-deformed (6, 6) and (12, 0) CNTs as a function of deformation angle θ (in degrees).

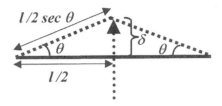

Figure 6.6: Schematic diagram representing deformation with an AFM tip. The tip-deformed tube undergoes a tensile strain. The deformation angle θ is related to the tip displacement (δ) and non-deformed tube-length (l) by $\theta = \tan^{-1}(2\delta/l)$.

the experiment of ref. [619], as well as for the metallic zigzag tube in Fig. 6.5? Also, why did the armchair tube display no significant drop in conductance even up to large angles of deformation? A simple explanation emerges if one zooms out from the middle of the tube and looks at the profile of the whole tube under AFM-deformation. One immediately discovers an overall stretching of the tube under AFM-deformation, as indicated schematically in Fig. 6.6. Fig. 6.7 compares drop in conductance in the (12, 0) tube subjected to: (1) AFM-deformation, and (2) uniform stretching. Such comparison makes it clear that tensile strain is the main reason behind the conductance drop in an AFM-deformed metallic zigzag tube. It was also verified that the (6, 6) armchair tube does not undergo any significant conductance drop upon stretching [649].

In order to explain the differences in conductance drops of the armchair (6, 6) and the zigzag (12, 0) tubes as a function of strain, it helps to re-visit the literature where a considerable amount of theoretical work already exists [650–654]. An

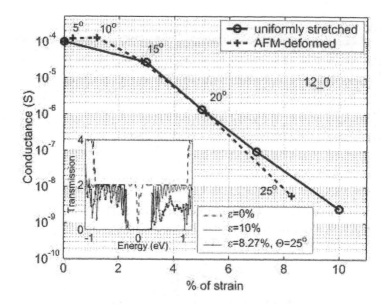

Figure 6.7: Conductance of an AFM-tip-deformed (12, 0) tube as compared to the same tube under a uniform stretch. Actual angles of tip-deformation are indicated. The % strain for the AFM-deformed tube is computed from the average C-C bond-stretch in the middle of the straight portion of the tube. The inset shows transmission in the vicinity of the Fermi surface (Energy = 0) for a uniform strain of 10 % and a tip-deformation angle of 25°, as compared to the non-deformed tube.

important result [653] is that the rate of change of bandgap as a function of strain depends on the CNT chiral angle θ, more precisely as proportional to $\cos(3\theta)$. Thus, stretched armchair tubes (θ= 30°) do not open any bandgap, and always remain metallic. On the other hand, a metallic $(3n, 0)$ zigzag tube ($\theta = 0$) can open a bandgap of $\sim$ 100 meV when stretched by only 1 %. This bandgap increases linearly with strain, thus transforming the CNT into a semiconductor at a strain of only a few percent. In general, all metallic tubes with $n_1 - n_2 = 3n$ will undergo the above metal-to-semiconductor transition, the effect being the most pronounced in metallic zigzag tubes. An experiment as in ref. [619] is thus expected to show a decrease in conductance upon AFM-deformation for all CNTs except the armchair tubes. Researchers are also beginning to explore the electromechanical response of a squashed CNT [655,656], where sp^3 coordination is a possibility.

In addition to the above results for metallic CNTs, theory also predicts that

[653] for semiconducting tubes ($n_1 - n_2 \neq 3n$), the bandgap can either increase (for $n_1 - n_2 = 3n - 2$) or decrease (for $n_1 - n_2 = 3n - 1$) with strain. These results have recently prompted more detailed experiments on a set of metallic and semiconducting CNTs deformed with an AFM-tip [657], as well as on CNTs under experimental tensile stretch [658]. Commercial applications from such work could lead to novel pressure sensors, transducers, amplifiers, and logic devices [659].

6.3 Nanowires and Nanoribbons

In spite of tremendous advances in carbon nanotube research, there remain some practical difficulties, which hinder many applications. Cheap mass-production remains one of the biggest hurdles. Other technological challenges involve controlling CNT-chirality, isolating/separating CNTs from bundles, alignment in nanocomposites, and so on. Chemical insertness of the CNT poses big problems in functionalization, sensor applications, and adhesion to structural materials. The above deficiencies have prompted researchers to explore other types of one-dimensional nanostructures. These efforts have recently led to the synthesis of nanowires and nanoribbons. Nanowires are typically solid (i. e., not hollow) cylindrical objects with a nearly uniform diameter of a few tens of nanometers or less. Most nanowires [543] have so far been synthesized from standard semiconductors: Si, Ge, GaAs, GaP, GaN, InAs, InP, ZnS, ZnSe, CdS, CdSe, and mixed compounds. Semiconducting nanowires have great potential in electronic and optoelectronic applications at the nanoscale. In addition, conducting nanowires made of transition and noble metals, silicides ($ErSi_2$), and polymeric materials have also been investigated in connection with interconnect applications.

Nanoribbons are a close cousin of the nanowires. As the name suggests, they possess a uniform rectangular cross-section with well-defined crystal structure, exposed planes, and growth direction (see Figs. 6.8 and 6.9). So far, nanoribbons have primarily been synthesized from the oxides of metals and semiconductors. In particular, SnO_2 and ZnO [660–662] nanoribbons have been materials systems of great current interest because of potential applications as catalysts, in optoelectronic devices, and as chemical sensors for pollutant gas species and biomolecules [663–665]. Although they grow to tens of microns long, the nanoribbons are remarkably single-crystalline and essentially free of dislocations. Thus they provide an ideal model for the systematic study of electrical, thermal, optical, and transport processes in one-dimensional semiconducting nanostructures, and their response to various external process conditions.

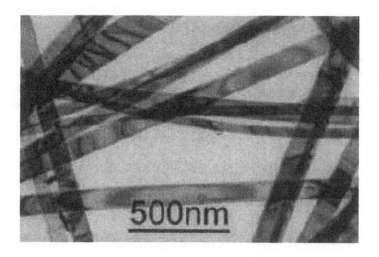

Figure 6.8: TEM bright-field image of SnO$_2$ nanoribbons displaying strain-contrast induced by bending of the ribbons. Each nanoribbon is single-crystalline without any dislocations. Courtesy of Dai et al., Ref. [660].

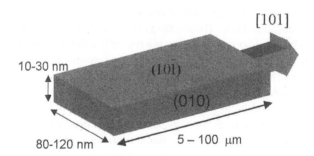

Figure 6.9: Schematic diagram of a SnO$_2$ nanoribbon, showing typical dimensions, exposed planes, and growth direction. Courtesy of Peidong Yang.

6.3.1 SnO$_2$ Nanoribbons

SnO$_2$ nanoribbons are usually synthesized by evaporating SnO or SnO$_2$ powder at high temperature, followed by deposition on an alumina substrate in the downstream of an Ar-gas flow [660]. Field-emission scanning electron microscopy (FE-SEM) and Transmission Electron microscopy revealed that the ribbons: (1) possess a highly crystalline rutile structure; (2) grow tens of microns long in the $\langle 1\,0\,1 \rangle$ direction; (3) display a uniform quasi-rectangular cross-section perpendicular to the growth direction; and (4) present the $(1\ 0\ \bar{1})$ and $(0\ 1\ 0)$ rutile planes as surface facets along the growth axis, with dimensions ranging from 80-120 nm by 10-30 nm (Fig. 6.9). Rutile SnO_2 is a wide-bandgap (3.6 eV) n-doped semiconductor, with the intrinsic carrier density determined by the deviation from stoichiometry, primarily in the form of oxygen vacancies [666].

6.3.2 SnO_2 Nanoribbons as Chemical Sensors

Recent experiments with SnO_2 nanoribbons [667] indicate that these are highly effective in detecting even very small amounts of harmful gases like NO_2. Upon adsorption of these gases, the electrical conductance of the sample decreases by several orders of magnitude. More interestingly, it is possible to get rid of the adsorbates by shining UV light, and the electrical conductance is completely restored to its original value. Such single-crystalline sensing elements have several advantages over conventional thin-film oxide sensors: low operating temperatures, no ill-defined coarse grain boundaries, and high active surface-to-volume ratio.

First-Principles DFT Simulations

Electron withdrawing groups like NO_2 and O_2 are expected to deplete the conduction electron population in the nanoribbon, thereby leading to a decrease in electrical conductance. To investigate this, first-principles DFT calculations of the adsorption process of NO_2, O_2, and CO on the exposed $(1\ 0\ \bar{1})$ and $(0\ 1\ 0)$ surfaces, as well as the edge atoms of a SnO_2 nanoribbon, were carried out. The DFT code used was DMol^3 [635], with the same basis set, parameter settings, and exchange-correlation functional as in section 6.2.4. Nanoribbon surfaces were represented in periodic supercells (Figs. 6.10 and 6.11), and accurate Brillouin Zone integration was performed by careful sampling of k-points chosen according to the Monkhorst–Pack scheme [479] with a k-point spacing of 0.1 Å^{-1}. In order to estimate charge transfer to adatoms, the partial charge on each atom was computed using the Mulliken population analysis [668]. In bulk rutile SnO_2, the Sn-atoms are octahedrally coordinated with six O-neighbors, while each O-atom is a threefold bridge between neighboring Sn-centers. At both $(1\ 0\ \bar{1})$ and $(0\ 1\ 0)$ surfaces the Sn-atoms lose an O-neighbor, thereby becoming fivefold coordinated (Fig. 6.10 (a, b)). The surface O-atoms become twofold-coordinated bridges connecting neighboring surface Sn-atoms (Fig. 6.10 (a, b)). Both surfaces were represented by three layers of Sn, each layer being sandwiched between two O-layers. The bottom SnO_2 layer was fixed in order to simulate the presence of several bulk-like layers in the actual sample. In order to reduce interaction with periodic images, the surface unit cell was doubled in the direction of the smaller surface lattice constant, and a vacuum of 15 Å was placed normal to the surface. To sim-

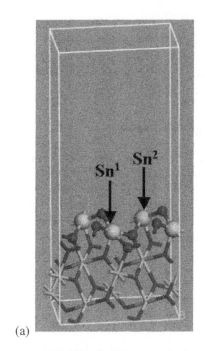 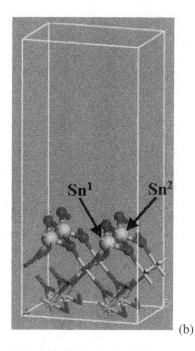

(a)   (b)

Figure 6.10: Simulation supercells representing exposed surfaces of a SnO_2
nanoribbon: (a) $(1\,0\,\bar{1})$ surface; (b) $(0\,1\,0)$ surface. Surface atoms are shown
in ball representation. White (larger) balls and Black (smaller) balls represent Sn
and O-atoms respectively. Sn^1 and Sn^2 are neighboring Sn-atoms connected with
a bridging O.

ulate nanoribbon edges (i. e., lines of intersection of $(1\,0\,\bar{1})$ and $(0\,1\,0)$ planes),
a structure as in Fig. 6.11 was embedded in a periodic supercell with the smallest repeat period (5.71 Å) along the length of the ribbon, and a vacuum of 15 Å
normal to both the $(1\,0\,\bar{1})$ surface (y-axis) and the $(0\,1\,0)$ surface (x-axis). At the
nanoribbon edges the Sn-atoms can be either threefold- or fourfold-coordinated
(Fig. 6.11).

Details of the results on binding energy and charge transfer are discussed else-
where [669]. The important results are summarized below:

1. All adsorbate structures involve one or more bonds to surface Sn-atoms.
   The binding energy on different surfaces and edges increases in the sequence $(010) < (10\bar{1}) <$ 3-fold edge $<$ 4-fold edge.

2. NO_2 adsorption displays a very rich chemistry because it can either form

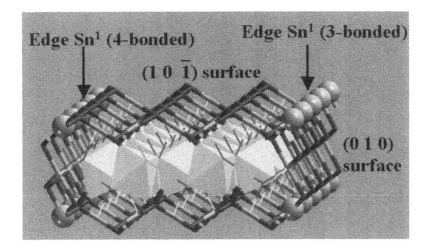

Figure 6.11: Simulation supercells representing SnO_2 nanoribbon edges. For clarity, the periodic cell is not shown, and the interior atoms are represented by polyhedra. The edge (Sn) atoms are shown in ball representation.

a single bond to a surface Sn, or can adsorb in the bidented form through two single bonds to neighboring Sn-atoms. The doubly bonded NO_2 is 2-3 kcal/mol more stable than the single-bonded NO_2, and the binding energies are in general 4-5 kcal/mol higher on the $(1\ 0\ \bar{1})$ surface than on the $(0\ 1\ 0)$. Activation barrier between the doubly-bonded and single-bonded structures is expected to be low, which should make the NO_2 species mobile on the exposed faces by performing a series of random walk steps along well-defined rows of Sn-atoms on the surface.

3. When two NO_2 molecules meet on the surface, either through random walk as described above, or through the incidence of a second NO_2 from gas phase in the vicinity of an already chemisorbed NO_2, there is a transfer of an O-atom from one NO_2 to the other, thus converting it to a surface NO_3 species. The net disproportionation reaction $NO_2 + NO_2 \rightarrow NO_3 + NO$ is well known in chemistry. The bidented NO_3 group has a substantially higher binding energy, especially on the $(1\ 0\ \bar{1})$ surface, and should not, therefore, be mobile. The resulting NO species is only weakly bound to the surface and should desorb easily. Synchrotron measurements using X-ray Absorption Near-Edge Spectroscopy (XANES) have recently confirmed the abundance of NO_3 species on the nanoribbon surface following NO_2 adsorption (Fig. 6.12) [669].

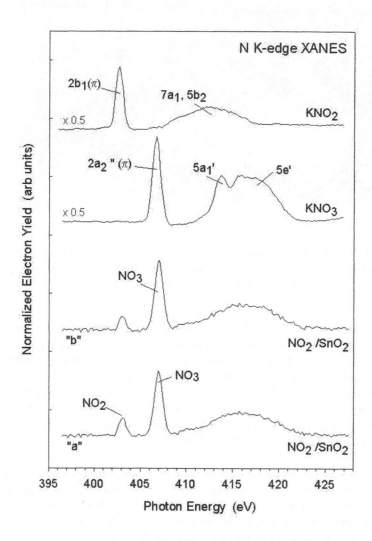

Figure 6.12: N K-edge XANES spectra acquired after exposing nanoribbons of SnO$_2$ to NO$_2$ at 300 K. Traces "a" and "b" correspond to two different nanoribbon samples, respectively. Included in the figure are the corresponding spectra for KNO$_2$ and KNO$_3$. The assignment of the features in these spectra is discussed in refs. [670].

4. A CO likes to adsorb in the following manner: the C forms two single bonds to the surface – one with a surface Sn and another with a bridging O, while the O of the CO forms a double bond to the C and sticks out of the surface. This way, the C-atom attains its preferred 4-valency and the O has its bivalency satisfied.

5. On a defect-free surface (i. e., surface with no O-vacancies), the O_2 molecules can only weakly physisorb. In this configuration, there is no charge transfer to the O_2, and therefore a nanoribbon surface without surface O-vacancies should be insensitive to atmospheric oxygen. However, at O-vacancy sites, the O_2 molecule has a strongly bound chemisorbed structure in the form of a peroxide bridge.

6. Both NO_3 groups and chemisorbed O_2 (at O-vacancy sites) accept significant amount of electronic charge from the surface. Therefore, such adsorbates should lead to the lowering of electrical conductance of the sample. CO, on the other hand, donates a moderate amount of electrons to the surface, and is therefore expected to increase the electrical conductance. All these results are consistent with direct experimental measurements of sample conductance [664,667,671]. Charge transfer between molecular species (donor or acceptor alike) and the nanoribbon surface could thus serve as a general mechanism for ultrasensitive chemical and biological sensing using single crystalline semiconductor nanowires.

Chapter 7

Theoretical Background

The physicists have made their universe,
and if you do not like it,
you must make your own.

E. R. Harrison

7.1 Quantum Chemistry

The topic of quantum chemistry is the application of quantum mechanics to the problem of the chemical bond. It considers molecules as a collection of particles (nuclei and electrons) which move according to the laws of quantum mechanics.

7.1.1 The Wave Function and the Schrödinger Equation

At the very basic of quantum mechanics is the wave function. Starting from experimentally observed phenomena (e. g., electron diffraction) de Broglie realized in 1924 that a (planar) wave can be assigned to each massive particle. Energy and impulse of the particle can be described by means of this wave

$$\Psi(\vec{r}; t) = c \exp\left[\frac{i}{\hbar}(\vec{p}\vec{r} - Et)\right] \qquad (7.1)$$

$\vec{r}$ here is the position and $\vec{p}$ the impulse vector of the particle, E the energy, t the time, c a normalization factor, i the imaginary unit, and $\hbar$ Planck's constant over 2π. The square of the wave function, or, more accurately, $|\Psi(\vec{r}; t)|^2 d^3r$ is the probability to find the particle in the volume element d^3r at the location $\vec{r}$ at time t.

An equation of motion has to be found for this wave function which fulfills the following criteria:

- plane waves have to be obtained in a force free state

- the superposition principle must be fulfilled (it must be a linear and homogeneous equation)

- it must yield $E = \vec{p}^2/2m$ (m – mass of the particle).

The following wave equation fulfills these criteria ($\nabla^2 = \partial^2/\partial x^2 + \partial^2/\partial y^2 + \partial^2/\partial z^2$)

$$i\hbar\frac{\partial}{\partial t}\Psi(\vec{r};t) = -\frac{\hbar^2}{2m}\nabla^2\Psi(\vec{r};t) = \hat{H}\Psi(\vec{r};t) \tag{7.2}$$

where $\hat{H}$ is the Hamilton operator. Equation (7.2) was generalized in 1926 by Schrödinger to the particle movement in a conservative force field when he extended the Hamilton operator by a term for the potential energy. The resulting Schrödinger equation

$$i\hbar\frac{\partial}{\partial t}\Psi(\vec{r};t) = \hat{H}\Psi(\vec{r};t) \tag{7.3}$$

with

$$\hat{H} = -\frac{\hbar^2}{2m}\nabla^2 + V(\vec{r})$$

is the basic equation in quantum chemistry [3].

The Schrödinger equation (7.3) can be separated into a time and a position dependent part by assuming that the wave function can be written as a product of a function of time $u(t)$ and a function of position $\psi(\vec{r})$

$$\Psi(\vec{r};t) = \psi(\vec{r})\,u(t) \tag{7.4}$$

After substituting in Equation (7.3) and some math one obtains

$$u(t) = \exp\left(\frac{-iEt}{\hbar}\right) \tag{7.5}$$

for the time dependent part (the integration over time has already been done here) and

$$\hat{H}\psi(\vec{r}) = E\psi(\vec{r}) \tag{7.6}$$

for the position dependent part [3]. Quantum chemistry usually only considers the position dependent part, and that is what we confine our attention to in the following.

The energy of a system described by the Schrödinger equation (7.6) can be obtained by multiplying Eq. (7.6) from the left with the complex conjugate of the wave function, ψ^*, followed by an integration over all space (from $-\infty$ to $+\infty$ in each of the x, y, and z directions indicated by $d\tau$)

$$\int \psi^* \hat{H} \psi d\tau = \int \psi^* E \psi d\tau \qquad (7.7)$$

Since the energy, E, is a constant with respect to the integration it can be taken out of the integration and the energy is obtained as

$$E = \frac{\int \psi^* \hat{H} \psi d\tau}{\int \psi^* \psi d\tau} = \frac{\langle \psi | \hat{H} | \psi \rangle}{\langle \psi | \psi \rangle} \qquad (7.8)$$

If the wave function is normalized the denominator of Eq. (7.8) will be equal to one. Eq. (7.8) is an expectation value of the operator $\hat{H}$. Expectation values are quantities calculated as in Eq. (7.8) for an operator. They can be considered an average of the quantity and play a fundamental role in quantum chemistry.

7.1.2 Many-Particle Systems

For many-particle systems such as molecules consisting of n electrons and N nuclei the wave function depends on four variables per particle. These are the three coordinates and the value of the spin, which also has to be considered here (both for nuclei as well as for electrons). The wave function contains all the information which is available for a system quantum mechanically. The Hamilton operator also takes on a more complicated form

$$\begin{aligned}
\hat{H} &= -\frac{\hbar^2}{2m_e} \sum_{i}^{n} \nabla^2 - e^2 \sum_{i}^{n} \sum_{k}^{N} \frac{Z_k}{r_{ik}} + e^2 \sum_{i}^{n} \sum_{j>i}^{n} \frac{1}{r_{ij}} \qquad (7.9) \\
&\quad - \frac{\hbar^2}{2} \sum_{k}^{N} \frac{1}{m_k} \nabla^2 + e^2 \sum_{k}^{N} \sum_{l>k}^{N} \frac{Z_k Z_l}{r_{kl}}
\end{aligned}$$

with m_e mass of the electron, e the elementary charge, Z_k the charge of a nucleus, and r_{ij} the distance between two particles. The summations run over all n electrons i and j and all N nuclei k and l.

A solution of the Schrödinger equation (7.6) for many-particle systems is not possible. It is therefore required to use certain simplifications and approximations to make the calculation of the electronic structure of many-particle systems feasible. The number of variables in the Schrödinger equation can be reduced by assuming a homogeneous and isotropic space. In this case the outer degrees of freedom (translation and rotation) can be separated and the position vectors in the wave function can be written by means of center-of-mass coordinates. This is a simplification which is still mathematically exact. Furthermore, due to the mass ratio between the atomic nuclei and the electrons the electronic problem for the many-particle system can be approximated by a problem in which the atomic nuclei are considered fixed in space and only the movement of the electrons is considered (Born–Oppenheimer approximation, [4]).

The Hamilton operator in the Born–Oppenheimer approximation does not contain the last two terms in Eq. (7.9) and is therefore reduced to its electronic part, which can be divided into an one-electron operator, $\hat{h}_i$, which represents the kinetic and potential energy of the ith electrons in the field of the nuclei (the first two terms in Eq. (7.9)) and a two-electron operator, $\hat{g}_{ij}$, which accounts for the electrostatic interaction of the electrons i and j (the third term in Eq. (7.9))

$$\hat{H} = \sum_i^n \hat{h}_i + \sum_{i<j}^n \hat{g}_{ij} \tag{7.10}$$

This Hamilton operator (7.10) can be applied on a modified wave function which only depends on the coordinates of the electrons. This way, one obtains an electronic Schrödinger equation.

To simplify the formulas we will use atomic units in the following discussion. Since certain constants such as the mass of an electron or the elementary charge occur quite often in the formulas they are included in the units. For a list of atomic units see the Appendix (p. 259).

7.1.3 Orbitals

To solve the electronic Schrödinger equation an ansatz for the wave function is necessary which makes the total wave function of the many-particle system constructable from one-electron wave functions. With such an ansatz the Schrödinger equation could be separated into one-electron equations. Since such an ansatz for the wave function has to have certain properties (e. g., it has to follow the Pauli principle [672]) one or more Slater determinants are usually used

$$\Phi = \frac{1}{\sqrt{n!}} \begin{vmatrix} \psi_1(\vec{r}_1;\sigma_1) & \psi_2(\vec{r}_1;\sigma_1) & \cdots & \psi_n(\vec{r}_1;\sigma_1) \\ \psi_1(\vec{r}_2;\sigma_2) & \psi_2(\vec{r}_2;\sigma_2) & \cdots & \psi_n(\vec{r}_2;\sigma_2) \\ \vdots & \vdots & \ddots & \vdots \\ \psi_1(\vec{r}_n;\sigma_n) & \psi_2(\vec{r}_n;\sigma_n) & \cdots & \psi_n(\vec{r}_n;\sigma_n) \end{vmatrix} \tag{7.11}$$

Its elements are the spin orbitals

$$\psi_i(\vec{r}_i;\sigma_i) = \phi_i(\vec{r}_i)\alpha_i(\sigma_i) \tag{7.12}$$

which can be written as a product of the function of position ϕ_i and of spin α_i.

A electron moves according to this ansatz statistically independent from all other electrons in the field which is created by all the other electrons. The electronic Hamilton operator (7.10), however, cannot be partitioned into parts depending only on the coordinates of one electron since it contains interaction terms of two electrons. Therefore, a variational calculation is required to determine the wave function.

7.1.4   The Hartree–Fock Equations

By substituting the Slater determinant (7.11) into the expectation value of the energy (Eq. 7.8) the energy of a many-particle system can be calculated. One obtains (for closed shell systems)

$$E = 2 \sum_i^{n/2} h_{ii} + \sum_i^{n/2} \sum_j^{n/2} (2J_{ij} - K_{ij}) \tag{7.13}$$

with

$$h_{ii} = \int \phi_i(1)^* \left(-\frac{1}{2}\nabla^2 - \sum_k^N \frac{Z_k}{r_{ik}} \right) \phi_i(1) d\tau_1 \tag{7.14}$$

$$J_{ij} = \iint \phi_i^*(1)\phi_j^*(2) \left(\frac{1}{r_{12}} \right) \phi_i(1)\phi_j(2) d\tau_1 d\tau_2 \tag{7.15}$$

$$K_{ij} = \iint \phi_i^*(1)\phi_j^*(2) \left(\frac{1}{r_{12}} \right) \phi_i(2)\phi_j(1) d\tau_1 d\tau_2 \tag{7.16}$$

The h_{ii} are called one-electron integrals, the J_{ij} are the Coulomb integrals, and the K_{ij} the exchange integrals. The Coulomb integrals result in the repulsive energy one would expect in the classical sense between the four electrons in the orbitals ϕ_i and ϕ_j. The exchange integrals, however, do not have a classical analogue. They are the result of the quantum mechanical treatment of a many-particle system.

In the quantum chemical literature the two-electron integrals of the Coulomb and exchange type are usually written using shorthand notations

$$\begin{aligned} J_{ij} &= \left\langle \phi_i^*\phi_j^* \left| \frac{1}{r_{12}} \right| \phi_i\phi_j \right\rangle \\ &= \left\langle ij \left| \frac{1}{r_{12}} \right| ij \right\rangle \\ &= (ij|ij) \end{aligned} \tag{7.17}$$

and

$$K_{ij} = \left\langle \phi_i^*\phi_j^* \left| \frac{1}{r_{12}} \right| \phi_j\phi_i \right\rangle$$

$$= \left\langle ij \left| \frac{1}{r_{12}} \right| ji \right\rangle$$

$$= (ij|ji) \tag{7.18}$$

The complex conjugate functions are written on the left-hand side, the real functions on the right-hand side both in the order of functions for electron 1 first followed by functions for electron 2.

The Slater determinant (7.11) is only an approximation for the wave function. The expectation value of the energy with such an approximated wave function is always larger than the exact energy [673]. To obtain the best possible, i. e., lowest energy, a variation of E in dependence of the variation of the orbitals $\delta\phi_i$ has to be performed. This will not result in the correct total wave function, but the best possible approximation for a wave function in form of a determinant which is constructed from orbitals. These orbitals are called *self-consistent* orbitals.

If the orbitals are determined such that they result in a stationary value of the energy and are orthogonal at the same time ($\langle \phi_i | \phi_j \rangle = \delta_{ij}$), one obtains

$$\left\{ \hat{h}_i + \sum_j^{n/2} \left[2\hat{J}_j - \hat{K}_j \right] \right\} \phi_i = \sum_j^{n/2} \epsilon_{ij} \phi_j \tag{7.19}$$

Here $\hat{J}_j$ and $\hat{K}_j$ are the Coulomb and exchange operators defined by the following equations

$$\hat{J}_j(1) = \left\langle \phi_j(2) \left| \frac{1}{r_{12}} \right| \phi_j(2) \right\rangle \tag{7.20}$$

$$\hat{K}_j(1)\phi_i(1) = \left\langle \phi_j(2) \left| \frac{1}{r_{12}} \right| \phi_i(2) \right\rangle \phi_j(1) \tag{7.21}$$

The term in curly brackets in Eq. (7.19) is called the Fock operator, $\hat{F}_i$. Using it, Eq. (7.19) can be re-written as

$$\hat{F}_i \phi_i = \sum_j^{n/2} \epsilon_{ij} \phi_j \tag{7.22}$$

The Fock operator in these equations is an effective one-electron Hamilton operator. These equations differ from a common one-electron wave equation in the way that they contain on the right hand side a set of constants ϵ_{ij} instead of a single eigenvalue. This is a result of the fact that the solutions for the wave equations are not unique since the Slater determinant used as ansatz cannot be changed in its value by a unitary transformation. If the orbitals ϕ_j are replaced by a new set ϕ'_j so that holds

$$\phi'_j = \sum_i T_{ji} \phi_i \tag{7.23}$$

where the T_{ji} are the elements of a unitary matrix it is possible to bring the matrix of Lagrange multiplicators ϵ_{ij} into diagonal shape. This way Equation (7.19) becomes a standard eigenvalue problem.

$$\hat{F}_i\phi_i = \epsilon_i\phi_i \qquad i = 1, n \qquad (7.24)$$

The equations (7.24) are called Hartree–Fock equations. With these equations we have achieved to convert the solution of the Schrödinger equation for the many-particle system to the solution of n equations for single particles. However, equations (7.24) can only be solved iteratively since the Fock operator contains the wave function (through $\hat{J}_j$ and $\hat{K}_j$).

7.1.5   The Roothaan–Hall Method

In order to numerically solve the Hartree–Fock equations the electronic orbitals ϕ_i have to be expressed analytically. To this end, one assumes that they can be represented as a linear combination of known functions χ_ν, called the basis functions

$$\phi_i = \sum_{\nu}^{m} c_{\nu i}\chi_\nu \qquad (7.25)$$

In principle, a complete (i. e., infinite) set of basis functions leads to exact orbitals within the Hartree–Fock limit. Since an infinite number of basis functions cannot be handled computationally one needs to make another approximation here by truncating the expansion in Eq. (7.25) at a finite number of basis functions. It is highly desirable to construct basis functions such that the error of truncation is small even with a reasonably small number of functions.

Historically, atomic orbitals (i. e., electronic wave functions for isolated atoms) have been used as basis functions in molecular calculations. The electronic orbitals ϕ_i for molecules, also known as molecular orbitals, are therefore obtained by a linear combination of atomic orbitals and the method is called the MO-LCAO approximation. Today atomic orbitals are no longer used as basis functions (see Section 7.1.6), but the name of the method is still used.

By substituting the linear combination of basis functions (7.25) into the Hartree–Fock equations (7.24) one obtains

$$\hat{F}_i\phi_i - \epsilon_i\phi_i = \hat{F}_i \sum_{\nu}^{m} c_{\nu i}\chi_\nu - \epsilon_i \sum_{\nu}^{m} c_{\nu i}\chi_\nu = 0 \qquad (7.26)$$

Multiplying from the left side by χ_μ^* and integrating results in the Roothaan equations

$$\sum_{\nu}^{m} c_{\nu i} \left[F_{\mu\nu} - \epsilon_i S_{\mu\nu}\right] = 0 \qquad (7.27)$$

where

$$F_{\mu\nu} = \langle \chi_\mu^* | \hat{F}_i | \chi_\nu \rangle \tag{7.28}$$

and

$$S_{\mu\nu} = \langle \chi_\mu^* | \chi_\nu \rangle \tag{7.29}$$

are the elements of the Fock matrix and of the overlap matrix, respectively. The elements of the Fock matrix can be expanded by substituting the definition of the Fock operator from Eq. (7.19). One obtains for a closed shell system

$$F_{\mu\nu} = H_{\mu\nu} + \sum_j^{N/2} \sum_\lambda^m \sum_\sigma^m c_{\lambda j} c_{\sigma j} [2(\mu\nu|\lambda\sigma) - (\mu\lambda|\nu\sigma)] \tag{7.30}$$

with

$$H_{\mu\nu} = \langle \chi_\mu^* | \hat{h}_i | \chi_\nu \rangle \tag{7.31}$$

This equation is usually simplified by introducing the density matrix P whose elements are

$$P_{\mu\nu} = 2 \sum_i^{N/2} c_{\mu i} c_{\nu i} \tag{7.32}$$

The Fock matrix elements then become

$$F_{\mu\nu} = H_{\mu\nu} + \sum_\lambda^m \sum_\sigma^m P_{\lambda\sigma} [(\mu\nu|\lambda\sigma) - 1/2(\mu\lambda|\nu\sigma)] \tag{7.33}$$

If one now collects the coefficients $c_{\nu i}$ in form of a matrix Eq. (7.27) can be recast in matrix form [674]

$$\boldsymbol{FC} = \boldsymbol{SC}\epsilon \tag{7.34}$$

This matrix equation represents the Self Consistent Field method (SCF), which meanwhile has become a standard method in quantum chemistry for the calculation of electronic properties of many-particle systems.

7.1.6   Basis Sets

Equation (7.25) shows that the molecular orbitals ϕ_i are constructed from basis functions, χ_ν. These basis functions are usually chosen to resemble the solutions of the Schrödinger equation for the hydrogen atom. If one solves the Schrödinger equation for the hydrogen atom the solution can be written as a product of a radial and an angular function. The radial solution has a distance dependence of $\exp(-\alpha r)$ and the angular function consists of spherical harmonics. Unfortunately, the calculation of four-center integrals over such functions (the so-called

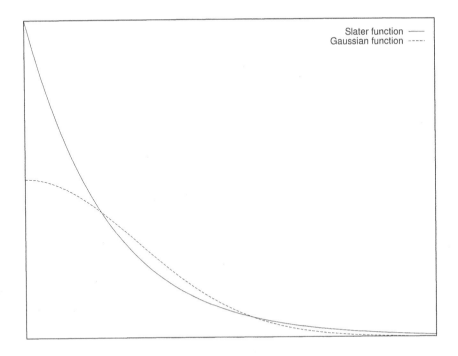

Figure 7.1: A Slater function, $2\exp(-r)$, and the Gaussian function which has maximum overlap with this Slater function, $\exp(-0.27095r^2)$.

Slater functions) is very difficult. As a result, Gaussian-type functions as introduced by Boys [675] are now commonly used in quantum chemical calculations. They have the following general mathematical form

$$g(\alpha_i) = x^a y^b z^c \exp(-\alpha_i r^2) \qquad (7.35)$$

where the exponents a, b, and c determine the shape of the function. For $a+b+c = 0$ a s function is obtained, for $a + b + c = 1$ a p function, etc.

Gaussian functions have the advantage that integrals over them can be calculated quickly. A disadvantage is that their distance dependence is different from Slater functions, as they approach zero much more rapidly. Figure 7.1 shows the Slater and Gaussian function which have the maximum overlap. A better description of a Slater function can be obtained by linear combination of Gaussian functions according to

$$\chi_\nu = \sum_i^L d_{i\nu} g(\alpha_i) \qquad (7.36)$$

where $g(\alpha_i)$ are the *primitive* Gaussian functions and d_i the *contraction coeffi-cients*. In commonly used basis sets the sum runs over less than ten primitives. The minimum number of primitives to give a reasonable approximation to a Slater function is three. A basis set which approximates a Slater function by three Gaussian functions is therefore called STO-3G. More commonly used basis sets use different contractions for the core and the valence shell of an atom. The very popular basis sets of Pople [676, 677], e. g., the 6-31G basis set, are made up from six primitives contracted to one basis function for the core shells and one function consisting of three primitives plus another function consisting of one primitive for the valence shell. For a good description of the bonding, in particular when electronegative elements are present, basis sets have to be augmented with basis functions which have a higher angular quantum number. An oxygen atom, e. g., in its ground state would have only s and p functions occupied. The augmentation functions for oxygen are then d functions. These functions are called polarization functions and have rather small exponents α making them extend far away from the nucleus. For calculations on anions additional functions, so-called diffuse functions, are required. Within the framework of Pople's basis sets polarization functions are marked with an asterisk while diffuse functions are marked by a plus sign. A single one of these symbols marks the presence of functions on all elements except hydrogen and helium while a second one is used for functions on hydrogen and helium. 6-31+G** would therefore be a 6-31G basis set with polarization functions on all elements and diffuse functions on all elements except hydrogen and helium. A more detailed explanation of basis set naming conventions can be found at page 258.

Basis sets are derived [678] by performing electronic structure calculations for atoms or, in some instances, for molecules. In these calculations the exponents and contraction coefficients of the basis functions are varied until the lowest energy has been found. Therefore, basis sets are strictly speaking only to be used with the electronic structure method they have been derived for. In practice basis sets are usually derived using Hartree–Fock calculations and then transferred to other electronic structure methods. This approach usually leads to good results. Employing atomic calculations for the derivation of basis sets also implies that the electronic structure of the atoms remains largely the same in the molecule as in the free atom. This is in fact the case.

7.1.7   The Direct SCF Method

The biggest computational problem with the SCF method is the large number of two-electron integrals which need to be handled. Since the Roothaan equations can only be solved iteratively these two-electron integrals are required for the construction of the Fock matrix in each iteration. Historically, in the so-called conventional SCF method this problem was solved by calculating all two-electron integrals at the beginning of the calculation and storing them on external storage (hard disk). They would then be read back from disk in each iteration to construct the Fock matrix. However, the number of two-electron integrals is huge even for small molecules, the capacity of hard disks is limited, and accessing external storage is by a few orders of magnitude slower compared to the speed of modern processors.

Therefore Almlöf et al. suggested the direct SCF method [679] where the two-electron integrals are recalculated in each iteration. Only with the direct SCF method calculations for large molecules have become feasible and nearly all modern SCF programs use the direct method. The number of two-electron integrals which have to be recalculated can be limited in several ways [680]

- very good integral estimates (only two-electron integrals which contribute significantly to the Fock matrix are really calculated)

- selective storage of the most often needed and the most difficult to calculate two-electron integrals, respectively

- minimizing the differential density matrix by linear combination

- use of symmetry for all finite point groups.

7.1.8 Potential Energy Hypersuperfaces

As we have seen in the previous sections it is possible to calculate the electronic energy of a molecule for a given set of coordinates of the atoms which are fixed in space. The knowledge of the electronic energy is, however, insufficient if one is interested in the structure (bond lengths, bond angles, etc.) of a molecule. How can the structure of a molecule be calculated?

The atomic nuclei of a molecule experience forces caused by the electrons surrounding them. They will move to the positions where these forces vanish. To calculate the structure of a molecule it is necessary to compute these force and to consider the movement of the nuclei in the force field created by the electrons. Generally, it is not necessary to solve the Schrödinger equation for the movement of the nuclei. It is often sufficient to solve the classical equations of movement for the nuclei only.

7.1.9 Forces

The force which acts on an atomic nuclei can be calculated as the first derivative of the energy with respect to the coordinates of the nucleus under consideration. This derivative can, in principle, be obtained by numerical differentiation of the energy. However, this approach is rather cumbersome and therefore the use of analytical derivatives is standard for the most common electronic structure methods. The

analytical calculation of derivatives has the advantage that all first derivatives of a
molecule can be calculated in about the same time as the energy of the molecule.
It holds

$$-f_{x_a} = \frac{\partial E}{\partial x_a} = \frac{\partial}{\partial x_a} \langle \Phi | \hat{H} | \Phi \rangle \qquad (7.37)$$

In this equation only the Hamilton operator depends directly on the positions of
the nuclei. The wave function Φ, however, also contains parameters which depend
on the position of the nuclei. The wave function, therefore, depends indirectly
on the positions of the nuclei, too. This dependency is caused on one hand by
the coefficients $c_{j\mu}$ for the linear combination of the atomic orbitals and, on the
other hand, by several in general non-linear parameters p which describe the basis
functions. The most important of these parameters are the positions of the basis
functions in space (r in Eq. (7.35)) and the exponents of the basis functions (cf.
Eq. 7.35).

The dependence of the parameters p on the coordinates of the nuclei is caused
by the dependence of the basis set centers on atomic positions in most of the
methods commonly used. The atomic part of the electron density moves with
the nuclei and the finer differences in the distribution of the electron density are
accounted for by a variation of the coefficients $c_{j\mu}$. That way the coefficients $c_{j\mu}$
become dependent on the positions of nuclei, too.

Thus,

$$E(x_a) = \left\langle \Phi(c_{j\mu}(x_a); p(x_a)) \left| \hat{H}(x_a) \right| \Phi(c_{j\mu}(x_a); p(x_a)) \right\rangle \qquad (7.38)$$

Performing the differentiations (as a simplification we write for $\partial Y / \partial x_a = Y^a$)
yields

- Hellmann–Feynman Force

$$\frac{\partial}{\partial x_a} \left[-e^2 \sum_k^N \sum_i^n \frac{Z_k}{r_{ki}} + e^2 \sum_{k<l}^N \sum_l^N \frac{Z_k Z_l}{r_{kl}} \right] \qquad (7.39)$$

$$= e^2 Z_a \left(\sum_i^n \frac{x_i - x_a}{r_{ai}^3} - \sum_l^N \frac{Z_l(x_l - x_a)}{r_{al}^3} \right)$$

The derivative of the Hamilton operator results in the negative Hellmann–
Feynman force [681].

- Density Force

$$\sum_{j\mu} \frac{\partial E}{\partial c_{j\mu}} c_{j\mu}^a = 2 \sum_{j\mu} (SC\epsilon)_{j\mu} c_{j\mu}^a = 2 \operatorname{Tr}(SC\epsilon C^{a\dagger}) \qquad (7.40)$$

Equation (7.40) can be further simplified by taking the derivative of the orthogonality condition $C^\dagger SC = 1$.

$$C^{a\dagger} SC + C^\dagger SC^a = -C^\dagger S^a C \qquad (7.41)$$

By multiplying Equation (7.41) from the left by ϵ and taking the trace one obtains

$$
\begin{aligned}
2\,\mathrm{Tr}(\epsilon C^{a\dagger} SC) &= 2\,\mathrm{Tr}(SC\epsilon C^{a\dagger}) = -\mathrm{Tr}(\epsilon C^\dagger S^a C) \quad (7.42)\\
&= -\mathrm{Tr}(C\epsilon C^\dagger S^a) = -\mathrm{Tr}(RS^a)
\end{aligned}
$$

with $R = C\epsilon C^\dagger$. It is therefore not necessary to calculate the derivatives of the orbital coefficients $c_{j\mu}$ to compute the first derivative of the energy. Rather, the much simpler to calculate matrices R and S^a are sufficient. Since this term depends on the density matrix its negative is called the density force.

- Integral Force
  The derivative of the energy with respect to the parameters p is called the integral force. To calculate it, it is necessary to know how the basis set depends on the coordinates of the nuclei. The most common way to use a simple, physically reasonable dependence (coupling of the basis set centers with the positions of the nuclei) leads to a relationship which can be used to compute the derivatives p^a. This is the most demanding part in the calculation of first derivatives.

It is also possible to optimize the parameters p. One obtains

$$\frac{\partial E}{\partial p_i} - \mathrm{Tr}\left(\epsilon C^\dagger \frac{\partial S}{\partial p_i} C\right) = \frac{\partial E}{\partial p_i} - \mathrm{Tr}\left(R \frac{\partial S}{\partial p_i}\right) = 0 \qquad (7.43)$$

The integral force then becomes

$$\sum_i \frac{\partial E}{\partial p_i} p_i^a = \sum_i \mathrm{Tr}\left(R \frac{\partial S}{\partial p_i}\right) \frac{\partial p_i}{\partial x_a} = \mathrm{Tr}(RS^a) \qquad (7.44)$$

In such a case the integral force and the density force cancel each other and the exact force is the Hellmann–Feynman force [681]. However, most common electronic structure methods do not optimize the parameters p and therefore need to calculate density and integral force.

Today, the analytical calculation of the first as well as the second derivatives for SCF wave functions has become a standard task [682, 683]. Even for correlated wave functions (e. g., second-order Møller–Plesset perturbation theory, or MP2) these derivatives are computed analytically [684, 685]. However, the working formulas for the derivation of the second derivatives are rather complex. A good summary can be found in [683].

7.1.10   Density Functional Theory

In the previous sections we have seen how the Schrödinger equation can be solved approximately for a molecular system. The Hartree–Fock method (HF) has been widely applied in ab initio quantum chemistry. The solutions to the electronic problem can be systematically improved upon by including electron correlation effects using perturbation theory (Møller–Plesset calculations) or a multi-determinant approach (Coupled Clusters or Configuration Interaction). We will not go into details for these kinds of calculations since the mathematical machinery to carry them out is rather complex and good summaries have already been provided [578, 686]. We will rather introduce a method which uses a slightly different approach to the whole problem. This method is density functional theory, which has found widespread use in recent years.

HF calculations and correlated methods based on them suffer from the enormous computational power required even for comparatively small molecules. HF calculations themselves scale theoretically with the size of the system as N^4, N being the number of basis functions (in practice, the scaling is somewhat better due to neglect of integrals smaller than a certain threshold). Correlated methods scale even worse; second order Møller–Plesset calculations (the simplest correlated method) scale as N^5. The inclusion of electron correlation is, however, often important. While methods based on HF permit a systematic improvement of the description of a molecular system, the computational costs involved might make them impractical for anything, but the smallest systems. It would be of benefit if correlated methods could be formulated such that the computational cost is similar to or lower than HF. This is indeed possible through what is known as Density Functional Theory (DFT).

In density functional theory the complicated N-electron wave function and the associated Schrödinger equation are replaced by the much simpler electron density and its associated calculational scheme. In the following we denote the electron density in the ground state Ψ by the notation $\rho(\vec{r}_1)$

$$\rho(\vec{r}_1) = N \int \cdots \int |\Psi(\vec{x}_1, \vec{x}_2, \ldots, \vec{x}_N)|^2 \, d\vec{x}_2 \cdots d\vec{x}_N \qquad (7.45)$$

As early as in the late 1920's Thomas and Fermi realized that statistical considerations can be used to approximate the distribution of electrons in an atom [687–689]. Thomas assumed that: "Electrons are distributed uniformly in the six-dimensional phase space for the motion of an electron at the rate of two for each h^3 of volume" and that there is an effective potential field that "is itself determined by the nuclear charge and this distribution of electrons" [687]. Using these assumptions it is possible to derive an approximation for the electronic kinetic energy in terms of the electron density [690]. If the exchange and correlation terms are also neglected and only the classical electrostatic energies of electron-nucleus attraction and electron-electron repulsion are included an energy formula for an atom in terms of electron density alone can be derived

$$E_{TF}[\rho(\vec{r})] = C_F \int \rho^{5/3}(\vec{r}) d\vec{r} - Z \int \frac{\rho(\vec{r})}{r} d\vec{r} + \frac{1}{2} \iint \frac{\rho(\vec{r}_1)\rho(\vec{r}_2)}{|\vec{r}_1 - \vec{r}_2|} d\vec{r}_1 d\vec{r}_2$$

(7.46)

with

$$C_F = \frac{3}{10}(3\pi^2)^{2/3}$$

(7.47)

If it is assumed that the electron density minimizes the energy functional Eq. (7.46) for the ground state of an atom under the constraint (N total number of electrons in the atom)

$$N = N[\rho(\vec{r})] = \int \rho(\vec{r}) d\vec{r}$$

(7.48)

the ground-state electron density must satisfy the variational principle

$$\delta \left\{ E_{TF}[\rho] - \mu_{TF} \left(\int \rho(\vec{r}) d\vec{r} - N \right) \right\} = 0$$

(7.49)

This yields the Euler-Lagrange equation

$$\mu_{TF} = \frac{\delta E_{TF}[\rho]}{\delta \rho(\vec{r})} = \frac{5}{3} C_F \rho^{2/3}(\vec{r}) - \phi(\vec{r})$$

(7.50)

with $\phi(\vec{r})$ being the electrostatic potential at point $\vec{r}$ due to the nucleus and the entire electron distribution

$$\phi(\vec{r}) = \frac{Z}{r} - \int \frac{\rho(\vec{r}_2)}{|\vec{r} - \vec{r}_2|} d\vec{r}_2$$

(7.51)

Eq. (7.50) can be solved in conjunction with the constraint Eq. (7.48) and the resulting electron density can then be inserted into Eq. (7.46) to give the total energy. This is the Thomas-Fermi theory of the atom, which is an elegantly simple model with one major flaw. It cannot predict binding in molecules [691]. Therefore, Thomas-Fermi theory was considered of not much real importance for quantitative predictions until Hohenberg and Kohn showed in 1964 [692] that the Thomas-Fermi model can be regarded as an approximation to an exact theory, the density functional theory.

It is easy to show that the energy of a many-electron system is a unique functional of the electron density, $\rho(\vec{r})$. Let us consider a collection of an arbitrary number of electrons, enclosed in a large box and moving under the influence of an external potential $v(\vec{r})$ and the mutual Coulomb repulsion. We can show that $v(\vec{r})$ is a *unique* functional of $\rho(\vec{r})$ by assuming that another potential $v'(\vec{r})$ with ground state Ψ' gives rise to the same density $\rho(\vec{r})$. Now Ψ' cannot be equal to

Ψ since they satisfy different Schrödinger equations. By denoting the Hamiltonian and ground-state energies associated with Ψ and Ψ' by $\hat{H}$, $\hat{H}'$ and E, E' the minimal property of the ground state requires

$$E' = \langle \Psi' | \hat{H}' | \Psi' \rangle < \langle \Psi | \hat{H}' | \Psi \rangle = \langle \Psi | \hat{H} + \hat{V}' - \hat{V} | \Psi \rangle \qquad (7.52)$$

so that

$$E' < E + \int [v'(\vec{r}) - v(\vec{r})] \rho(\vec{r}) \mathrm{d}\vec{r} \qquad (7.53)$$

By exchanging primed and unprimed quantities we find in exactly the same way that

$$E < E' + \int [v(\vec{r}) - v'(\vec{r})] \rho(\vec{r}) \mathrm{d}\vec{r} \qquad (7.54)$$

Adding Eqs. (7.53) and (7.54) leads to the inconsistency

$$E + E' < E + E' \qquad (7.55)$$

Thus, the assumption that there is another potential $v'(\vec{r})$ which gives rises to the same density $\rho(\vec{r})$ cannot be valid and $v(\vec{r})$ must be a unique functional of $\rho(\vec{r})$.

However, this is one of the most curious theorems of physics. As shown above it can be proven that such a functional exists, but it is equally clear that it is impossible to determine this functional. The only criterion for the quality of an *approximated* functional is how well it performs in calculations.

A second problem arises from the need to calculate the kinetic energy as a functional of the density. This is the problem which stopped Thomas-Fermi theory. However, it is possible to avoid this problem by replacing the density functional formulation of the kinetic energy by the exact kinetic energy of a noninteracting model system. This can be accomplished by introducing orbitals into the problem in such a way that the kinetic energy can be computed to good accuracy, leaving a small residual correction that is handled separately. The exact formula for the ground state kinetic energy is given by

$$T = \sum_{i}^{N} n_i \left\langle \psi_i \left| -\frac{1}{2} \nabla^2 \right| \psi_i \right\rangle \qquad (7.56)$$

where the ψ_i and n_i are orbitals and their occupation numbers. The Pauli principle requires that $0 \le n_i \le 1$. The Hohenberg-Kohn theory assures that this is a functional of the total electron density

$$\rho(\vec{r}) = \sum_{i}^{N} n_i |\psi_i(\vec{r})|^2 \qquad (7.57)$$

For any interacting system of interest, there are an infinite number of terms in Eq. (7.56) and (7.57), which is of not much help. But it is possible to go to the special case $n_i = 1$ for N orbitals and $n_i = 0$ for the rest, therefore describing a system of N non-interacting electrons. Eq. (7.56) and (7.57) thus become

$$T_s[\rho] = \sum_i^N \left\langle \psi_i \left| -\frac{1}{2}\nabla^2 \right| \psi_i \right\rangle \tag{7.58}$$

(the subscript s denotes that it is the kinetic energy calculated from a Slater determinant) and

$$\rho(\vec{r}) = \sum_i^N |\psi_i(\vec{r})|^2 \tag{7.59}$$

To obtain a unique decomposition of $\rho(\vec{r})$ in terms of orbitals a non-interacting reference system with the Hamiltonian

$$\hat{H}_s = \sum_i^N \left(-\frac{1}{2}\nabla_i^2 \right) + \sum_i^N v_s(\vec{r}_i) \tag{7.60}$$

needs to be introduced. In this Hamiltonian there are no electron-electron repulsion terms and its ground state electron density is exactly ρ. For this system there will be an exact determinantal ground state wave function

$$\Psi_s = \frac{1}{\sqrt{N!}} \det[\psi_1 \psi_2 \cdots \psi_N] \tag{7.61}$$

where the ψ_i are the N lowest eigenstates of the one-electron Hamiltonian $\hat{h}_s$

$$\hat{h}_s \psi_i = \left[-\frac{1}{2}\nabla^2 + v_s(\vec{r}) \right] \psi_i = \epsilon_i \psi_i \tag{7.62}$$

The kinetic energy is given by Eq. (7.58)

$$T_s[\rho] = \left\langle \Psi_s \left| \sum_i^N \left(-\frac{1}{2}\nabla_i^2 \right) \right| \Psi_s \right\rangle = \sum_i^N \left\langle \psi_i \left| -\frac{1}{2}\nabla^2 \right| \psi_i \right\rangle \tag{7.63}$$

and the density is decomposed. This quantity, although uniquely defined for any density, is still not the exact kinetic energy functional $T[\rho]$, but if one sets up a problem of interest in such a way that $T_s[\rho]$ is its exact kinetic energy component a separation is possible

$$F[\rho] = J[\rho] + E_{xc}[\rho] + T_s[\rho] \tag{7.64}$$

where

$$E_{xc}[\rho] \equiv T[\rho] - T_s[\rho] + V_{ee}[\rho] - J[\rho] \tag{7.65}$$

The quantity $E_{xc}[\rho]$ is called the exchange-correlation energy and accounts for the, presumably small, difference between T and T_s. This was the breakthrough for density functional theory accomplished by Kohn and Sham in 1965 [693].

Therefore, to obtain the energy it is only necessary to know the one-particle density instead of the many-particle wave function. Kohn and Sham have further shown [693] that the density that yields the minimum energy of a given system can be found by solving a single-particle equation with an effective "exchange-correlation" potential. The ground state electron density is the density that minimizes the energy functional

$$E[\rho] = \int \rho(\vec{r})v(\vec{r})\mathrm{d}\vec{r} + F[\rho] \tag{7.66}$$

and hence satisfies the Euler equation

$$\mu = v(\vec{r}) + \frac{\delta F[\rho]}{\delta \rho(\vec{r})} \tag{7.67}$$

By substituting Eq. (7.64) in the Euler equation one obtains

$$\mu = v(\vec{r}) + \frac{\delta J[\rho]}{\delta \rho(\vec{r})} + \frac{\delta E_{xc}[\rho]}{\delta \rho(\vec{r})} + \frac{\delta T_s[\rho]}{\delta \rho(\vec{r})} \tag{7.68}$$

where the first three terms define the Kohn–Sham effective potential

$$
\begin{aligned}
v_{eff}(\vec{r}) &= v(\vec{r}) + \frac{\delta J[\rho]}{\delta \rho(\vec{r})} + \frac{\delta E_{xc}[\rho]}{\delta \rho(\vec{r})} \\
&= v(\vec{r}) + \int \frac{\rho(\vec{r}')}{|\vec{r} - \vec{r}'|}\mathrm{d}\vec{r}' + v_{xc}(\vec{r})
\end{aligned} \tag{7.69}
$$

and $v_{xc}(\vec{r})$ being the exchange-correlation potential. For a given $v_{eff}(\vec{r})$ one can obtain the $\rho(\vec{r})$ that satisfies the Euler equation (7.68) by solving the N one-electron equations

$$\left[-\frac{1}{2}\nabla^2 + v_{eff}(\vec{r}) \right]\psi_i = \epsilon_i\psi_i \tag{7.70}$$

and setting

$$\rho(\vec{r}) = \sum_i^N |\psi_i(\vec{r})|^2 \tag{7.71}$$

Since v_{eff} depends on $\rho(\vec{r})$ these equations can only be solved self-consistently. An initial guess of $\rho(\vec{r})$ allows the construction of $v_{eff}(\vec{r})$, which can be used to find a new $\rho(\vec{r})$. The total energy can be computed directly from Eq. (7.66) with Eq. (7.64). Equations (7.69) to (7.71) are known as the Kohn–Sham equations.

These equations closely resemble the Hartree–Fock equations, Eq. (7.24), where the non-local Coulomb and exchange operators have been replaced by the Kohn–Sham effective potential v_{eff}. Unfortunately, the exact form of the functional of the exchange-correlation energy and, hence, also of the exchange-correlation potential, v_{xc}, which is needed to construct the Kohn–Sham effective potential, are unknown in general. The search for an accurate functional of the exchange-correlation energy has encountered tremendous difficulty and is the greatest challenge in density functional theory. The following sections describe what approximations can be made for this functional.

The exchange-correlation energy is customarily divided into two parts, a pure exchange, E_x, and a correlation part, E_c. Each of the energies is often written in terms of the energy per particle, ϵ_x and ϵ_c, also known as energy density

$$E_{xc}[\rho] = E_x[\rho] + E_c[\rho] = \int \rho(\vec{r})\epsilon_x[\rho(\vec{r})]\mathrm{d}\vec{r} + \int \rho(\vec{r})\epsilon_c[\rho(\vec{r})]\mathrm{d}\vec{r} \quad (7.72)$$

and the different density functionals in use are expressed in terms of those. It should be noted that the definitions of exchange and correlation energies are different for Hartree–Fock and density functional theory. The correlation energy in Hartree–Fock theory is defined as the difference between the exact energy and the energy calculated within the Hartree–Fock approximation. The exchange energy is the total electron-electron repulsion minus the Coulomb energy. Both of these energies have a short-range and a long-range part. The long-range part of the exchange energy effectively cancels the long-range part of the correlation energy. The definitions of exchange and correlation in density functional theory are short-range (local) and depend only on the density at a given point and in the immediate vicinity. The cancellation at long range is implicitly built into the exchange-correlation functional. It is therefore impossible to calculate the exchange energy from Hartree–Fock theory and combine it with a correlation energy calculated from density functional theory.

The correlation between electrons of parallel spin is different from that between electrons of opposite spin. The exchange energy is given as a sum of contributions from the α and β spin densities, as the exchange energy only involves electrons of the same spin. The total density is the sum of the α and β contributions, $\rho = \rho_\alpha + \rho_\beta$, and for a closed shell singlet these are identical. Functionals for the exchange and correlation energies may be formulated in terms of separate spin-densities. However, they are often given instead as functions of the spin polarization ζ and the radius of the effective volume containing one electron, r_S

$$\zeta = \frac{\rho_\alpha - \rho_\beta}{\rho_\alpha + \rho_\beta} \quad (7.73)$$

$$\frac{4}{3}\pi r_S^3 = \frac{1}{\rho} \quad (7.74)$$

In the formulas below it is implicitly assumed that the exchange energy is a sum over both the α and β densities.

Local Density Methods

The local density approximation (LDA) assumes that the density can locally be treated as a uniform electron gas or, equivalently, that the density is a slowly varying function. The exchange energy for a uniform electron gas is given by

$$\epsilon_x^{LDA}[\rho] = -C_x \rho^{1/3} \tag{7.75}$$

with

$$C_x = \frac{3}{4} \left(\frac{3}{\pi}\right)^{1/3} \tag{7.76}$$

In the more general case, where the α and β spin densities are not equal, LDA has to be replaced by the local spin density approximation (LSDA)

$$\epsilon_x^{LSDA}[\rho] = -2^{1/3} C_x \left(\rho_\alpha^{1/3} + \rho_\beta^{1/3}\right) \tag{7.77}$$

For closed shell systems LDA and LSDA are equal and since this is the most common case the terms LDA and LSDA are often used interchangeably, although this is not true in the general case.

Slater's X_α method [694] is a special case of an LDA method where the correlation energy is neglected and the exchange term is given by

$$\epsilon_x^{X_\alpha}[\rho] = -\frac{3}{2}\alpha C_x \rho^{1/3} \tag{7.78}$$

With $\alpha = 2/3$ it becomes identical to the expression for a uniform electron gas above. Originally, the value of $\alpha = 1$ was used, but it was shown that a value of 3/4 gives better agreement for atomic and molecular systems. The name Slater is often used as a synonym for the L(S)DA exchange energy involving the energy density raised to the power of 1/3.

The correlation energy of a uniform electron gas has been determined by Monte Carlo methods for a number of different densities. A suitable analytic interpolation formula for these results has been constructed by Vosko, Wilk, and Nusair (VWN) [644]. This very accurate fit interpolates between the unpolarized (ζ=0) and spin polarized (ζ=1) limits by the functional

$$\epsilon_c^{VWN}(r_S, \zeta) = \epsilon_c(r_S, 0) + \epsilon_a(r_S) \left[\frac{f(\zeta)}{f''(0)}\right] [1 - \zeta^4] + [\epsilon_c(r_S, 1) - \epsilon_c(r_S, 0)] f(\zeta)\zeta^4 \tag{7.79}$$

where

$$f(\zeta) = \frac{(1+\zeta)^{4/3} + (1-\zeta)^{4/3} - 2}{2(2^{1/3} - 1)}$$

The $\epsilon_c(r_S, \zeta)$ and $\epsilon_a(r_S)$ functionals are parameterized

$$
\begin{aligned}
\epsilon_{c/a}(x) = \ &A \left[\ln \frac{x^2}{X(x)} + \frac{2b}{Q} \tan^{-1} \left(\frac{Q}{2x+b} \right) - \frac{bx_0}{X(x_0)} \left[\ln \frac{(x-x_0)^2}{X(x)} \right. \right. \\
&+ \left. \left. \frac{2(b+2x_0)}{Q} \tan^{-1} \left(\frac{Q}{2x+b} \right) \right] \right]
\end{aligned}
\tag{7.80}
$$

where

$$
\begin{aligned}
x &= \sqrt{r_S} \\
X(x) &= x^2 + bx + c \\
Q &= \sqrt{4c - b^2}
\end{aligned}
$$

The parameters A, x_0, b, and c are fitting constants, different for $\epsilon_c(r_S, 0)$, $\epsilon_c(r_S, 1)$, and $\epsilon_a(r_S)$.

The local density approximation works fine for solids, but has its shortcomings when it comes to the calculation of properties of isolated molecules. The exchange energy is usually underestimated by approximately 10 %; the correlation energy is furthermore overestimated by as much a factor as 2. As a consequence, bond strengths are commonly overestimated. Moreover, the use of an exchange-correlation functional based on the electron gas model is not a very good solution. As a result, a lot of effort has gone and is still going into finding an exchange-correlation functional that gives better results both for solids and isolated molecules.

Gradient Corrected Methods

A major breakthrough was the use of gradient-corrected functionals [695–697] where the exchange-correlation energy depends not only on the density itself, but also on the first derivative (gradient) of the density. These methods are known as gradient-corrected or generalized gradient approximation (GGA) methods. Another, misleading, term used in the literature is non-local methods, which is not strictly true since these functionals depend only on the density and their derivatives at a given point and not on a space volume. Thus, Eq. (7.72) becomes

$$E_{xc} = \int \rho(\vec{r}) \epsilon_{xc} \left[\rho(\vec{r}), |\nabla \rho(\vec{r})| \right] d\vec{r} \tag{7.81}$$

The modification of the LSDA exchange expression proposed by Perdew and Wang (PW86) is

$$\epsilon_x^{PW86} = \epsilon_x^{LDA}(1 + ax^2 + bx^4 + cx^6)^{1/15} \tag{7.82}$$

where

$$x = \frac{|\nabla\rho|}{\rho^{4/3}} \tag{7.83}$$

and a, b, and c are constants [698]. Another widely used modification of LDA is that of Becke (B88)

$$\epsilon_x^{B88} = \epsilon_x^{LDA} - \beta\rho^{1/3}\frac{x^2}{1 + 6\beta x \sinh^{-1} x} \tag{7.84}$$

where the parameter β is determined by fitting to known atomic data and x is defined by Eq. (7.83) [504]. An exchange functional similar to B88 has been proposed by Perdew and Wang (PW91)

$$\epsilon_x^{PW91} = \epsilon_x^{LDA}\left(\frac{1 + xa_1 \sinh^{-1}(xa_2) + (a_3 + a_4 e^{-bx^2})x^2}{1 + xa_1 \sinh^{-1}(xa_2) + a_5 x^2}\right) \tag{7.85}$$

where a_1 to a_5 and b are constants and x is given by Eq. (7.83).

The gradient-corrected functionals shown so far are approximating the exchange part of the exchange-correlation energy. A popular correlation functional is that of Lee, Yang, and Parr (LYP) [505]

$$\begin{aligned}
\epsilon_c^{LYP} &= -a\frac{\gamma}{1 + d\rho^{-1/3}} - ab\frac{\gamma e^{-c\rho^{-1/3}}}{9(1 + d\rho^{-1/3})\rho^{8/3}} \\
&\times \left[18\left(2^{2/3}\right) C_F \left(\rho_\alpha^{8/3} + \rho_\beta^{8/3}\right) - 18\rho t_W\right. \\
&+ \left.\rho_\alpha(2t_W^\alpha + \nabla^2\rho_\alpha) + \rho_\beta(2t_W^\beta + \nabla^2\rho_\beta)\right]
\end{aligned} \tag{7.86}$$

where

$$\gamma = 2\left[1 - \frac{\rho_\alpha^2 + \rho_\beta^2}{\rho^2}\right] \tag{7.87}$$

and

$$t_W^\sigma = \frac{1}{8}\left(\frac{|\nabla\rho_\sigma|^2}{\rho_\sigma} - \nabla^2\rho_\sigma\right) \tag{7.88}$$

The parameters a, b, c, and d are determined by fitting to data for the helium atom. The second derivative of the density in Eq. (7.86) can be removed by partial integration [578, 699].

A gradient correction to LSDA was proposed by Perdew (P86) [698]

$$\epsilon_c^{P86} = \epsilon_c^{LDA} + \frac{e^\Phi C(\rho)|\nabla\rho|^2}{f(\zeta)\rho^{7/3}} \tag{7.89}$$

where

$$f(\zeta) = 2^{1/3}\sqrt{\left(\frac{1+\zeta}{2}\right)^{5/3} + \left(\frac{1-\zeta}{2}\right)^{5/3}} \tag{7.90}$$

$$\Phi = a\frac{C(\infty)|\nabla\rho|}{C(\rho)\rho^{7/6}} \tag{7.91}$$

and

$$C(\rho) = b_1 + \frac{b_2 + b_3 r_S + b_4 r_S^2}{1 + b_5 r_S + b_6 r_S^2 + b_7 r_S^3} \tag{7.92}$$

The a and b_1 to b_7 are numerical constants.

Perdew and Wang improved the P86 correction in 1991 (P91 or PW91) [697, 700]

$$\epsilon_c^{PW91} = \epsilon_c^{LDA} + \rho\left(H_0(t, r_S, \zeta) + H_1(t, r_S, \zeta)\right) \tag{7.93}$$

where

$$
\begin{aligned}
H_0(t, r_S, \zeta) &= \frac{f(\zeta)^3}{b}\ln\left[1 + a\frac{t^2 + At^4}{1 + At^2 + A^2t^4}\right] \\
H_1(t, r_S, \zeta) &= \left(\frac{16}{\pi}\right)(3\pi^2)^{1/3}[C(\rho) - c]f(\zeta)^3 t^2 e^{-dx^2/f(\zeta)^2} \\
f(\zeta) &= \frac{1}{2}((1+\zeta)^{2/3} + (1-\zeta)^{2/3}) \\
t &= \left(\frac{192}{\pi^2}\right)^{1/6}\frac{|\nabla\rho|}{2f(\zeta)\rho^{7/6}} \\
A &= a[e^{-b\epsilon_c(r_S,\zeta)/f(\zeta)^3} - a]^{-1}
\end{aligned}
$$

$\epsilon_c(r_S, \zeta)$ is a modified form of Eq. (7.80) (cf. Ref. [701]), x and $C(\rho)$ have the same meaning as shown above and a, b, c, and d are constants.

Most of these functionals violate fundamental restrictions. They fail, for example, to have the exchange energy cancel the Coulomb self-repulsion. The P86 and PW91 functionals cannot predict the correlation energy for one-electron systems. Recent work in this area has focused on creating functionals which do not suffer from these problems.

Hybrid Methods

An exact connection can be made between the exchange-correlation energy and
the corresponding potential connecting the non-interacting reference and the actual system

$$E_{xc} = \int_0^1 \langle \Psi_\lambda | \hat{V}_{xc}(\lambda) | \Psi_\lambda \rangle d\lambda \tag{7.94}$$

This is known as the *adiabatic connection formula*. The integral can be approximated by assuming $\hat{V}_{xc}$ to be linear in λ as the average of the values as the two
end points

$$E_{xc} \approx \frac{1}{2} \left(\langle \Psi_0 | \hat{V}_{xc}(0) | \Psi_0 \rangle + \langle \Psi_1 | \hat{V}_{xc}(1) | \Psi_1 \rangle \right) \tag{7.95}$$

For $\lambda = 0$ the electrons are not interacting. Therefore, there is no correlation
energy, but only exchange energy. This exchange energy is exactly the exchange
energy calculated by Hartree–Fock theory as long as the Kohn–Sham orbitals are
identical to the Hartree–Fock orbitals. If the last term in Eq. (7.95) is approximated by the LSDA result the Half-and-Half (H+H) method is obtained [702]

$$E_{xc}^{H+H} = \frac{1}{2} \left(E_x^{HF} + E_{xc}^{LSDA} \right) \tag{7.96}$$

This approach can be generalized by including gradient-correction and one ar-
rives, for example, at the Becke three parameter functional (B3) [703, 704]

$$E_{xc}^{B3} = (1 - a)E_x^{LSDA} + aE_x^{HF} + b\Delta E_x^{B88} + E_c^{LSDA} + c\Delta E_c^{GGA} \tag{7.97}$$

The a, b, and c are parameters which are determined by fitting to experimental data. Methods based on the adiabatic connection formula are called hybrid
methods or adiabatic connection model (ACM) since they mix Hartree–Fock and
density functional theories. These methods are substantially more costly in calculations than plain density functional methods since the calculation of the Hartree–
Fock exchange energy requires the evaluation of all the same two-electron integrals as in Hartree–Fock calculations. However, these methods have been proven
to give much more accurate results than plain density functional methods.

Commonly Used Functionals

The most important and most widely used functionals are

- BP86
  The combination of Becke's gradient-corrected exchange functional and
  Perdew's gradient-corrected correlation functional from 1986.

- BLYP
  The combination of Becke's gradient-corrected exchange functional and
  Lee–Yang–Parr's gradient-corrected correlation functional Eq. (7.86).

- BPW91
  The combination of Becke's gradient-corrected exchange functional and Perdew-Wang's gradient-corrected correlation functional.

- B3LYP
  The combination of Becke's three-parameter hybrid functional Eq. (7.97) and Lee–Yang–Parr's gradient-corrected correlation functional (E_c^{GGA} in Eq. (7.97) becomes equal to Eq. (7.86)).

- ACM
  A hybrid functional similar to B3LYP, but with slightly different parameters.

Density functional methods offer a number of ways to improve the numerical performance. In HF calculations analytical basis functions are commonly used (in particular Gaussian functions) since it is comparatively easy to evaluate the integrals with them. Gaussian functions are, however, only an approximation to the shape of the electron distribution and a number of them have to be used together to reach suitable accuracy. Density functional methods can be used with analytical basis functions as well. Since the integrations required to calculate the Fock matrix elements in density functional theory have to be carried out numerically anyway, it is also possible to use numerical basis functions, which can be adapted much better to the shape of the electron distribution. Density functional methods have also been the first to achieve better scaling by using a method called resolution of identity. With this method the one-electron density is fitted to an auxiliary basis set of Gaussian type functions [705–708] and the calculation of four-center two-electron integrals is completely avoided. Therefore, density functional methods can reach a theoretical scaling of N^3, which makes them, despite including correlation, even faster than HF calculations.

Compared to Hartree–Fock theory there is one major difference in density functional theory. If the exact functional of the exchange-correlation energy would be known density functional theory would provide the exact total energy including electron correlation. Density functional methods have therefore the potential to include electron correlation at a computational cost similar to that of calculating the uncorrelated Hartree–Fock energy. This is the case for the approximations to the functional in use today; however, this might not be true for the exact functional.

7.1.11 Application to Solids – Bloch's Theorem

The previous sections have provided a short introduction into various methods to calculate electronic structures. These methods can readily be applied to molecules. However, solids play a very important role in materials science. To extend electronic structure calculations to solids a few additional problems have to be solved.

First, the potential felt by an electron in a solid is periodic. We can account for the presence of other electrons by simply assuming that we know an effective potential, $V(\vec{r})$, for our electron. The electron will experience the same potential in the crystal at all positions related by translational symmetry. Thus

$$V(\vec{r} + \vec{R}_j) = V(\vec{r}) \tag{7.98}$$

where

$$\vec{R}_j = l_j \vec{a} + m_j \vec{b} + n_j \vec{c} \tag{7.99}$$

is a translation vector in the lattice defined by the lattice vectors $\vec{a}$, $\vec{b}$, and $\vec{c}$. It is convenient to define a set of translation operators, $\hat{T}_j$, which have the property

$$\hat{T}_j f(\vec{r}) = f(\vec{r} + \vec{R}_j) \tag{7.100}$$

These operators commute[1] with each other and with the Hamilton operator. As a consequence the wave function can be chosen to be an eigenfunction of the energy and all translations

$$\hat{T}_j \psi_i(\vec{r}) = \psi_i(\vec{r} + \vec{R}_j) = \lambda_j \psi_i(\vec{r}) \tag{7.101}$$

where λ_i is the eigenvalue that describes the effect of the operator $\hat{T}_j$ on the function $\psi_i(\vec{r})$. It is, of course, required that the translations do not change the norm of the wave function, which is accomplished if λ_i is a complex number of modulus unity written as

$$\lambda_j = \exp(i\theta_j) \tag{7.102}$$

where θ_j is real. If we now consider two translation operators, $\hat{T}_j$ and $\hat{T}_l$, acting in succession we obtain

$$\hat{T}_j \hat{T}_l \psi_i(\vec{r}) = \psi_i(\vec{r} + \vec{R}_j + \vec{R}_l) = \lambda_j \lambda_l \psi_i(\vec{r}) = \lambda_{j+l} \psi_i(\vec{r}) \tag{7.103}$$

Therefore it is required that

$$\theta_j + \theta_l = \theta_{j+l} \tag{7.104}$$

which can be fulfilled if

$$\theta_j = \vec{k} \cdot \vec{R}_j \tag{7.105}$$

Here $\vec{k}$ is an arbitrary vector which is the same for each operation. It characterizes the particular wave function $\psi_i(\vec{r})$, which can be written more precisely as $\psi_{i\vec{k}}(\vec{r})$. By allowing for other quantum numbers, n, we obtain Bloch's theorem [709]

$$\psi_{n\vec{k}}(\vec{r} + \vec{R}_j) = e^{i\vec{k} \cdot \vec{R}_j} \psi_{n\vec{k}}(\vec{r}) \tag{7.106}$$

---

[1] Operators are said to commute with each other if the outcome of their successive application to a function does not depend on the order in which they are applied. This is often written using a commutator $[\hat{O}_1, \hat{O}_2] = \hat{O}_1 \hat{O}_2 - \hat{O}_2 \hat{O}_1 = 0$.

which says that we can obtain the wave function for the whole periodic solid from the wave function of a unit cell. The vector $\vec{k}$ is of central importance in describing Bloch functions. It is a vector in reciprocal space (since it has the dimension of an inverse length).

Bloch functions can be expanded in terms of a suitable set of basis functions the same way as in molecular calculations (cf. p. 167). The expansion coefficients in this case depend on the wave vector, $\vec{k}$.

By applying Bloch's theorem we can reduce the wave function of the solid to the wave function of the unit cell. In principle, it is possible to use the same atom-centered basis sets in periodic electronic structure calculations, as used for molecular calculations. This leads to crystal Hartree–Fock or periodic density functional theory calculations. Such calculations can be very demanding because a sufficient number of wave vectors has to be included, especially for metallic systems. However, systematic schemes of k-point sampling and other algorithmic advances have recently enabled practical calculations on systems of a few hundred atoms. Periodic solids offer other possibilities for basis functions, which are less oriented on a chemist's atomic orbital picture. A valence electron moving through a solid will experience a potential which varies strongly. Close to a nucleus (within the core region) it will feel a strong electrostatic force. This force weakens when the electron leaves the core region since the core electrons shield the nucleus. In between two atoms (in the interstitial region) the potential becomes nearly constant due to periodicity. As a consequence we can consider two opposite classical limits: tight binding theory and nearly-free-electron theory. Tight binding theory applies if the potential in the neighborhood of some or all ions is very strong compared to that in the interstitial regions between them (e. g., in alkali halide). In this case the uncoupled isolated atoms provide a good first approximation to the solid. The eigenfunctions of the isolated atom are suitable basis functions for the description of the electronic structure of the solid. Nearly-free-electron theory can be applied if the individual atomic potentials overlap so much that a better approximation is to assume that the potential is constant (e. g., in non-transition metals). A natural choice of basis functions in this case is plane waves. In reality solids occupy a spectrum between these two extremes. Table 7.1 provides an overview over electronic structure methods suitable for calculations on solids.

7.1.12 Tight Binding Theory

Tight binding theory can be viewed as a semi-empirical theory for periodic solids. It takes the point of view that a solid is a collection of individual atoms, which have been brought close together from infinite separation. When the atoms are far apart, its constituent electrons feel only the potential from the atomic core plus other electrons in that atom. Thus the electronic wave functions are simply the

Table 7.1: Electronic structure methods for solids, surfaces, interfaces, molecules, and clusters [710]

Method	Systems	Atom Types	Properties Obtainable	Methodological Characteristics
Pseudopotential plane wave	Periodic structures, metals, semiconductors, insulators	All main group elements and transition metals; not well tested for rare-earth and actinides	Crystallographic structures, binding energies, phonon spectra, elastic constants, band gaps, band offsets	Only valence electrons treated explicitly, systematic convergence of basis set, computationally efficient
Full-potential linearized augmented plane wave (FLAPW)	Periodic structures and thin films; all bonding types	All atoms of the periodic table including rare earth atoms and actinides	Crystallographic structures, binding energies, elastic constants, energy band structures, work functions, core level shifts, magnetic moments, magneto-optical properties, hyperfine fields, field gradients	All-electron method, includes relativistic effects explicitly, highly flexible basis set, no shape approximation to the potential ("full-potential" method)

Table 7.1, continued

Method	Systems	Atom Types	Properties Obtainable	Methodological Characteristics
Linearized muffin-tin orbital (LMTO)	Close-packed periodic structures	All atoms	Energy band structures, substitutional energies, electrical, magnetic, and optical properties	Often used in connection with atomic-sphere-approximation (ASA), computationally fast, but restricted to close-packed solids
Full-potential linearized muffin-tin orbital (FPLMTO)	Close-packed and open periodic structures	All atoms	Same as LMTO and crystallographic structures	More accurate than LMTO-ASA, but computationally more demanding
Augmented spherical wave (ASW)	Same as LMTO	All atoms	Same as LMTO	Similar to LMTO
Linear combination of atomic orbitals (LCAO)	Periodic structures, molecules, and clusters; molecules on surfaces	All atoms	Crystal and molecular structures, binding energies, energy band structures, band gaps, core level shifts	Implemented with Gaussian, Slater-type, or numerical atomic orbitals; predominant method for molecular systems

atomic orbitals $\chi(\vec{r})$. As the atoms are brought close together to form the solid, the electrons begin to feel the potential from surrounding atoms, and its wave function starts to become delocalized. Mathematically, this can be expressed as a linear combination of orbitals centered on various atoms, just as in the LCAO formalism for molecules (see section 7.1.5). The only difference is that, in a periodic solid, the electronic wave function has to satisfy the Bloch theorem (see section 7.1.11), which means that the linear combination has to be of the form

$$\psi_{\vec{k}}(\vec{r}) = \frac{1}{\sqrt{N}} \sum_{\vec{R}} e^{i\vec{k}\cdot\vec{R}} \chi(\vec{r} - \vec{R}) \tag{7.107}$$

In other words, the wave function is a linear superposition of atomic wave functions with a phase shift between the functions on neighboring sites. If the Bloch functions are created for atoms that are far apart then the energy is that of the atomic state and does not depend on $\vec{k}$. However, as the atoms come closer, the potential energy is a function of the phase shift and is proportional to the extent of overlap of the wavefunctions and to the extent of change from the free atom potential.

To a first-order perturbation, the total energy, E, can be written as

$$E = E_{atomic} + \int \psi^* \Delta V(\vec{r}) \psi d\vec{r} \tag{7.108}$$

where E_{atomic} is the energy of the original atomic orbital and $\Delta V(\vec{r})$ is the difference in the potential energy between the isolated atom and the atom in a crystal lattice.

Substituting for the wavefunction in Eq. (7.107) and using the lattice invariance under translation through $\vec{R}$, a simplified form of the above equation is as follows

$$E(\vec{k}) = E_{atomic} + \sum_{\vec{R}} e^{i\vec{k}\cdot\vec{R}} A(\vec{R}) \tag{7.109}$$

where the summation runs over all the lattice sites in the crystal and

$$A(\vec{R}) = \int \chi^*(\vec{r}) \Delta V(\vec{r}) \chi^*(\vec{r} - \vec{R}) d\vec{r} \tag{7.110}$$

When $\vec{R} = 0$, this function corresponds to the overlap of the orbital with itself. In other cases, when $\vec{R} \neq 0$, this function corresponds to the overlap of orbitals that are centered on neighboring sites. Thus, the equation for the energy can be written as follows

$$E(\vec{k}) = E_{atomic} + \alpha + \sum_{\vec{R} \neq 0} e^{i\vec{k}\cdot\vec{R}} A(\vec{R}) \tag{7.111}$$

where α is the change in energy of the atomic orbital due to the environment in the solid and the last term is the change in energy of the atomic orbital due to overlap of that orbital centered on one site with the same orbital centered on a neighboring site.

If the overlap is zero or if the crystal potential is the same as the atomic potential, then $\Delta V(\vec{r}) = 0$ and $A(\vec{R}) = 0$, which result in zero bandwidth as in the case of deep core states.

Tight-binding theory is mostly used to describe electrons in narrow bands, e. g., partially filled d bands of transition metals, or the electronic structure of insulators. For more delocalized electrons, e. g., valence s electrons in alkali metals, it is more appropriate to start from a free-electron picture, as described in the following.

7.1.13 Nearly-Free-Electron Theory – Plane Waves and Pseudopotentials

Nearly-free-electron theory uses a Fourier series as basis expansion

$$\chi_j(\vec{r}) = \exp[i(\vec{k} + \vec{K}_s) \cdot \vec{r}] \qquad (7.112)$$

where $\vec{K}_s$ is a reciprocal lattice vector. We can therefore build the wave function from these "plane waves"

$$\psi_{n\vec{k}}(\vec{r}) = \frac{1}{\sqrt{\Omega}} \sum_s a_{ns}(\vec{k}) e^{i(\vec{k}+\vec{K}_s)\cdot\vec{r}} \qquad (7.113)$$

where Ω is the volume of the crystal used to ensure normalization of $\psi_{n\vec{k}}(\vec{r})$.

To obtain the expansion coefficients, $a_{ns}(\vec{k})$, and the eigenvalues, $\epsilon_{n\vec{k}}$, the matrix of the Hamiltonian can be set up in the same way as in the Roothaan–Hall method described above using plane waves as basis set. The resulting secular equation can be written as

$$\sum_s \left[\left[\left(\vec{k} + \vec{K}_s \right)^2 - \epsilon_{\vec{k}} \right] \delta_{st} + V(\vec{K}_t - \vec{K}_s) \right] a_s(\vec{k}) = 0 \qquad (7.114)$$

and the eigenvalues and eigenvectors obtained by matrix diagonalization.

A plane-wave expansion in a real crystal would require a very large number of plane waves to describe an electron since the potential near a nucleus is significantly different from the interstitial region. In practice a pure plane-wave ansatz fails for real crystals. Two methods have been developed which modify the plane waves so that the problem becomes tractable: pseudopotentials and augmented plane waves.

Pseudopotentials can be used to eliminate the necessity that plane waves have to describe core states. While pseudopotentials are more general we will follow the historical development and derive them from orthogonalized plane waves (OPW). To treat the core of an atom differently we denote a plane wave by $|\vec{K}\rangle$ and a core state by $|c\rangle$. We can then write an orthogonalized plane wave, $|\chi_{\vec{K}}\rangle$, by

$$|\chi_{\vec{K}}\rangle = |\vec{K}\rangle - \sum_c |c\rangle\langle c|\vec{K}\rangle \tag{7.115}$$

where the sum runs over all core states. Since the different core states are orthogonal to each other it follows

$$\langle c|\chi_{\vec{K}}\rangle = 0 \tag{7.116}$$

which means that $|\chi_{\vec{K}}\rangle$ is orthogonal to all core states. If we now use Bloch functions of the form

$$|\psi_{\vec{K}}\rangle = \sum_s a_s(\vec{k})|\chi_{\vec{k}+\vec{K}_s}\rangle \tag{7.117}$$

we have a wave function which consists of a smooth part, $|\vec{K}\rangle$, and a part that possesses small-wavelength oscillations over the core region, $|c\rangle\langle c|\vec{K}\rangle$.

Using this wave function we can obtain the secular equation in analogy to the plane wave case, Eq. (7.114),

$$\sum_s \left[\left[\left(\vec{k}+\vec{K}_s\right)^2 - \epsilon_{\vec{k}} \right] \delta_{st} + V^{OPW}(\vec{K}_t - \vec{K}_s) \right] a_s(\vec{k}) = 0 \tag{7.118}$$

which differs from Eq. (7.114) only in that it contains an orthogonalized plane wave Fourier coefficient, $V^{OPW}(\vec{K}_t - \vec{K}_s)$. We can therefore interpret Eq. (7.118) as a plane-wave representation of [711, 712]

$$(\hat{T} + \hat{V}_{ps})|\phi_{\vec{k}}\rangle = \epsilon_{\vec{k}}|\phi_{\vec{k}}\rangle \tag{7.119}$$

where $\hat{V}_{ps}$ is an operator which results in matrix elements equal to the matrix elements obtained from V^{OPW}. This operator is

$$\hat{V}_{ps} = v + \sum_c (\epsilon_{\vec{k}} - \epsilon_c)|c\rangle\langle c| \tag{7.120}$$

The quantity $\hat{V}_{ps}$ is the pseudopotential which has the essential property that its eigenvalues are identical to those of the real potential for the valence bands. Therefore the core states are eliminated from $|\phi_{\vec{k}}\rangle$ [713, 714]. The pseudopotential is weaker than the real potential since the real potential is attractive and the terms added on the right-hand side are positive. The determination of a practical pseudopotential is difficult and will not be demonstrated here [714].

Plane waves can also be augmented by dividing the solid into non-overlapping spheres surrounding the atoms (chosen as large as possible) and interstitial regions between them. If we now assume that the potential inside the spheres is spherical symmetric and constant outside we obtain a so-called "muffin-tin" potential, which simplifies the problem greatly. The potential must, of course, be continuous at the surface of the spheres. Basis functions in the interstitial region are plane waves

$$\langle \vec{r} | \chi_{\vec{k}+\vec{K}_s} \rangle = e^{i(\vec{k}+\vec{K}_s)\cdot\vec{r}} \tag{7.121}$$

while inside the muffin-tin spheres they are

$$\langle \vec{r} | \chi_{\vec{k}+\vec{K}_s} \rangle = \sum_L C_{L\nu}(\vec{k}+\vec{K}_s) R_{l\nu}(\epsilon, r) Y_L(\vec{r}) \tag{7.122}$$

where $Y_L(\vec{r})$ are spherical harmonics and the functions $R_{l\nu}(\epsilon, r)$ are solutions of the radial Schrödinger equation regular at the origin. The coefficients $C_{L\nu}(\vec{k}+\vec{K}_s)$ have to be obtained so that the plane waves match the basis functions inside the muffin-tin spheres at the sphere's surface continuously. The combination of the basis functions Eq. (7.121) and (7.122) together with the matching condition defines an augmented plane wave.

In the linearized augmented plane wave method (LAPW) [715, 716] the radial part of the basis functions in Eq. (7.122) is recalculated during each iteration of the self-consistency cycle. Therefore, they adapt to the actual shape of the potential inside the spheres. In the full-potential linearized augmented plane wave (FLAPW) method [496, 497] the radial parts for both the valence electrons and the core electrons are recalculated during each iteration. This allows for interpretation of x-ray photoelectron spectral data since core level shifts can be calculated. FLAPW calculations can also be used for predicting hyperfine splitting.

7.1.14 Semi-Empirical Methods

Solving the Schrödinger equation using the Hartree–Fock or density functional methods described previously requires large computational resources. It has therefore been of interest to find ways to reduce the computational complexity of these methods. One possibility is to use empirical data from experiments to adjust certain quantities which would have to be computed otherwise. Therefore, these methods have been named semi-empirical.

Semi-empirical methods avoid the calculation of certain difficult to calculate two-electron integrals by either neglecting them or replacing them with quantities fitted to experimental data. Then they use these quantities to set up the Fock matrix. Over the years a number of different semi-empirical methods has been developed, which are distinguished by the way integrals are neglected and by the fit of the remaining quantities to experimental data. These methods have been designed primarily for first row atoms and are particularly well established in organic chemistry. Traditionally, semi-empirical methods use a restricted basis set of one s and three p functions per atom. Though there are extensions to heavier atoms which include d functions, semi-empirical methods have seen comparatively little application in the materials science area. However, since a number of important materials such as polymers can be studied using semi-empirical methods we will provide a summary here.

MNDO, AM1, and PM3

The most widely used semi-empirical methods today are Modified Neglect of Diatomic Overlap (MNDO), Austin Model 1 (AM1), and Modified Neglect of Diatomic Overlap, Parametric Method Number 3 (PM3). These methods are based on the neglect of diatomic differential overlap model (NDDO), which was introduced by Pople and co-workers [717]. In this model the differential overlap between atomic orbitals on different atoms is neglected. Thus, the overlap matrix is reduced to a unit matrix. All two-electron, two-center integrals of the form $(\mu\nu|\lambda\sigma)$ where μ and ν are on the same atom and λ and σ are also on the same atom are retained. The one-center two-electron integrals are derived from experimental data on isolated atoms (atomic spectra) for the MNDO and AM1 methods. For PM3 these integrals were optimized to reproduce experimental molecular properties.

The Fock matrix elements in MNDO are

$$F_{\mu\mu} = H_{\mu\mu} + \sum_{\nu\,on\,A} \left[P_{\nu\nu}(\mu\mu|\nu\nu) - \frac{1}{2}P_{\nu\nu}(\mu\nu|\mu\nu) \right] + \sum_{B \neq A} \sum_{\lambda\,on\,B} \sum_{\sigma\,on\,B} P_{\lambda\sigma}(\mu\mu|\lambda\sigma)$$

(7.123)

where

$$H_{\mu\mu} = U_{\mu\mu} - \sum_{B \neq A} V_{\mu\mu B}$$

(7.124)

If μ and ν are basis functions which are both located on atom A

$$F_{\mu\nu} = H_{\mu\nu} + \frac{3}{2}P_{\mu\nu}(\mu\nu|\mu\nu) - \frac{1}{2}P_{\mu\nu}(\mu\mu|\nu\nu) + \sum_{B \neq A} \sum_{\lambda\,on\,B} \sum_{\sigma\,on\,B} P_{\lambda\sigma}(\mu\nu|\lambda\sigma)$$

(7.125)

where

$$H_{\mu\nu} = - \sum_{B \neq A} V_{\mu\nu B}$$

(7.126)

and if μ is a basis function located on atom A and ν one on atom B

$$F_{\mu\nu} = H_{\mu\nu} + \frac{1}{2} \sum_{\lambda\,on\,B} \sum_{\sigma\,on\,A} P_{\lambda\sigma}(\mu\sigma|\nu\lambda)$$

(7.127)

where

$$H_{\mu\nu} = \frac{1}{2}S_{\mu\nu}(\beta_\mu + \beta_\nu) \tag{7.128}$$

The one-center one-electron integrals in MNDO have a value corresponding to the energy of a single electron experiencing the full nuclear charge plus terms from the potential due to all the other nuclei in the system. These terms are parameterized using reduced nuclear charges and a two-electron integral.

$$U_{\mu\mu} = \left\langle \mu_A \left| -\frac{1}{2}\nabla^2 - \frac{Z_A}{|\vec{r}_1 - \vec{r}_A|} \right| \mu_A \right\rangle \tag{7.129}$$

$$V_{\mu\mu B} = -Z'_B(\mu_A\mu_A|s_Bs_B) \tag{7.130}$$

$$V_{\mu\nu B} = -Z'_B(\mu_A\nu_A|s_Bs_B) \tag{7.131}$$

where μ_A and ν_A are either s or p type functions on atom A. Two-center one-electron integrals are written as a product of the corresponding overlap integral times the average of two atomic "resonance" parameters, Eq. (7.128).

There are a maximum of five one-center two-electron integrals per atom with a sp-basis.

$$\begin{aligned}
\langle ss|ss \rangle &= G_{ss} \\
\langle ss|pp \rangle &= G_{sp} \\
\langle sp|sp \rangle &= H_{sp} \\
\langle pp|pp \rangle &= G_{pp} \\
\langle pp|p'p' \rangle &= G_{p2}
\end{aligned} \tag{7.132}$$

where p and p' are two different p type atomic orbitals. G type parameters are Coulomb integrals, H type parameters are exchange integrals. The two-center two-electron integrals are modeled as interactions between multipoles. There are a total of 22 of these integrals in a sp-basis, which can be written in terms of orbital exponents and the one-center two-electron parameters given in Eq. (7.132).

The core-core repulsion between the nuclear charges is reduced by the number of core electrons. For this purpose the product of the charges is divided by their distance. Due to the approximations in the NDDO model this term is not canceled by electron-electron terms at long distances, resulting in a net repulsion between uncharged molecules. Therefore, the core-core repulsion needs to be modified to generate the proper limiting behavior which involves two-electron integrals. The following terms are used for the core-core repulsion

• MNDO

$$V_{nn}^{MNDO}(AB) = Z'_A Z'_B \langle s_As_B|s_As_B \rangle (1 + e^{-\alpha_A r_{AB}} + e^{-\alpha_B r_{AB}}) \tag{7.133}$$

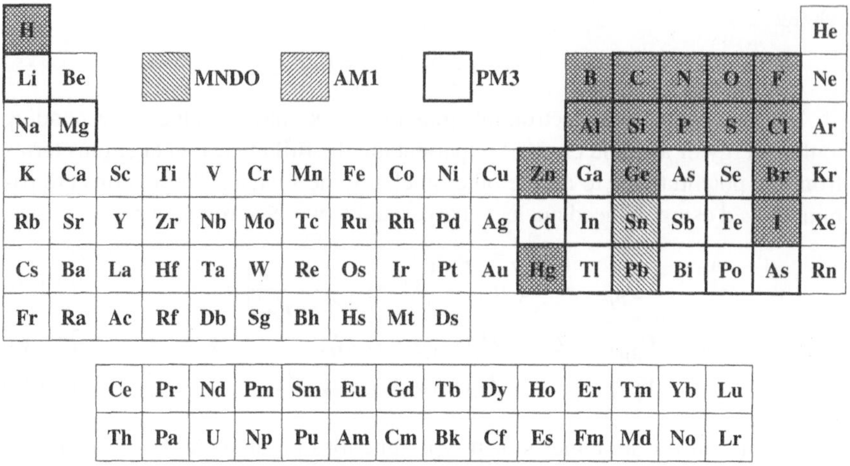

H																	He
Li	Be		MNDO		AM1		PM3					B	C	N	O	F	Ne
Na	Mg											Al	Si	P	S	Cl	Ar
K	Ca	Sc	Ti	V	Cr	Mn	Fe	Co	Ni	Cu	Zn	Ga	Ge	As	Se	Br	Kr
Rb	Sr	Y	Zr	Nb	Mo	Tc	Ru	Rh	Pd	Ag	Cd	In	Sn	Sb	Te	I	Xe
Cs	Ba	La	Hf	Ta	W	Re	Os	Ir	Pt	Au	Hg	Tl	Pb	Bi	Po	As	Rn
Fr	Ra	Ac	Rf	Db	Sg	Bh	Hs	Mt	Ds								

Ce	Pr	Nd	Pm	Sm	Eu	Gd	Tb	Dy	Ho	Er	Tm	Yb	Lu
Th	Pa	U	Np	Pu	Am	Cm	Bk	Cf	Es	Fm	Md	No	Lr

Figure 7.2: Elements parameterized in the MNDO, AM1, and PM3 semi-empirical methods.

- AM1

$$V_{nn}^{AM1}(AB) = V_{nn}^{MNDO}(AB) + \frac{Z'_A Z'_B}{r_{AB}} \left(\sum_k a_{kA} e^{-b_{kA}(r_{AB}-c_{kA})^2} \right.$$
$$\left. + \sum_k a_{kB} e^{-b_{kB}(r_{AB}-c_{kB})^2} \right) \quad (7.134)$$

with $k = 2 \cdots 4$ depending on atom

- PM3

$$V_{nn}^{PM3}(AB) = V_{nn}^{AM1}(AB) \quad (7.135)$$

with $k = 2$ for all atoms.

Each of the NDDO model based methods uses at least twelve parameters per atom. These are the orbital exponents, $\zeta_{s/p}$, the one-electron terms, $U_{s/p}$ and $\beta_{s/p}$, the two-electron terms, G_{ss}, G_{sp}, G_{pp}, G_{p2}, and H_{sp}, as well as the parameters used in the core-core repulsion, α, and the constants a, b, and c for AM1 and PM3. Figure 7.2 shows which elements have been parameterized for each method.

MNDO/d

Since MNDO, AM1, and PM3 are based on a sp-basis only a small part of the periodic table can be handled by these methods. The inclusion of d functions significantly complicates the NDDO model. Instead of the five one-center two-electron integrals with a sp-basis there are 17 with a spd-basis. Similarly, the number of two-center two-electron integrals raises from 22 to 491. Nevertheless, a workable NDDO model which includes d functions for use with MNDO has been constructed by Thiel and Voityuk [718, 719]. This method is called MNDO/d. New parameters include the basis function exponent, ζ_d, and the one-electron terms, U_d and β_d. Only the one-center two-electron integral G_{dd} is taken as a freely varied parameter. The other eleven new one-center two-electron integrals are calculated analytically based on pseudo-orbital exponents, which are assigned so that the analytical formulas regenerate G_{ss}, G_{pp}, and G_{dd}.

Semi-ab initio Method 1

The philosophy behind the semi-ab initio method 1 (SAM1) differs from the philosophy behind MNDO, AM1, and PM3 insofar as the one- and two-center two-electron integrals are no longer replaced by parameters [720]. Due to the increase in available computational power it was considered justified to calculate these integrals directly from the atomic basis functions. These integrals are then scaled by a function containing adjustable parameters to reproduce experimental data

$$\langle \mu_A \nu_B | \mu_A \nu_B \rangle = f(r_{AB}) \langle \mu_A \nu_B | \mu_A \nu_B \rangle \tag{7.136}$$

The SAM1 model uses the STO-3G basis set, but in principle other basis sets can be used with this model as well. Due to the inclusion of basis sets with d functions the SAM1 model can be used for a larger fraction of the periodic table of the elements.

7.1.15   The Basis Set Superposition Error

All quantum chemical calculations which use atom centered basis functions and try to study interactions between different molecules are affected by the basis set superposition error. This error arises because of the necessary truncation of the basis set expansion in practical calculations. If we consider the energy calculated for a single molecule with two different basis sets, X and Y, where Y is larger than X, the basis set Y will give a lower energy than X because the basis set Y is more flexible. To study interactions between two molecules, say A and B, one calculates the energy of the molecules alone followed by the energy for the molecules interacting (the complex). By putting molecule B into interaction with molecule A, the basis functions of molecule B become available to molecule A and vice versa. The energies for both interacting molecules will therefore be artificially lowered due to the larger basis available and the interaction energy calculated will be wrong.

It is, however, possible to obtain an estimate of this basis set superposition error using the counterpoise correction method of Boys and Bernardi [215]. This method requires the evaluation of two more energies, namely the energies of the molecules A and B alone in the basis set of the complex. Practically, this is accomplished by adding molecule B to A, but setting all the nuclear charges in molecule B to zero. Since this leaves us with basis functions without any nuclei, these functions are also called "ghost functions". If we write the energies as $E_A(A)$, $E_B(B)$, $E_{AB}(A)$, $E_{AB}(B)$, and $E_{AB}(AB)$ where the index refers to the basis set and the value in parentheses to the species we can calculate the interaction energy as

$$\Delta E = E_{AB}(AB) - E_A(A) - E_B(B) + E_{BSSE} \tag{7.137}$$

where

$$E_{BSSE} = E_A(A) - E_{AB}(A) + E_B(B) - E_{AB}(B) \tag{7.138}$$

Thus, by substituting Equation (7.138) into Equation (7.137) one obtains the working formula

$$\Delta E = E_{AB}(AB) - E_{AB}(A) - E_{AB}(B) \tag{7.139}$$

which may be misleading in the way that ΔE is exact since it is calculated in the same basis set. This is, however, not the case. It is an (ingenious and often accurate, but no ways exact or mathematically rigorous) *a posteriori* correction scheme.

7.1.16 Nuclear Magnetic Resonance Spectra

Nuclear Magnetic Resonance (NMR) spectroscopy is a fundamental tool in today's analytical chemistry. The interpretation of NMR spectra is not always straightforward and the application of theoretical methods to obtain chemical shifts for atoms in molecules has drawn large interest. The calculation of chemical shifts has, however, become a routine task only recently. A NMR chemical shift for a given nucleus is the difference between the shielding of an external magnetic field, $\vec{B}$, of that nucleus and the shielding of a nucleus of the same element in a standard compound, such as tetramethylsilane for ^{1}H NMR. The shielding differs depending on the electron density around the nucleus and the nucleus therefore experiences only an effective local field. The external field induces a current density in the electronic charge distribution, which itself produces a small magnetic field that opposes $\vec{B}$. The nuclear spin levels are split by

$$\Delta E = \hbar B (1 - \sigma) \gamma \Delta I_z \tag{7.140}$$

where σ is the shielding constant, γ the gyromagnetic ratio of the nucleus, and ΔI_z the change in the spin quantum number. If the electronic energy, E, is expanded with respect to $\vec{B}$ and the nuclear magnetic moment, $\vec{m}_N$,

$$
\begin{aligned}
E \;=\; & E_0 + \sum_i \frac{\partial E}{\partial B_i} B_i + \sum_N \sum_i \frac{\partial E}{\partial m_{Ni}} m_{Ni} + \frac{1}{2} \sum_{i,j} \frac{\partial^2 E}{\partial B_i \partial B_j} B_i B_j \\
& + \sum_N \sum_{i,j} \frac{\partial^2 E}{\partial m_{Ni} \partial B_j} m_{Ni} B_j + \frac{1}{2} \sum_{N,N'} \sum_{i,j} \frac{\partial^2 E}{\partial m_{Ni} \partial m_{N'j}} m_{Ni} m_{N'j} + \cdots
\end{aligned}
$$

$$(7.141)$$

it can be seen that the shielding tensor is given by the second derivative of the electronic energy with respect to components of $\vec{B}$ and $\vec{m}_N$ (since $\vec{m}_N = \hbar \gamma \vec{I}$)

$$
\sigma_{ij}^N = \frac{\partial^2 E}{\partial m_{Nj} \partial B_j} \tag{7.142}
$$

The last equation forms the theoretical basis for the determination of NMR chemical shifts. The reason that the calculation of NMR chemical shifts has become routine only recently lies in the problem of the gauge origin. To describe the interaction of an electron with an external magnetic field a vector potential, $\vec{A}$, can be used which adds the following terms to the Hamilton operator

$$
\hat{H}^{inter} = -\frac{e}{mc} \vec{A} \cdot \vec{p} + \frac{e^2}{2mc^2} \vec{A}^2 \tag{7.143}
$$

These additional terms are obtained by using the generalized momentum

$$
\vec{\pi} = \vec{p} - \frac{e}{c} \vec{A} \tag{7.144}
$$

in the kinetic energy term of the Hamilton operator. There is, however, no unique choice of $\vec{A}$ for a given magnetic field $\vec{B}$. For a static homogeneous field there is an infinite number of different vector potentials

$$
\vec{A}(\vec{r}) = \frac{1}{2} \vec{B} \times (\vec{r} - \vec{R}_G) \tag{7.145}
$$

where $\vec{R}_G$ is an arbitrary parameter, which is usually called the gauge origin. Results from exact calculations would be invariant with respect to the gauge origin, but the use of finite atomic orbital basis sets for the representation of the molecular orbitals results in the loss of gauge invariance.

The gauge-origin problem can be overcome in two ways. First, it is possible to assign local (separate) gauge origins to various "building blocks" of the wave function in such a way that a unique description is guaranteed. This is the approach used by Kutzelnigg and Schindler [721, 722] in their IGLO (individual gauge for localized orbitals) method. Since the notation of local gauge origins is only meaningful for localized quantities this approach is limited to localized molecular orbitals. This is no obstacle in the case of SCF wave functions, but it is difficult to generalize. The second, more general, approach is the use of gauge-including atomic orbitals (GIAO) [723, 724]. Instead of using field-independent basis functions, $\chi_\mu(0)$, the basis functions are made field-dependent

$$\chi_\mu(\vec{B}) = \exp\left(-\frac{i}{2c}\vec{B} \times \left[\vec{R}_\mu - \vec{R}_G\right] \cdot \vec{r}\right) \chi_\mu(0) \qquad (7.146)$$

Gauge-origin independent results are guaranteed by using Equation (7.146). More importantly, the perturbed wave function is described in a balanced way, since the GIAO's are the correct first-order wave functions in a magnetic field assuming that $\chi_\mu(0)$ represents the exact solution in the field-free case. The use of GIAO's requires the evaluation of additional, differentiated two-electron integrals, but this has been implemented efficiently [724]. The GIAO method can and has been straightforwardly extended to correlated wave functions [725–731]. The computational costs for the evaluation of chemical shifts do not explicitly depend on the size of the system under study (only implicitly through the number of basis functions) and are four to eight times the cost of the corresponding energy calculation [732].

7.2 Vibrational Spectra

In addition to NMR techniques vibrational spectroscopy is routinely applied today. Vibrational spectroscopic techniques are the infra-red (IR) and Raman spectroscopies, which differ only in selection rules and can therefore be handled computationally in the same manner.

The calculation of vibrational spectra requires knowledge of the potential energy surface near the equilibrium geometry of a molecule. Since every method which can be used to obtain such knowledge can be used to calculate IR and Raman spectra the following discussion will not be limited to electronic structure methods as in the case of the calculation of NMR spectra.

The potential energy of the system can be expanded into a Taylor series around the equilibrium geometry

$$
\begin{aligned}
V &= V_e + \sum_i \frac{\partial V}{\partial x_i} x_i + \frac{1}{2}\sum_{i,j}\frac{\partial^2 V}{\partial x_i \partial x_j}x_i x_j + \frac{1}{6}\sum_{i,j,k}\frac{\partial^3 V}{\partial x_i \partial x_j \partial x_k}x_i x_j x_k + \\
&\quad + \cdots + \frac{1}{n!}\sum_{i,\cdots,n}\frac{\partial^n V}{\partial x_i \cdots \partial x_n}x_i \cdots x_n
\end{aligned}
\qquad (7.147)
$$

The first term can be set to zero since it only determines the origin of the energy scale. The second term will be zero since $\partial V/\partial x_i$ is zero at the equilibrium geometry (the forces disappear at the equilibrium geometry). Therefore the Taylor expansion becomes

$$
\begin{aligned}
V &= \sum_{i,j} F_{ij} x_i x_j + \sum_{i,j,k} F_{ijk} x_i x_j x_k + \sum_{i,j,k,l} F_{ijkl} x_i x_j x_k x_l + \cdots \\
&+ \sum_{i,\cdots,n} F_{i\cdots n} x_i \cdots x_n
\end{aligned}
\tag{7.148}
$$

The F_{ij}, F_{ijk}, etc. are the quadratic (or harmonic), cubic, quartic, etc. force constants. In general, the Taylor expansion is aborted after the harmonic term and the vibrational analysis is therefore done in the harmonic approximation. The F_{ij} can be calculated from any method which provides the energy of a system as function of the coordinates. The easiest, but most time-consuming way is to distort the molecule along each coordinate and calculate the energy for each distortion. These energies can then be used in numerical differentiation schemes to obtain the force constants. Most electronic structure and force field methods in use today allow the direct analytical calculation of all harmonic force constants. For this purpose the energy expression underlying the particular method is differentiated. The calculation of the complete set of harmonic force constants is significantly more time-consuming than the calculation of a single energy or force vector even with this method. Nevertheless, strictly speaking the calculation of the force constants is necessary to identify whether a given stationary point on the potential energy surface is a minimum or a transition state.

The kinetic energy of the molecule is equal to the sum of the kinetic energies of its atoms

$$
T = \frac{1}{2} \sum_{i,j} v_i M_{ij} v_j
\tag{7.149}
$$

where v_i and v_j are the derivatives of Cartesian coordinates with respect to time (the velocities) and M_{ij} is the generalized mass matrix, which is diagonal in Cartesian coordinates

$$
M_{ij} = m_i \delta_{ij}
\tag{7.150}
$$

The Lagrange function for the molecule becomes

$$
L = T - V = \frac{1}{2} \sum_{i,j} v_i M_{ij} v_j - \frac{1}{2} \sum_{i,j} x_i F_{ij} x_j
\tag{7.151}
$$

and the molecule's movement is described by the Lagrange equations

$$
\frac{\mathrm{d}}{\mathrm{d}t} \frac{\partial L}{\partial v_i} - \frac{\partial L}{\partial x_i} = 0
\tag{7.152}
$$

Using Eq. (7.151) one obtains a system of coupled linear homogeneous differential equations

$$\sum_j \left(M_{ij} \frac{dv_j}{dt} + F_{ij} x_j \right) = 0 \tag{7.153}$$

This system of differential equations can be transformed into a system of algebraic equations by using the ansatz

$$
\begin{aligned}
x_j &= L_j \exp(i\omega t) \\
\frac{dv_j}{dt} &= \frac{d^2 x_j}{dt^2} = -\omega^2 L_j \exp(i\omega t)
\end{aligned}
\tag{7.154}
$$

One obtains

$$\sum_j (F_{ij} - \omega^2 M_{ij}) L_j = 0 \tag{7.155}$$

where $\omega = 2\pi c \bar{\nu}$ and $\bar{\nu}$ is the wave number. This system of algebraic equations has $3N - 6$ non-trivial solutions for a non-linear molecule

$$x_{jr}(t) = L_{jr} \exp(i\omega_r t) \tag{7.156}$$

which correspond to vibrations. The general solution is obtained by superimposing the solutions using coefficients c_r, which are determined by the initial conditions

$$x_j(t) = \sum_r L_{jr} c_r \exp(i\omega_r t) = \sum_r L_{jr} q_r(t) \tag{7.157}$$

The $q_r(t)$ are called normal coordinates and describe the vibrational movements of the atoms of the molecule. In practice, the wave numbers and normal coordinates are obtained using matrix algebra employing the GF matrix method [733–735]. To account for anharmonicities higher-order force constants have to be considered. This usually requires the use of numerical derivatives and such calculations are therefore very expensive.

7.3   Statistical Mechanics

7.3.1   Partition Functions

In the previous sections we have seen how it is possible to calculate electronic energies for molecules and solids. However, chemists usually do not deal with single molecules, but with macroscopic quantities of a chemical. To be able to predict thermodynamic properties at the macroscopic level one needs to consider how a large number of molecules behaves.

Thermodynamics teaches that the thermodynamic probability needs to be maximized. The thermodynamic probability is the number of possible microstates which can realize a macrostate and is defined as

$$w_i = \binom{g_i}{n_i} = \frac{g_i!}{(g_i - n_i)! \, n_i!} \tag{7.158}$$

where g_i is the degeneracy and n_i is the occupation number of microstate i. The total thermodynamic probability for the macrostate is given as the product of the probabilities for the microstates

$$W = \prod_i w_i \tag{7.159}$$

and needs to be maximized. Therefore the first derivative needs to vanish

$$d \ln W = \sum_i \frac{\partial \ln W}{\partial n_i} dn_i = 0 \tag{7.160}$$

subject to the constraints

$$\sum_i n_i = N \tag{7.161}$$

$$\sum_i n_i \epsilon_i = E \tag{7.162}$$

where N is the total number of particles, ϵ_i the energy of microstate i, and E the total energy of the system. These constraints specify that the occupation numbers need to sum up to the total number of particles in the system and that the total energy of the system needs to be conserved. Differentiating the constraints yields

$$dN = \sum_i \frac{\partial N}{\partial n_i} dn_i = 0 \tag{7.163}$$

$$dE = \sum_i \frac{\partial E}{\partial n_i} dn_i = 0 \tag{7.164}$$

and adding them to Eq. (7.160) using Lagrangian multipliers α and β gives

$$\sum_i \left(\frac{\partial \ln W}{\partial n_i} + \alpha \frac{\partial N}{\partial n_i} + \beta \frac{\partial E}{\partial n_i} \right) dn_i = 0 \tag{7.165}$$

For simplification we remember Stirling's formula $\ln x! \approx x \ln x - x$ for large x and calculate from Eq. (7.158)

$$\frac{\partial \ln W}{\partial n_i} = \ln(g_i - n_i) - \ln n_i \approx \ln \frac{g_i}{n_i} \tag{7.166}$$

for $n_i \ll g_i$. Since $\partial N/\partial n_i = 1$ and $\partial E/\partial n_i = \epsilon_i$ we obtain

$$\sum_i (\ln \frac{g_i}{n_i} + \alpha + \beta\epsilon_i)dn_i = 0 \tag{7.167}$$

Solving for n_i gives

$$n_i = g_i e^\alpha e^{\beta\epsilon_i} \tag{7.168}$$

By using the constraint from Eq. (7.161) it can be found that

$$e^\alpha = \frac{N}{\sum_i g_i e^{\beta\epsilon_i}} \tag{7.169}$$

and we obtain the Maxwell–Boltzmann energy distribution law

$$\frac{n_i}{N} = \frac{g_i e^{\beta\epsilon_i}}{\sum_i g_i e^{\beta\epsilon_i}} \tag{7.170}$$

The denominator in Eq. (7.170) is called a partition function and is the fundamental quantity in statistical thermodynamics. By calculating the average kinetic energy of a particle it is possible to identify the second Lagrangian multiplier β as $-1/k_B T$ where k_B is the Boltzmann constant and T the temperature.

Thus, the partition function, q, for a single molecule is defined as a sum over all distinct energy levels, N_e,

$$q = \sum_i^{N_e} g_i e^{-\epsilon_i/k_B T} \tag{7.171}$$

The partition function, Q, for N molecules is defined as

$$Q = q^N \quad \text{(for different particles)} \tag{7.172}$$

$$Q = \frac{q^N}{N!} \quad \text{(for identical particles)} \tag{7.173}$$

Thermodynamic functions such as the internal energy, U, and the Helmholtz free energy, A, can be calculated from the partition function

$$U = k_B T^2 \left(\frac{\partial \ln Q}{\partial T}\right)_V \tag{7.174}$$

$$A = -k_B T \ln Q \tag{7.175}$$

Other thermodynamic functions can be constructed from these relations

$$H = U + pV = k_B T^2 \left(\frac{\partial \ln Q}{\partial T}\right)_V + k_B TV \left(\frac{\partial \ln Q}{\partial V}\right)_T \tag{7.176}$$

$$S = \frac{U - A}{T} = k_B T \left(\frac{\partial \ln Q}{\partial T}\right)_V + k_B \ln Q \tag{7.177}$$

$$G = H - TS = k_B TV \left(\frac{\partial \ln Q}{\partial V}\right)_T - k_B T \ln Q \tag{7.178}$$

Macroscopic observables, such as pressure, p, or heat capacity at constant volume, C_v, can be calculated as derivatives of the thermodynamic functions

$$p = -\left(\frac{\partial A}{\partial V}\right)_T = k_B T \left(\frac{\partial \ln Q}{\partial V}\right)_T \tag{7.179}$$

$$C_v = \left(\frac{\partial U}{\partial T}\right)_V = 2k_B T \left(\frac{\partial \ln Q}{\partial T}\right)_V + k_B T^2 \left(\frac{\partial^2 \ln Q}{\partial T^2}\right)_V \tag{7.180}$$

7.3.2 Calculation of Thermodynamic Functions

All possible quantum states need to be known to calculate the partition function. To do this one usually assumes that the energy of a molecule can be approximated as a sum of terms involving translational, rotational, vibrational, and electronic states. This is a good assumption as long as the Born–Oppenheimer approximation holds. The total partition function then becomes a product of the partition function for each type of states

$$q_{total} = q_{trans}\, q_{rot}\, q_{vib}\, q_{elec} \tag{7.181}$$

The partition function for translational states can be calculated by solving the Schrödinger equation for a particle in a box. One obtains

$$q_{trans} = \left(\frac{2\pi M k_B T}{h^2}\right)^{3/2} V \tag{7.182}$$

where M is the molecular mass of the particle, h Planck's constant, and V the volume of one mole of gas of the molecule.

The partition function for rotational states can be calculated by solving the Schrödinger equation for a rigid rotor. This assumes that the rotation does not change the geometry of the molecule due to centrifugal forces. One obtains for linear molecules

$$q_{rot} = \frac{8\pi^2 I k_B T}{h^2 \sigma} \tag{7.183}$$

where I is the moment of inertia of the molecule and σ the symmetry number. For non-linear molecules the partition function becomes

$$q_{rot} = \frac{\sqrt{\pi}}{\sigma} \left(\frac{8\pi^2 k_B T}{h^2}\right)^{3/2} \sqrt{I_1 I_2 I_3} \tag{7.184}$$

where I_1, I_2, and I_3 are the moments of inertia of the molecule in its principal axes coordinate system. The symmetry number σ is the order of the rotational subgroup in the molecular point group. Table 7.2 lists possible values.

Table 7.2: Symmetry number for a given point group

Point group	σ	Point group	σ	Point group	σ
C_1	1	S_n	$n/2$	$C_{\infty v}$	1
C_n	n	C_{nv}	n	$D_{\infty h}$	2
D_n	$2n$	C_{nh}	n	T_d	12
C_s	1	D_{nd}	$2n$	O_h	24
C_i	1	D_{nh}	$2n$	I_h	60

The partition function for vibrational states can be calculated by solving the Schrödinger equation for a one-dimensional harmonic oscillator. One obtains

$$q_{vib} = \frac{e^{-h\nu/2k_bT}}{1 - e^{-h\nu/k_bT}} \tag{7.185}$$

where ν is the vibrational frequency. For polyatomic molecules the coordinate system can be transformed to the normal mode coordinate system (cf. Section 7.2) and the partition function becomes the product of the partition functions for each normal mode i

$$q_{vib} = \prod_i^{N_f} \frac{e^{-h\nu_i/2k_bT}}{1 - e^{-h\nu_i/k_bT}} \tag{7.186}$$

where N_f is the number of vibrational degrees of freedom ($3N - 5$ for a linear molecule, $3N - 6$ for a non-linear molecule, $3N - 7$ for a transition state).

The electronic partition function involves a sum over electronic quantum states. These are the solutions to the electronic Schrödinger equation (ground state and all possible excited states). Since the energy difference between the ground state and the first excited state usually is much larger than k_BT only the first term in Eq. (7.171) is important. If the zero point for the energy is defined as the electronic energy of the reactant, the electronic partition functions for the reactant and the transition state become

$$q_{elec}^{reactant} = g \tag{7.187}$$

$$q_{elec}^{ts} = ge^{-\Delta E^{\neq}/k_BT} \tag{7.188}$$

where $\Delta E^{\neq}$ is the difference in electronic energy between the reactant and the transition state and g is the electronic degeneracy of the wave function. The degeneracy may be either in the spin part ($g = 1$ for singlet, 2 for doublet, 3 for triplet, etc.) or in the spatial part ($g = 1$ for wave functions belonging to an A, B, or Σ representation, 2 for an E, π, or Δ representation, 3 for a T representation, etc.). Since most stable molecules usually have non-degenerate ground state wave functions g is 1 for them.

Enthalpy and entropy can now be calculated using these partition functions. One obtains for a non-linear molecule

$$H_{trans} = \frac{5}{2}RT \tag{7.189}$$

$$S_{trans} = \frac{5}{2}R + R\ln\left(\frac{V}{N_A}\left(\frac{2\pi M k_B T}{h^2}\right)^{3/2}\right) \tag{7.190}$$

$$H_{rot} = \frac{3}{2}RT \tag{7.191}$$

$$S_{rot} = \frac{1}{2}R\left[3 + \ln\left(\frac{\sqrt{\pi}}{\sigma}\left(\frac{8\pi^2 k_B T}{h^2}\right)^{3/2}\sqrt{I_1 I_2 I_3}\right)\right] \tag{7.192}$$

$$H_{vib} = R\sum_i^{N_f}\left(\frac{h\nu_i}{2k_B} + \frac{h\nu_i}{k_B}\frac{1}{e^{h\nu_i/k_B T} - 1}\right) \tag{7.193}$$

$$S_{vib} = R\sum_i^{N_f}\left(\frac{h\nu_i}{k_B T}\frac{1}{e^{\frac{h\nu_i}{k_B T}} - 1} - \ln(1 - e^{-h\nu_i/k_B T})\right) \tag{7.194}$$

$$H_{elec}^{reactant} = 0 \tag{7.195}$$

$$H_{elec}^{ts} = \Delta E^{\neq} \tag{7.196}$$

$$S_{elec}^{reactant} = S_{elec}^{ts} = R\ln(g) \tag{7.197}$$

For a linear molecule only the rotational contributions are different

$$H_{rot} = RT \tag{7.198}$$

$$S_{rot} = R\left[1 + \ln\left(\frac{8\pi^2 I k_B T}{\sigma h^2}\right)\right] \tag{7.199}$$

The summation for the vibrational contributions runs over the vibrational degrees of freedom as outlined above.

7.4 Molecular Mechanics

The exact solution of the equations of motion for a molecule based on quantum chemistry is — as we have seen in the previous section — a rather complex problem. The large amount of numbers which have to be processed requires elaborate programming algorithms and very fast computers with large memory. The size of molecules which can be handled by non-empirical quantum mechanical methods is, therefore, limited.[2] Molecules of the size of interest in the chemical industry are often so large that they are not accessible to quantum mechanical methods (e. g., polymers or heterogeneous catalysts).

If the electronic properties are not of direct concern, but structures, conformations, vibrational spectra, etc. are the properties of interest molecular mechanics methods can be an extremely useful alternative. Such methods involve a classical description of a molecular system through interatomic potentials or force fields. The idea behind molecular mechanics is that the character, e. g., of a bond, is largely determined by its immediate environment and therefore the properties of a bond (e. g., its length and strength) are nearly constant in different molecules if the same atoms are bond together. For this concept to work it is necessary to define the term *atom* more narrowly. It is insufficient, as experience has proven, to only understand a certain element as *atom* (as quantum mechanical methods do). It is also necessary to include other properties as, e. g., the hybridization, into the term *atom*. Molecular mechanics methods usually use *atom types* to distinguish parameters for different bonding situations. As a concrete example, one could assume that the properties of a N–H bond in NH_3 are transferable to a N–H bond in methylamine, but not in NH_4^+. The nitrogen in NH_4^+ will therefore have a different *atom type* compared to the nitrogen in NH_3 and methylamine. Transferability of properties is the basic assumption of molecular mechanics methods. This approach has its basis in vibrational spectroscopy where certain functional groups possess characteristic vibrational frequencies. A functional group can therefore be characterized by a set of force constants.

Molecular mechanics methods use a set of parameterized functions aimed at representing the potential energy hypersurface of a molecule as accurately as possible. Parameters in these functions are force constant like quantities, which also lead to the name force field (or potential) for the whole set of parameters and functions. Calculating the energy of a conformation of a molecule is achieved by simply summing up the values of the different functions in dependence of the molecule's geometry. One does not obtain a physically meaningful energy this way,[3] but stable conformations can be obtained by energy minimization in the same way as in quantum chemistry. The stability of different conformers can, e. g., be compared based on the molecular mechanics (steric) energy.

The description of a molecular system using force fields speeds up the calculations considerably. Molecular mechanics calculations can therefore be used with molecules much bigger than what is possible for quantum mechanical methods.

---

[2]The parameter which determines the size of the molecule is considered here to be the number of atoms in the molecule. For quantum mechanical calculations more important is the number of basis functions used to describe a molecule. There is, however, no strict relation between the number of atoms and the number of basis functions in a molecule.

[3]This energy is called steric energy, an energy which only depends on the arrangement of the atoms in space.

However, the need for force field parameters presents a problem. Force field parameters are often only transferable between closely related molecules and great care has to be taken to assure that the parameters give the appropriate results. Force fields should always be checked to predict the desired properties accurately for known molecules closely related to the systems under study. It is usually also not possible to combine force field parameters from different sources without sacrificing accuracy.

7.4.1 Force Fields

An oversimplified classical treatment of a molecule might assume that the atoms in the molecule are held together purely by electrostatic forces. Therefore, it would seem reasonable to use Coulomb's law to represent the energy of a molecular system. Coulomb's law alone, however, is insufficient for this purpose because oppositely charged atoms would collapse. In real molecules the strong repulsion of the electron clouds of the atoms approaching one another prevents the collapse. This repulsion can be modeled in a force field using the repulsive part of a Van der Waals term. The simplest workable force fields therefore consist only of a Coulomb and a Van der Waals term. Such force fields are suitable for modeling of ionic crystals and are called *ion pair potentials*. Usually, the Van der Waals term includes an attractive part as well, because neutral atoms experience attractive forces due to dispersion.

There exists another possibility for describing a molecular system in the framework of classical mechanics. This approach assumes that the potential energy of a molecular system can be expanded in a Taylor series around its equilibrium geometry. It is based on the notion of covalent bonds as they are found mainly in organic chemistry and provides a mathematical description of the chemical bond. By modifying this approach to make it transferable between different systems, molecular mechanics force fields are obtained.[4]

Common among all force fields is that they provide parameters for certain atom types, which are combined according to the bonding situation in a molecule. A force field for organic molecules, e. g., might provide the atom types C (sp^3 hybridized), C (sp^2 hybridized), and C (sp hybridized) for carbon and an aliphatic and an alcoholic atom type for hydrogen. Parameters for all (meaningful) bonds among all the provided atom types have to be available in a force field.

---

[4]It should be noted that the use of the term *force field* in the literature is somewhat arbitrary. It is usually applied to both the set of potential energy functions and their corresponding parameters, but it is also common for some authors to refer to just a certain functional form as the force field.

7.4.2 Ion Pair and Shell Model Potentials

The development of ion pair potentials has its origin in the analysis of the cohesive energy of ionic crystals by Madelung and Born [736, 737]. These workers, and later Ewald and Evjen, showed that by making the assumption that a crystal is composed of simple positive and negative ions, one can obtain a surprisingly effective description of many inorganic systems. Indeed, the accuracy of the predicted energies and the insights obtained with this simple assumption stimulated the extension of such models to systems containing point defects within their lattices in addition to the examination of transport properties.

The fundamental assumption in Born's description of the solid is that the crystal is composed of spherically symmetric positive and negative ions. These ions interact with one another according to their separation (except at short distances where atoms begin to collide). The value of their energetic interaction is given by Coulomb's law

$$V(r_{ij}) = \frac{q_i q_j}{\epsilon r_{ij}} \tag{7.200}$$

where $V(r_{ij})$ is the potential energy of interacting species i and j at separation r_{ij} and ϵ is an appropriate constant related to the permittivity of a vacuum.

Because a crystalline solid is composed of a three dimensional array or lattice of ions, the total electrostatic energy of interaction, E_{el}, must account for all the resulting pairwise interactions,

$$E_{el} = \frac{1}{2} \sum_{i,j} \frac{q_i q_j}{\epsilon r_{ij}} \tag{7.201}$$

The 1/2 before the summation is a result of the fact that the complete summation counts each electrostatic interaction twice. Alternatively, one could sum over $i < j$ and remove the factor 1/2.

However, as Madelung and Born originally pointed out, the summation of the totality of ion-ion interactions can be reduced to a simple electrostatic term and a structure dependent constant, the value of which is uniquely determined by the lattice in question. The resulting "Madelung constants" captured the limiting values of the conditionally convergent sums of electrostatic interactions,

$$E_{Madelung} = -A \frac{e^2}{\epsilon r} \tag{7.202}$$

Here $E_{Madelung}$ is the electrostatic energy of the crystal, e the electronic charge multiplied by the valence of the ions of the structure, r a measure of the lattice constants of the crystal, and A is the Madelung constant which depends only on the crystal structure.

This expression for the electrostatic interaction energy as a function of lattice constant tends toward infinitely negative values with decreasing crystal size. The model in its purely electrostatic form, thus, fails because it does not account for the repulsion of the ions at close separation. To account for this, Born introduced the simple assumption that the repulsive energy of the ions could be expressed as a power law of the type Br^{-n}, where the values of B and n are constants for a given system at its equilibrium crystal structure. It was shown that the lattice energy of the system depends only upon the exponent, n. Born used compressibility data to obtain values for the repulsive exponent, employing the following argument. The total energy of the system, being the sum of repulsive and electrostatic interactions, is

$$E_{Total} = -A\frac{e^2}{\epsilon r} + \frac{B}{r^n} \tag{7.203}$$

At equilibrium separation of the ions, r_0, there will be no net force on the system and the derivative of Eq. (7.203) with respect to r can therefore be set to zero, yielding

$$\frac{dE_{Total}}{dr}\bigg|_{r=r_0} = A\frac{e^2}{\epsilon r_0^2} - n\frac{B}{r_0^{n+1}} = 0 \tag{7.204}$$

Rearranging gives

$$B = \frac{Ae^2}{\epsilon n}r_0^{n-1} \tag{7.205}$$

and Eq. (7.203) becomes

$$E_{Total} = -A\frac{e^2}{\epsilon r_0}\left(1 - \frac{1}{n}\right) \tag{7.206}$$

The compressibility K_0 at absolute zero is related to the energy and volume of the crystal

$$\frac{1}{K_0 V_0} = \left(\frac{d^2 E}{dV^2}\right)_{V=V_0} \tag{7.207}$$

Because the volume at equilibrium, V_0, is a simple function of the equilibrium geometry, r_0, Eq. (7.206) above may be differentiated and substituted into Eq. (7.207) to give

$$n = 1 + \frac{18r_0^4}{K_0 e^2 A} \tag{7.208}$$

yielding n directly from the known compressibility of the material. The values of n obtained this way range from 5.9 to 10 [738]. Because of the $(1-\frac{1}{n})$ dependence of the energy on n (Eq. (7.206)) one finds only a small effect on the overall lattice energy when n is varied.

It is interesting to note that this work predates the discovery of quantum mechanics. The model's success for a variety of systems provided an early indication of the ability of force field methods to describe complex phenomena.

With an improved understanding of the origin of interatomic forces based on the developments in quantum mechanics, the Born model description was refined. Quantum mechanical calculations of the forces between simple ions indicated that the power law used for describing the repulsive interaction of closed-shell species could be improved by implementing an exponential form, as in Eq. (7.209)

$$E_{rep} = B \exp\left(\frac{-r}{\rho}\right) \tag{7.209}$$

Despite the presence of the still adjustable constants, B and ρ, the functional form has a reasonable theoretical footing and has indeed formed the basis of many potential based investigations [311, 739–741].

Quantum mechanical calculations were also applied to evaluate the nature of the weak interactions between atoms. These weak interactions are responsible for the induced dipole cohesion of neutral atoms. It was shown that the dispersion interaction energy of two atoms or ions with closed electronic configurations varied according to their polarizability and ionization energies. London used perturbation theory to obtain a simplified expression for the dispersion energy, E_{dis}, of two interacting species i and j, at separation r,

$$E_{dis} = \frac{3}{2}\left(\frac{\alpha_i \alpha_j I_1 I_2}{I_1 + I_2}\right)\frac{1}{r_{ij}^6} = C_{ij}\frac{1}{r_{ij}^6} \tag{7.210}$$

with atomic polarizability volumes α_i and α_j for interacting atoms and first ionization potentials I_1 and I_2 (Slater–Kirkwood formula). For identical atoms Eq. (7.210) becomes

$$E_{dis} = \frac{3}{4}\left(\alpha^2 I\right)\frac{1}{r_{ij}^6} \tag{7.211}$$

The dispersion energy is greatest for species with high polarizability and ionization potential values.

We note that more sophisticated schemes for deducing parameters for dispersive interactions from atomic data, e. g., for adsorbate/adsorbent interactions in zeolites, have been developed and applied in recent years [742–745].

The potential model that was thus derived for interaction based upon quantum mechanical results together with the original electrostatic description of ionic materials has the following form

$$E_{Total} = \frac{1}{2}\sum_{i,j}\frac{q_i q_j}{\epsilon r_{ij}} + \frac{1}{2}\sum_{i,j}B_{ij}\exp\left(\frac{-r_{ij}}{\rho_{ij}}\right) + \frac{1}{2}\sum_{i,j}C_{ij}\frac{1}{r_{ij}^6} \tag{7.212}$$

The summations are over all atoms i and j with separations, r_{ij}, and B_{ij}, C_{ij}, and ρ_{ij} are the adjustable parameters of the model. The interaction energy of the system as a whole is then the sum of the Coulombic energies, short range repulsive energies, and the weakly attractive energy components for all constituents. As we shall see below, the individual components are typically the atomic centers and the points representing the polarization centers of the system. The successes of the simple ionic models introduced by Born and co-workers have been well documented and cover a wide range of applications. For many systems, however, the directionality of chemical bonding requires treatment. Accordingly, a more general way of treating the energetic interaction of a system containing many atomic species has the following form

$$V = V_0 + \sum_{i>j} \frac{q_i q_j}{\epsilon r_{ij}} + \sum_{i>j} \Psi_{ij}(r_{ij}) + \sum_{i>j>k} \Psi_{ijk}(r_{ijk}) + \sum_{i>j>k>l} \Psi_{ijkl}(r_{ijkl}) + \cdots$$

$$(7.213)$$

where successive terms on the right hand side of the equality describe two, three, four, and higher body component interactions. What the early potentials therefore suggest is that often the potential energy of the system can be accounted for by truncation of Eq. (7.213) before triad and higher order terms. This approximation, that we may factorize the total potential energy of the system into a sum of two-body interactions, has proved effective for many systems. However, it fails for metallic systems [746, 747], systems with high degrees of covalence, and for systems where forces are weak.

To describe the higher order terms of Eq. (7.213) a variety of multi-body potentials have been introduced. In the molecular mechanics method developed for organic molecules, bond angle and torsion deformation terms are used to describe the energetics of the cooperative displacement of groups of atoms thereby incorporating multibody dependencies (see the following section). A contrasting approach has been employed for the study of noble gases where the triple-dipole formula of Axilrod and Teller [748], Eq. (7.214), is used with success

$$V_{Three\,Body} = \frac{v(1 + 3\cos\theta_i \cos\theta_j \cos\theta_k)}{r_{ij}^3 r_{jk}^3 r_{ik}^3}$$

$$(7.214)$$

where v is a constant and r_{ij}, r_{jk}, r_{ik}, θ_i, θ_j and θ_k are the interatomic distances and angles of a triad of interacting atoms.

The molecular mechanics description of deformations about a target geometric variable and the triple-dipole interaction of assemblies represent somewhat different descriptions of multibody energetic interactions. The triple-dipole expression describes the physical interaction of a triad of polarizable species. The bond angle bending potential of a force field represents a more phenomenological approach, one which incorporates the physics of the situation beyond the initially assumed equilibrium geometry through parameterization of force constants. Multibody potential forms have also been used by Kollman and co-workers in the simulation of liquids [749].

The most extensive use of multibody potentials for the simulation of materials using ionic pair potentials has been the addition of bond angle bending terms to two-body potentials used to describe silicate and framework structured materials [266, 750, 751]. In addition, harmonic planarity restraining terms have been employed in the simulation of polyatomic anions in, e. g., inorganic carbonates [752]. Here, as in the organic molecular mechanics methodology described later, structural distortions about an expected geometry are realized at an energetic cost.

Charge Models

A fundamental component of all early potential models is the charge model and this remains the situation for systems studied today using pair-wise potential models [176, 184]. Born originally suggested the use of formal oxidation states for the study of ionic materials, a choice prompted by the need to accommodate defects where the constraint of charge neutrality renders the use of partial charges troublesome. Formal or oxidation-state charge models have been developed, and they have found growing applications for halides and oxides and a variety of other types of materials [93, 753, 754].

Despite the early successes of formal charge models, quantum mechanical calculations revealed complete electron transfer to be an extreme model even for the most ionic of inorganic materials. As far as partial atomic charges can reliably be obtained from the electron density of well defined systems (a feat not without its own uncertainties [755–757]), it appears that charges that are somewhat lower in magnitude than the formal valence ones are often more appropriate [755]. Indeed, potentials based on such charge models have enjoyed success, thus underscoring the validity of this approach. However, just as Born anticipated, there exist difficulties with non-formal charge models. For example, if Al replaces Si in a silicate lattice, the formal charge model easily provides a route to the restoration of charge neutrality (one mono-valent cation per Al^{3+} ion in this example returns such a system to an overall zero charge). Contrarily, the situation is far more complex with a partial charge model. If the charge difference between Al and Si is constrained to be $1e$, charge compensating cations must have integral charges only. This situation is satisfactory for a mono-valent cation case, but a divalent cation such as calcium must then be given a charge $2e$, a charge which may be higher than the charge on the framework Al species itself. Such a non-physical charge distribution will inevitably have an effect on subsequent simulations, adversely affecting the location of the other cations or the location of sorbates within the lattice as well as other undesirable influences. The charge distribution most suited to the description of the framework under study is then of limited applicability to the analysis of extra-framework cation binding. In effect, the approximations or compromises of one chemical environment are of limited compatibility within a new environment.

Table 7.3: Non-bonded interaction potentials employed in materials simulation

Functional Form	References
$E = D_0((r_0/r)^{12} - 2(r_0/r)^6)$	[758]
$E = D_0(2(r_0/r)^9 - 3(r_0/r)^6)$	[759–763]
$E = D_0(6/(y-6)e^{y(1-r/r_0)} - y/(y-6)(r_0/r)^6)$	[764, 765]
$E = D_0(X^2 - 2X)$	[766]
where $X = e^{-(y/2)(r/r_0-1)}$	
$E = D_0(5(r_0/r)^{12} - 6(r_0/r)^{10})$	[767]
$E = D_0 e^{y(1-r/r_0)} + a_1/(1 + e^{b_1(r-c_1)}) + a_2/(1 + e^{b_2(r-c_2)}) +$	[768]
$a_3/(1 + e^{b_3(r-c_3)})$	
$E = Be^{-\rho/r} - C_6/r^6$	[769]
$E = D_0[X^2 - 2X]/(1 + e^{20(r-r_c)}) + (C_6/r^6)/(1 + e^{20(r_c-r)})$	[770]
where $X = e^{-(y/2)(r/r_0-1)}$	

D_0 is the well depth, r and r_0 the current and equilibrium interatomic distances respectively, y is a scaling factor, and B, ρ, a_n, b_n c_n, C_4, and C_6 are potential parameters. These non-bonded interaction forms are employed by Cerius^2 version 3.9 [771].

Functional Forms

The repulsion of fully occupied orbitals in model systems received attention in the earliest application of quantum mechanical methods. From those studies an exponential representation of the energy-distance curve was obtained [772]. This functional form has been used extensively in the simulation of both solids and molecular systems. Also, derived from early quantum mechanical results are potentials using inverse power repulsive forms (see, e. g., Refs. [773,774]). Such potentials have also been employed with success in the simulation of liquids, molecular solids, and ionic systems. As described above, pair potentials characterized by use of both Coulombic and short range interaction terms were adopted early for the description of condensed materials. However, these were not the only potential forms employed. Of relevance in practical simulations is the fact that the exponential form, though theoretically well founded, possesses a numerical instability at short interatomic separation. The attractive sixth power term overwhelms the repulsive exponential contribution leading to a departure to energetic negative infinity with attendant energy conservation problems. Molecular dynamics simulations with long time steps and Monte Carlo simulations will sample this physically unreasonable region of potential description unless appropriate avoidance strategies are employed. Accordingly, a variety of modified potential forms are employed in practical simulations. Several representative examples with references to illustrative applications are collected in Table 7.3.

It is also important to recognize that a particular potential form may impose a significant computational or algorithmic burden in practical calculations. Trigonometric and exponential functions are evaluated slowly and the use of look up tables or splined potentials (see, e. g., Ref. [775]) needed to speed up some computations requires development effort on the part of the programmer. These efforts typically reduce the generality of the resulting software. Thus, a careful balance between CPU costs and the physical reality of the functions used is desirable.

Polarizability

When placed in an electrostatic field, an atom or ion's electrons are polarized, as shown schematically in Figure 7.3. The polarized species then exerts an altered effect on its surroundings in comparison with its unpolarized form. Accounting for the substantial polarizability of atoms and ions that exist in many solids has elicited the development of additional potential models. Most notable are the point polarizable ion model and the shell model.

The point polarizable ion model [776] ascribes a discrete atom-based polarizability to each interacting center. For any given electrostatic configuration this polarizability yields a set of atom based-dipoles. The resulting dipoles themselves affect the electrostatic environment of the material during the simulation, and in practice iteration is used to reach a self-consistent set of atom-centered dipole moments. This methodology allows for atomic polarization and permits the component species of the system to respond to the electrostatic field exerted by an instantaneous configuration of neighbor atoms.

The shell model, originally proposed by Dick and Overhauser [16], describes each atom of the system as being composed of two centers, as illustrated in the schematic of Figure 7.4. The first center is the core that represents the nucleus of the atom. The core generally has a positive charge. The second center (the shell) represents the valence electrons of the atom, possesses no mass, bears a negative charge, and is attached to the core by a harmonic spring. There is no Coulombic interaction between the core and the shell. When this two component representation of an atom is placed in an electric field, the core and shell move in opposite directions with a degree of separation that is governed by the field, their charges, and the spring strength which connects them. The core-shell energy is given by

$$V(r_{ij}) = K_{ij}r_{ij}^2 \qquad (7.215)$$

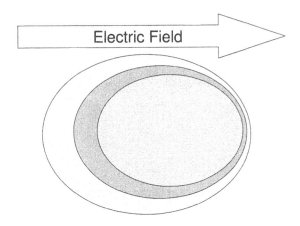

Figure 7.3: Schematic representation of atomic electron density (represented by contours of differing shade) in an applied electric field.

where r_{ij} is the core-shell separation and K_{ij} the spring constant. The total energy (cf. Eq. (7.212)) of the system is then given by

$$E_{Total} = \frac{1}{2}\sum_{ij}\frac{q_iq_j}{\epsilon r_{ij}} + \frac{1}{2}\sum_{ij}B_{ij}\exp\left(\frac{-r_{ij}}{\rho_{ij}}\right) + \frac{1}{2}\sum_{ij}C_{ij}\frac{1}{r_{ij}^6} + \frac{1}{2}\sum_{ij}K_{ij}r_{ij}^2$$

(7.216)

The atom interacts with other species in the system through two-body forces centered on the shell component. Note that in the shell model the electronic polarizability, α, is given by

$$\alpha = Y^2/K_{ij} \qquad (7.217)$$

where Y is the shell charge and, as before, K_{ij}, the harmonic spring constant [769]. In this way, the model is able to connect the coupling of electronic polarizability and short range interactions. There are clear parallels between the shell model, which allows charge separation from atom centers, and the use of non-atom centered partial charges in the TIPS potentials of Jorgensen and co-workers [777] and in the EPEN force field developed by the Scheraga group [778].

The shell model does, however, possess an inherent instability if the local electric field is of sufficient strength to detach the shell from the core [779]. Such a situation may occur in the simulation of inorganic surfaces, particularly for surfaces possessing dipoles and with a tendency to reconstruct. To compensate for

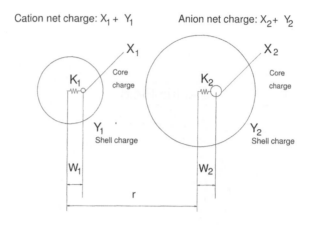

Figure 7.4: Schematic representation of the shell model for a cation-anion pair. Negatively charged shells (with charges Y_1 and Y_2 e) are attached to positively charged cores (with charges X_1 and X_2 e) by harmonic springs, K_1 and K_2. The interionic separation between shells is r and the separation between core and shell, for cation and anion respectively, W_1 and W_2. Although shells are shown with finite radii, to recognize their relation to the atom's valence charge density, energetic interactions are computed at the point at the shell's center. Electrostatic interactions between core and shell of the same species are excluded in potential evaluation.

this deficiency of the shell model a number of authors have employed an additional quartic term to supplement the original harmonic spring [780].

7.4.3   Molecular Mechanics Force Fields

In contrast to the ion pair potentials discussed in the previous section, a different approach has been used successfully in organic chemistry. Organic molecules are mainly covalently bonded. A covalent bond is, in contrast to an ionic bond, directed, so it is insufficient to use the simple ion pair model, which works so well for inorganic solids. Rather, the directed character of the bond has to be accounted for in the force field. A convenient approach for accomplishing this has its origin in vibrational spectroscopy.

To describe the vibrations of a molecule, spectroscopists expand the molecule's potential energy in a Taylor series around its equilibrium geometry (which is considered to be a fixed arrangement of atoms in space according to the Born–

Oppenheimer approximation). Hence the potential is expressed as

$$V = V_0 + \sum_i \frac{\partial V}{\partial x_i}(x_i - x_i^0) + \frac{1}{2}\sum_{i,j}\frac{\partial^2 V}{\partial x_i \partial x_j}(x_i - x_i^0)(x_j - x_j^0)$$

$$+ \frac{1}{6}\sum_{i,j,k}\frac{\partial^3 V}{\partial x_i \partial x_j \partial x_k}(x_i - x_i^0)(x_j - x_j^0)(x_k - x_k^0) + \cdots \quad (7.218)$$

V_0 is a constant which determines the origin of the energy scale. It can be set to zero without loss of generality. $\partial V/\partial x_i$ is zero since the expansion is done around the equilibrium where the first derivatives vanish by definition. The remaining derivatives are the quadratic (or harmonic), cubic, quartic, etc. force constants

$$F_{ij} = \frac{\partial^2 V}{\partial x_i \partial x_j} \quad (7.219)$$

$$F_{ijk} = \frac{\partial^3 V}{\partial x_i \partial x_j \partial x_k} \quad (7.220)$$

$$F_{ijkl} = \frac{\partial^4 V}{\partial x_i \partial x_j \partial x_k \partial x_l} \quad (7.221)$$

$$\cdots$$

where all quantities in Eq. (7.218) with index 0 are equilibrium values. In principle, it would be possible to use just this Taylor series expansion directly in force field calculations. However, we cannot do this because the spectroscopic force constants of Eqs. (7.219)–(7.221) are not unique. Consider a simple molecule such as ethane, which has eight atoms and therefore 18 internal degrees of freedom. Since ethane has seven bonds, twelve bond angles, and nine torsional angles it has a total of 28 internal coordinates. One of the bond angles in each methyl group is always determined by the other five angles – it is linearly dependent on the others. A set of internal coordinates made up from *all* bond lengths and bond angles would therefore be redundant. We can neglect any one of the bond angles (it does not matter which one) and still obtain a complete description of the molecule. However, that creates a problem because we cannot change one of the CCH angles without affecting the HCH angles. How can we isolate the effect a single CCH bond angle has on the energy of ethane? To do this, spectroscopists use linear, independent combinations of internal coordinates in the Taylor series described above, but unfortunately these combinations are in general not unique because we could combine the CCH and HCH angles with each other or we could mix CCH and HCH angles. Consequently, the spectroscopic force constants are themselves not unique either, and therefore not transferable between different molecules.

More importantly, to obtain the Taylor series expansion one must know the equilibrium geometry of the molecule, which is one of the properties we wish to derive from our force field calculations in the first place! The Taylor series expansion is also just a mathematical way of describing the potential energy surface near an equilibrium. It does not capture the physics of interatomic interactions.

To overcome these problems, we must find a generalization of the Taylor series expansion. By looking at different molecules containing the same functional group, we observe from vibrational spectroscopy that each functional group has its characteristic vibrational frequencies (that is the reason that vibrational spectroscopy can be used as a tool for structure determination). This means that functional groups have something in common even in different molecules. This commonality is their characteristic topology and the strength of the bonds between their atoms. But if the bond strengths are the same, the bond lengths have to be the same, too, since they depend on the bond strength. So for each bond, we can define a common, most probable bond length and a constant bond strength. If we now reconsider our Taylor series expansion of the potential energy, Eq. (7.218), we see that we could get two parameters for each bond if we expand the potential energy up to the quadratic term (with x_i now being the bond length). The bond strength would then be described by the (quadratic) force constant and our "standard" bond length will be x_i^0. Characteristic vibrational frequencies are not limited to bond stretching. Angle bending and changes in torsional angles can also be seen in vibrational spectra. Therefore we can identify x_i in Eq. (7.218) with bond lengths, bond angles, or torsional angles. In addition, we see that there are couplings (or cross terms) between all of these internal motions. Of course, we are not limited to a quadratic force field; we can also include higher order terms which just require more parameters to model, e. g., bond stretching more accurately. Thus a simple and transferable force field can look like

$$
\begin{aligned}
V \quad &= \quad V_{bonds} + V_{angles} + V_{torsions} + V_{out-of-plane} + V_{bond-bond} \\
&+ \quad V_{angle-angle} + V_{bond-angle} + V_{bond-torsion} + V_{angle-torsion} \\
&+ \quad V_{angle-angle-torsion} + V_{non-bond}
\end{aligned}
$$

$$(7.222)$$

with

$$
V_{bonds} \quad = \quad \sum_{bonds} [K_2 (b - b_0)^2 + K_3 (b - b_0)^3 + K_4 (b - b_0)^4]
$$

$$
V_{angles} \quad = \quad \sum_{angles} [H_2 (\theta - \theta_0)^2 + H_3 (\theta - \theta_0)^3 + H_4 (\theta - \theta_0)^4]
$$

$$
V_{torsions} \quad = \quad \sum_{torsions} [V_1 [1 - \cos \phi] + V_2 [1 - \cos(2\phi)]
$$
$$
+ V_3 [1 - \cos(3\phi)]]
$$

$$V_{out-of-plane} = \sum K_\chi \chi^2$$

$$V_{bond-bond} = \sum_b \sum_{b'} F_{bb'} (b - b_0)(b' - b'_0)$$

$$V_{angle-angle} = \sum_\theta \sum_{\theta'} F_{\theta\theta'} (\theta - \theta_0)(\theta' - \theta'_0)$$

$$V_{bond-angle} = \sum_b \sum_\theta F_{b\theta} (b - b_0)(\theta - \theta_0)$$

$$V_{bond-torsion} = \sum_b \sum_\phi (b - b_0)[V_1 \cos\phi + V_2 \cos 2\phi + V_3 \cos 3\phi]$$

$$V_{angle-torsion} = \sum_\theta \sum_\phi (\theta - \theta_0)[V_1 \cos\phi + V_2 \cos 2\phi + V_3 \cos 3\phi]$$

$$V_{angle-angle-torsion} = \sum_\theta \sum_{\theta'} \sum_\phi K_{\theta\theta'\phi} \cos\phi\, (\theta - \theta_0)(\theta' - \theta'_0)$$

$$V_{non-bond} = \sum_{i>j} \frac{q_i q_j}{\epsilon r_{ij}} + \sum_{i>j} \left[\frac{B_{ij}}{r_{ij}^9} + \frac{C_{ij}}{r_{ij}^6} \right] \tag{7.223}$$

which is the functional form of the Consistent Force Field (CFF) [191, 192].

Excluding the terms for torsions and non-bond interactions, this force field looks similar to our Taylor series expansion, Eq. (7.218), where different terms have been expanded to a different order. The question to ask now is: How does this force field differ from the Taylor series expansion? First, the energy depends on all (redundant) internal coordinates and not just on a particular combination. Second, each of the terms has a physical meaning that describes the change in total energy caused by a change in a particular internal coordinate. The use of redundant coordinates will cause some problems during the parameterization of the force field (as we shall explain in more detail in Section 7.4.5), but these problems can be overcome. Third, the Consistent Force Field contains terms for torsions and non-bonded interactions, which are not found in the Taylor series expansion. Torsional potentials adopted in many force fields are usually described by a Fourier series because torsions often have more than a single "equilibrium" value, they can change over a large range, and they are periodic. Fourier series are well suited for describing such behavior. In addition to the terms from the Taylor and Fourier series, we see that the familiar Coulombic and Van der Waals terms also appear here. A force field devoid of these latter terms would not be able to describe intermolecular interactions. Coulombic and Van der Waals terms can also be used to fine tune the remainder of the force field. Non-bonded interactions also exist within different parts of the same molecule (intramolecular effects). It should be noted that the introduction of Coulombic and Van der Waals terms is somewhat artificial because they are not strictly necessary (as seen from the Taylor expansion). The introduction of Coulombic and Van der Waals terms can cause problems in the parameterization because the parameters for these terms correlate with parameters of other terms (especially torsional terms). Therefore, charges used in this type of force field are often much smaller than formal charges.

To distinguish a force field like the one in Eq. (7.222) from ion pair potentials, we shall call the former a *molecular mechanics force field*, a notation adopted in Allinger's pioneering work on predicting structures of hydrocarbons [781, 782] and used for his series of force fields [311, 314–316, 783]. In contrast to ion pair potentials, molecular mechanics force fields include higher order terms from Eq. (7.213), usually up to fourth order for torsions.

Equation (7.222) also brings to light the largest problem of molecular mechanics force fields – the large number of parameters associated with such a complex potential function. As we will see later, molecular mechanics force field parameters are today often derived using quantum mechanical calculations. The larger number of parameters compared to ion pair potentials does not significantly slow molecular mechanics force fields calculations once the parameters are in hand. Most of the parameters are for localized interactions such as bonds and angles, and the number of these localized interactions grows only linearly with the size of the system. The most time consuming part of molecular mechanics force field calculations is the evaluation of the Coulombic interaction, a process which scales with the square of the number of atoms in the system and must be evaluated over long distances to ensure convergence. There are methods for speeding up the calculation of the Coulombic interactions such as Ewald summation [466], cell multipole methods [784], or spherically truncated, pairwise r^{-1} summation [785]. These methods are also employed with ion pair potentials.

Molecular mechanics force fields have been employed in a number of areas of materials science. The most prominent use has been in polymer simulations since polymers are often organic molecules, which are well suited for treatment by molecular mechanics force fields. Zeolites represent an intermediate case in this regard, because they possess some covalent character in their bonding [745] and so molecular mechanics force fields are able to capture many of their properties effectively. Shell model potentials have also been employed with success for zeolites. The interaction of the zeolite framework with organic sorbates is also of great industrial significance. Here molecular mechanics force fields are better suited to the description of the organic component. Nonetheless, further investigation of such methods for framework descriptions is justified. There are other areas in materials science where interfaces between inorganic and organic systems are important including the study of corrosion inhibitors, coatings, and electronic displays. There exist attempts to create force fields that can handle inorganic and organic compounds equally well [640, 786–788], but these approaches have yet to find widespread application. The development and refinement of force fields for these types of systems is an area of active research.

7.4.4 Comparison of Ion Pair and Molecular Mechanics Force Fields

In the previous two sections we provided the theoretical underpinning for the ion pair and molecular mechanics force fields. The most prominent difference between these force fields arises from the fact that they are attempting to describe very different types of systems. Ion pair potentials describe ionic systems where bonds are not directed, whereas molecular mechanics force fields require the presence of directed bonds because they describe covalently bonded systems. There are materials that clearly fall into one or the other of the two categories: oxidic catalysts, e. g., are largely ionic in nature whereas many polymers are covalent. But not all materials can be categorized so conveniently as illustrated by zeolites, which contain ionic as well as covalent character in their bonding. Moreover, for materials design, it is often necessary to consider the interaction of one type of material with another, e. g., how does an organic corrosion inhibitor behave on a metal surface? Consequently, there exists a need for force fields which can be used reliably in these kinds of heterogeneous systems.

Table 7.4 compares advantages and disadvantages of both force field types. Ion pair potentials are much simpler and are therefore easier to parameterize. The parameterization is often done based on experimental data of solids alone. Because bond lengths and bond angles in experimental structures are usually limited to a rather narrow range around their equilibrium values, the sampling of the potential energy surface is not very good for these potentials. As a result the transferability of the potential and its capability to correctly predict bond lengths and angles are often limited as well. An example of where this becomes problematic is on surfaces where, because of lack of strain or constraints from the remainder of the lattice, significant bond elongation can occur. Sometimes it is even necessary to add additional molecular mechanics type terms to an ion pair potential. For example, the best performing ion pair potentials for zeolites contain a harmonic angle bending term for the OSiO and OAlO angles. Experimentally, these angles are found to be close to the tetrahedral angle, and the angle bending term is required to reproduce this structural feature.

In contrast, molecular mechanics force fields usually predict bond lengths and bond angles well because their terms explicitly treat those bonds and angles. The impediment to their use, however, is the large number of parameters required, many of which are obtained by using quantum mechanical calculations. These molecular mechanics force fields also have the advantage that they can be extended consistently if interactions with organic molecules are to be studied. In principle, it should be possible to combine a molecular mechanics force field for an inorganic host with one used for an organic guest. In practice this generally demands careful derivation because the non-bonding parameters in different force fields are often tuned to compensate for some other effect, such as polarization,

Table 7.4: Advantages and disadvantages of ion pair potentials and molecular mechanics force fields

	Ion pair and shell model potentials	Molecular mechanics potentials
Mathematical form	simple	complicated
Number of parameters	small	large
Parameter derivation	from experimental data as well as from ab initio calculations	mostly from ab initio calculations
Reproduction of lattice constants	excellent	excellent
Reproduction of bond lengths and angles	good if derived from ab initio	good
Reproduction of vibrational spectra	good if derived from ab initio	good
Extensibility to interaction with organics	inconsistent	consistent
Applicable to change of coordination	yes	no

that was not explicitly included in the force field at the outset.

Finally, ion pair potentials are able to describe changes in coordination since they do not consider bonds. In some areas, notably geophysical and glass simulations, this can be important because it allows the study of phase transitions.

If we look at the functional forms of both ion pair potentials and molecular mechanics force fields we see that both force fields share Coulomb and Van der Waals terms. It should therefore be possible to combine both types of force fields in studies involving inorganic as well as organic components. In so doing, one may make use of the particular strengths of each force field.

7.4.5 Force Field Parameterization

In addition to the functional form of a force field one needs the corresponding set of force field parameters. In the parlance of force fields, the term *atom* is defined in a somewhat more narrow way than in quantum mechanics. In quantum mechanical calculations, it is sufficient to specify the element for an atom and the "parameters" (basis set) will be usually the same for all atoms of the same element. Force fields require the specification of additional properties for an atom such as its connectivity, hybridization, oxidation state, and so on. The element and these additional properties are commonly referred to as *atom types*. In a molecular mechanics force field for hydrocarbons there might be atom types for sp^3, sp^2, and sp hybridized carbon atoms. Sometimes, it is also necessary to distinguish atoms according to their bonding situation, e. g., there might be different atom types for hydrogen atoms in an alkyl group compared to those in a hydroxyl group or carboxylic acid.

To parameterize a force field, the molecular modeler must usually rely on chemical intuition to establish atom types for the molecular system under study from the beginning. Parameters can then be derived using the standard fitting procedure outlined below. With these parameters, properties of the system known experimentally or derived from quantum mechanical calculations can then be calculated and compared. If significant deviations exist between computed and known values it is either necessary to introduce new atom types or to use different functional forms for the force field. This fitting process has to be repeated until sufficiently small deviations between computed and observed values are obtained and can be protracted, but it is necessary to ensure accurate predictions from a given force field. To guarantee the quality and transferability of the parameterized force field, it is advisable to divide the available experimental or quantum mechanical data into two sets, a *training set* which is used to fit the force field and a *test set* which is used only to validate the quality of the derived force field. The training and test set should contain different molecules with the same functional groups, but in different configurations and/or conformations. If quantum mechanical data are used in the fit, the training set could consist of the smaller molecules where extensive sampling of the potential energy surface is possible. The test set could then contain larger molecules, which are more demanding in quantum mechanical calculations and the number of data points obtainable is therefore more limited (see, e. g., Refs. [143, 144]).

Special attention has to be paid to the sampling of the potential energy surface. To obtain meaningful transferable parameters, it is necessary to sample properties of the test systems at a variety of bond lengths and angles. This issue is, of course, a problem when only limited experimental data is to be used in a fit. A solution to this dilemma is the use of diverse systems together during fitting [789]. However, the use of quantum mechanical data which allow a systematic variation of bond lengths, bond angles, and other features is increasingly possible. Because the latter approach is often more convenient for a computational chemist we discuss it first and then we consider the derivation of force fields from experimental data.

Ab Initio Based Force Fields

The use of quantum mechanical calculations, and in particular ab initio calculations, to derive force fields has seen much application in the last decade [144, 175, 182, 191, 317, 790]. This is due to the dramatic improvement of computer capabilities allowing ab initio calculations to be performed for reasonably sized molecules. Ab initio calculations also have the advantage that they allow much better sampling of the potential energy surface than can usually be assessed by experimental methods. Other than CPU time, there are practically no limits on the number of conformations which can be sampled. However, in the materials science area, force fields are often employed to study solids. Ab initio calculations on periodic systems like solids are currently still time consuming, when feasible at all. Furthermore, force field parameterization demands multiple conformations be computed. Smaller units, therefore, have to be found that best represent the solid. Preferably, these smaller units are molecules treatable by standard molecular ab initio codes. Hence, *models* of the solid are selected to be small enough to be computed by ab initio calculations, but also large enough to represent the structural features of the solid. Parameters derived for the models can in many cases be transferred to the solid without any loss of generality.

Quantum mechanical calculations provide energies, first and second derivatives of the energy with respect to atomic coordinates, and dipole moments for the models. All these *observables* can also be calculated from a force field. The force field parameters are then determined by minimizing the sum in the following equation

$$S = \sum_a^{nmol} \sum_b^{nconf} W_{ab} \left[W_{ab}^E (E_{ab} - E_{ab}^{\mathrm{qm}})^2 + W_{ab}^{1st} \sum_i^{3N} \left(\frac{\partial E_{ab}}{\partial x_i} - \frac{\partial E_{ab}^{\mathrm{qm}}}{\partial x_i} \right)^2 \right.$$

$$+ \quad W_{ab}^{2nd} \sum_{i \geq j}^{3N} \left(\frac{\partial^2 E_{ab}}{\partial x_i \partial x_j} - \frac{\partial^2 E_{ab}^{\mathrm{qm}}}{\partial x_i \partial x_j} \right)^2 + W_{ab}^{\mu} \sum_{i}^{3} (\mu_{abi} - \mu_{abi}^{\mathrm{qm}})^2 \Bigg] \quad (7.224)$$

where the quantities with superscript qm are calculated quantum mechanically, whereas the others are obtained from the force field. The sums a and b run over all molecules and all conformations, respectively. The W represents the weights that are used to increase or decrease the importance of certain quantities, e. g., it may be desirable to obtain a force field which predicts relative energies more correctly than, say, vibrational frequencies or dipole moments. The relative energy for a conformation is only one number whereas for N atoms there exist $3N(3N - 1)/2$ Cartesian second derivatives for one conformation. It is therefore common to use weights to increase the importance of the relative energy, e. g., $W_{ab}^E = 10000, W_{ab}^{1st} = 100, W_{ab}^{2nd} = 1$. Although it is theoretically possible to obtain charges for the Coulomb term in a force field by fitting to dipole moments [791, 792], in practice it is very difficult to get physically reasonable charges this way. Therefore, charges and Van der Waals parameters are usually obtained separately, before the remaining parameters are fitted [175, 191, 793–798].

There are two methods commonly used to obtain atomic charges. For ionic systems, it is not uncommon to use simple formal charges, which generally guarantees charge neutrality. The other possibility is to use charges obtained from a population analysis in quantum mechanical calculations [799] or by fitting the electrostatic potential obtained from a quantum mechanical calculation [800,801]. These charges are usually smaller than formal charges, and so it can be difficult to guarantee charge neutrality when they are transferred to arbitrary systems. The Van der Waals parameters have usually been obtained by fitting to crystal structure data and to heats of sublimation of molecular crystals [192, 311, 316, 793–795, 797, 798, 802–804], but it is also possible to obtain parameters from quantum mechanical calculations [805]. Because Van der Waals interactions include dispersion, it would be necessary to use a correlated quantum mechanical method. Unfortunately, density functional methods do not work for dispersive interactions [806] and post-Hartree–Fock methods, capable of capturing this physical effect, are computationally expensive [807]. The dispersion interactions are therefore often ignored. The repulsive part of the Van der Waals interaction, on the other hand, can be determined from test particle calculations [808–813] using the Hartree–Fock method.

As mentioned previously, molecular mechanics force fields use redundant internal coordinates. In a tetrahedrally coordinated center such as SiO_4, all six bond angles $\Delta\alpha_1 \cdots \Delta\alpha_6$ are used, but only five of them are independent. Thus correlations exist between the force constants and the reference values of the coordinates, and this poses a problem in the fitting process because there is no unambiguously optimal set of parameters. Hence, parameters fitted to redundant coordinates may not be transferable and have a well-defined physical meaning.

This correlation problem between force constants and reference values can be avoided by adding a linear term to each diagonal term of the force field. For instance, for bond angles

$$E = H_1 (\theta - \theta_t) + H_2 (\theta - \theta_t)^2 + H_3 (\theta - \theta_t)^3 + H_4 (\theta - \theta_t)^4 \quad (7.225)$$

In this equation the linear term added is the first term, and the linear force constant H_1 is optimized instead of the trial reference value θ_t. The optimum reference value θ_0 can then be computed from the trial reference value θ_t and the optimized force constants by solving the following equation

$$\left. \frac{\partial E}{\partial \theta} \right|_{\theta=\theta_0} = H_1 + 2H_2 (\theta_0 - \theta_t) + 3H_3 (\theta_0 - \theta_t)^2 + 4H_4 (\theta_0 - \theta_t)^3 = 0 \quad (7.226)$$

To eliminate the correlations between the linear force constants an explicit relationship between them can be used. The energy is expressed once as a Taylor series in local symmetry coordinates, Δs_i,

$$E = \sum_i h^{(i)} \Delta s_i + \cdots \quad (7.227)$$

which are linear combinations of the redundant set of internal coordinates, Δr_j,

$$\Delta s_i = \sum_j c_{ij} \Delta r_j \quad (7.228)$$

The higher-order terms have been avoided here since they have no effect on the derivation. Substitution of Eq. (7.228) into Eq. (7.227) yields

$$E = \sum_i h^{(i)} \sum_j c_{ij} \Delta r_j + \cdots \quad (7.229)$$

The energy can, equivalently, be expanded in a Taylor series that explicitly depends on the redundant coordinates Δr_j, giving

$$E = \sum_j H_1^{(j)} \Delta r_j + \cdots \quad (7.230)$$

By equating terms in Eqs. (7.229) and (7.230), the relationship among the linear force constants can be derived:

$$\sum_j H_1^{(j)} = \sum_i h^{(i)} \sum_j c_{ij} \quad (7.231)$$

Since the local symmetry coordinates normally used obey $\sum_j c_{ij} = 0$, it follows from Eq. (7.231) that

$$\sum_j H_1^{(j)} = 0 \quad (7.232)$$

This relationship has to be used explicitly to avoid correlation between the linear force constants [814].

In addition to avoiding parameter correlations this way, a few more constraints need to be imposed on the parameters in some of the terms. Usually, bond stretching is quite well described by a Morse function. However, a Morse function has a very small slope if a bond is stretched far from equilibrium, and thus the force driving this bond back to its equilibrium value is also very small. This situation can cause problems in geometry optimizations. The use of a quadratic, cubic, and quartic force constant in the bond stretching term assures that there will be a large force for extended bond lengths, but this in turn introduces the possibility that either more than one minimum exists for this force field term (if the quadratic force constant becomes negative) or the energy goes to $-\infty$ for very short and very long bond lengths (if the quartic force constant becomes negative, Fig. 7.5). To avoid this unphysical situation the quartic force constant can be obtained from the quadratic and cubic values using

$$K_4 = \frac{7K_3^2}{12K_2} \tag{7.233}$$

This relation can be derived by expanding the Morse function in a Taylor series and comparing coefficients. By applying this relation the correct shape of the potential is guaranteed (Fig. 7.5), which also assures that the energy for very short and very long bond lengths remains positive.

Another problem associated with parameterization comes from the requirement that the angle bending term has to be symmetric. Clearly, an angle of, say, 170° has to have the same energy as when it is 190°. In other words, the angle bending term has to have a slope of zero at 0° and 180°. To maintain this symmetry, the cubic and quartic force constant can be calculated from the quadratic force constant and the reference angle,

$$H_3 = H_2 \frac{2(\pi - 2\theta_0)}{3\theta_0(\pi - \theta_0)} \tag{7.234}$$

$$H_4 = -\frac{H_2}{2\theta_0(\pi - \theta_0)} \tag{7.235}$$

instead of being fitted. This procedure also ensures that there is only one minimum between 0° and 180°.

The process of deriving force fields from ab initio calculations should follow some essential general guidelines to ensure maximum accuracy of the force field parameters. These can be summarized as follows:

- Select molecular models that are suitable for ab initio calculations

- Check the suitability of these models with respect to a correct description of properties of the systems to be modeled

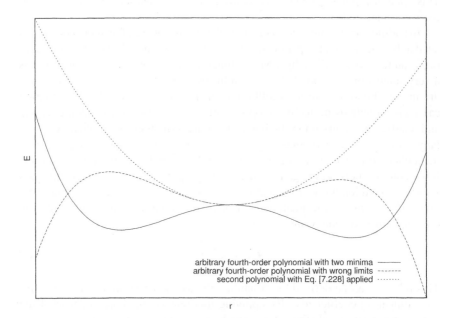

E

arbitrary fourth-order polynomial with two minima ———
arbitrary fourth-order polynomial with wrong limits ------
second polynomial with Eq. [7.228] applied ········

r

Figure 7.5: An arbitrary fourth-order polynomial might either have two minima or the wrong shape (the energy goes toward $-\infty$ for very short and very long bond length r). In contrast, by applying Eq. (7.233) the correct shape is guaranteed.

- Use the ab initio data for these models to derive potential parameters and then check the parameters using the models

- Check the transferability of the derived parameters by carrying out force field calculations on larger, more complex models which are still amenable to ab initio calculations

- Test the quality of the parameters by doing calculations on the real system and comparing those results with available experimental data.

Empirical Force Fields

Early force field parameter derivations were completely empirical owing to the lack of reliable first-principles, quantum mechanical data, plus a relative abundance of experimental data. As described in preceding sections, ab initio methods are capable of providing a reasonable description of many molecular systems and these methods have become an effective point of reference for derivation of potential functions. However, in practice, empirical and theoretical parameter derivation are often combined. For example, the $O^{2-} \cdots O^{2-}$ potential employed in the simulation of many inorganic oxides was originally derived through ab initio calculations [815], and this starting point has been combined in many cases with empirical fitting for cation-anion interaction parameters needed to obtain a complete potential for describing a wide range of oxide materials [93, 789]. This resulting hybridized approach to potential derivation has been demonstrated to be effective in reproducing structure and properties of known systems as well as prediction of unknown systems as diverse as organic polymers and inorganic silicates [144, 816].

Empirical fitting is achieved in practice by calculation of observables for a given system using an initial guess of the potential parameters. Then the weighted sum of deviations between calculation and observation are computed and the potential parameters are adjusted to minimize this sum. The observable parameters employed are generally drawn from the following set:

1. lattice parameters

2. structural parameters (including atomic coordinates, bond lengths, and angles)

3. dielectric properties (static and dynamic dielectric constants)

4. elastic constants

5. vibrational properties (modes of vibration, phonons), and

6. other properties such as heat capacities, densities, etc.

As with any data fitting process the ratio of observations to adjustable parameters should be kept as high as possible. A schematic of the empirical potential fitting process is shown in Figure 7.6.

It is worth noting that the incorporation of structural information in empirical fitting can be achieved in two contrasting manners: (1) the overall force on a particular structural component can be minimized with respect to the potential parameters or (2) at every fitting step the structure can be relaxed to equilibrium prior to the calculation of physical properties [789]. The latter method may often be superior to the former, [124], though it has not always been employed, because energy minimization of a large system can be prohibitively time consuming. Furthermore it is only possible where a reasonable initial potential model can be provided.

An interesting approach for the empirical fitting of potential parameters is provided by the work of Gale and Bush [789] who fitted a range of oxide parameters to physical characteristics for a diverse set of materials simultaneously.

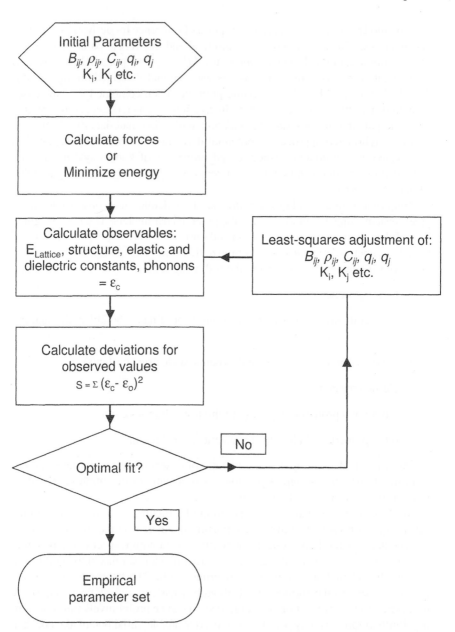

Figure 7.6: A schematic representation of the empirical method for deriving short-range potential energy function parameters. ϵ_c are the computed properties for the system, ϵ_o observed properties, and S the sum of squared deviations from target values. B_{ij}, ρ_{ij}, and C_{ij} are potential parameters of Eq. (7.212); K_i and K_j shell model parameters and q_i and q_j the charges of component species.

Here, $O^{2-} \cdots O^{2-}$ interaction parameters, among others, were derived by reference to the complete set of materials. The sampling of a large range of possible interatomic separations and configurations is thus representative of the whole set. It is reasonable to suppose the parameters obtained through such a fitting procedure will retain a greater degree of transferability than those fitted to isolated components of the training set.

Transferability

Force field transferability is an important consideration in practical simulations. When parameters have been carefully refined to reproduce observation for only a model system one needs to ask: how will they perform when transfered to new chemical environments? For example, if potentials are available and are known to describe adequately Al_2O_3 and Li_2O, how effective will these potential models be in the description of Li^+ migration in Al_2O_3? How effective will these potentials be in describing the surfaces of mixed oxides containing Li^+ and Al^{3+}? (In such a case the surface environment alone will be a test of the transferability for the potential model.) Decades of experience in the application of shell model potentials [93,776,817] and the arrival of reliable first principles methods indicate that such calculations are often surprisingly accurate [818,819]. The trust one can place in pair potentials depends on the range of interatomic separations sampled in their derivation and intended use. Figure 7.7 depicts the interatomic separations of relevance for different types of calculations one might attempt to carry out, along with their respective sources of experimental support. Quantum mechanical calculations can in principle access any desired interatomic separation but difficulty in their use lies in providing a reasonable chemical environment with which to probe the requisite interatomic separations. Here "reasonableness" is assessed by both computational tractability as well as realism of the chemical environment employed. It is, however, interesting to note that several highly successful potentials have been obtained by using accurate quantum chemistry in combination with only small models of the local environment about a given atom [176,184]. One can expect that when systematic quantum mechanical calculations are possible, and thorough sampling of interatomic separations can be effected [819], greater transferability of potential parameters will result. Set against this expectation is the fact that for many physically interesting properties, experimental values are not only readily accessible but they also exist with accuracy far exceeding those from present quantum mechanical methods. However, given the success of quantum mechanics in accounting for the geometries and the properties of molecular systems, it is also reasonable to expect that computationally derived fitting procedures will gain popularity in the future especially because empirical fitting provides sampling of a comparatively limited region of the potential as depicted in Figure 7.7 [820,821].

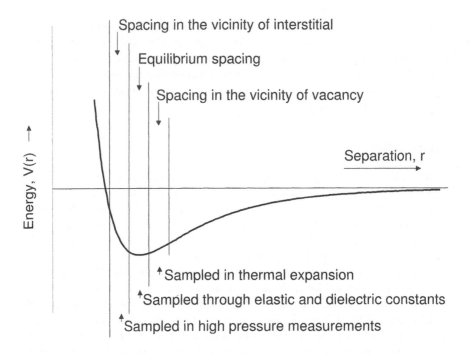

Figure 7.7: A schematic diagram showing the differing regions of interatomic short-range potential interactions. Regions of relevance in defect calculations are highlighted above the axis. Where ionic species are obliged to occupy locations between normal lattice sites (interstitial positions) shorter interatomic separations are sampled. Conversely, when lattice site vacancies are present larger interatomic separations are present. The areas where these regions are sampled by experimental observation are indicated below the axis.

Force field methodologies for materials simulation involve an assortment of assumptions and approximation. Within a given class of systems, the approximations, although sweeping, may effectively cancel, if one is fortunate. For example, in silicate systems, both experimental measurement and accurate quantum chemical calculations show that formal oxidation states are an overstatement of the local charges on framework cations. However, a model employing such an approximation may provide a highly accurate description of local geometries when compared to crystallographic data and may also describe adequately amorphous systems where experimental evidence is less readily available [131, 822]. Problems ensue when different classes of material meet, each having their own unique properties embodied in the computer model with their own independent approximations. An example would be when an organic molecule, typically largely described by harmonic potentials and low partial atomic charges, is modeled in the vicinity of a highly charged amorphous silicate surface. However, there is no intrinsic reason why a hybrid experimental system cannot be described by a hybrid potential function with appropriate local descriptions and a special interaction treatment for the two differing regions. Furthermore, there have been many simulations of organic molecule-microporous material interactions which demonstrate that even existing generation potentials are effective in the description of such systems [823–827]. Indeed, the subject of water interacting at metal surfaces has been treated in the literature [828].

Combining Rules For Non-bonded Interactions

When organic force fields are developed it is common practice to assign only like center based potential parameters for non-bonded interactions. To generate parameters for heterogeneous interactions between species i and j (as opposed homogeneous, i-i or j-j interactions) combining rules can be specified to approximate the needed i-j parameters. Thus, for example, Lorentz-Bertholet mixing rules yield Lennard-Jones σ and ϵ parameters as follows:

$$\sigma_{ij} = \frac{1}{2}[\sigma_i + \sigma_j] \tag{7.236}$$

$$\epsilon_{ij} = [\epsilon_i \epsilon_j]^{\frac{1}{2}} \tag{7.237}$$

Here σ_i, σ_j, and σ_{ij} are the Lennard-Jones contact radii (the radii at which the interaction potential has a value of zero) for atoms pairs i-i, j-j, and i-j respectively; and ϵ_i, ϵ_j, and ϵ_{ij} are the Lennard-Jones well depths of the respective component interactions. The Lennard-Jones potential then has the following form

$$V(r_{ij}) = 4\epsilon \left[\left(\frac{\sigma_{ij}}{r_{ij}} \right)^{12} - \left(\frac{\sigma_{ij}}{r_{ij}} \right)^{6} \right] \tag{7.238}$$

Also frequently employed for organic force fields are geometric mean combining rules [312] and also more sophisticated mixing schemes [808, 829].

For inorganic systems, it has been more common to specify pair-wise interaction parameters individually (see, e. g., Ref. [789]). This is a consequence of the historical emphasis on the simulation of a limited number of compounds within a particular study. However, one notable exception is provided by the ionic potential parameters derived by Vessal [830, 831]. In his approach, combining rules were a design constraint in the derivation of the parameter set as a whole. The resulting force field is effective for both the structural and energetic description of a wide range of materials and, in particular, the incorporation of non-framework cations in microporous materials, which can prove an exacting test of more locally derived and less transferable force fields.

7.4.6   Rule Based Force Fields

The previous sections have shown that the derivation of force field parameters is a formidable task. Therefore, force fields derived in a strict manner based on ab initio calculations or experimental data are limited to rather narrow classes of compounds (certain functional groups in organic molecules, zeolites, metal oxides, etc.) and it might be impossible to find force field parameters for a particular system under study. Within the last decade there have been approaches to remedy this situation by providing force fields which cover the entire periodic table of elements. Since a strict derivation of parameters based on ab initio calculations or experimental data for all possible combinations of elements is impossible, another way of obtaining the required parameters was needed. These force fields use general rules based on element and hybridization to calculate the required parameters [640, 832–834].

For example, the Universal Force Field (UFF) [640], which is a molecular mechanics force field, uses the sum of atom type specific single bond radii with some corrections for bond order and electronegativity as the reference bond length and derives the force constant for this bond based on a generalization of Badger's rules [835, 836]. Bond angle distortions are described using a cosine Fourier expansion

$$V_{angles} = \sum_{angles} H \sum_{n=0}^{m} C_n \cos n\theta \qquad (7.239)$$

where the coefficients, C_n, can be selected so that the potential has minima at the correct angles for different coordination environments. Reference angles are based on the angles occurring in the hydride of an element and the force constant is also derived based on a generalization of Badger's rules. Torsional parameters are based on the hybridization and are fitted to experimental torsional barriers for the element's hydrides. Combination rules as in some other force fields are used for Van der Waals parameters and charges are obtained by a charge equilibration scheme.

Such an approach has the advantage that it is generally applicable; however the quality of such a force field is often lower than that of a specialized force field in particular with respect to conformational energies. Casewit et al. have compared their UFF with MM2/3 [786] and find that, e. g., the root mean square error for CC bonds is 0.021 Å for UFF, but 0.012 Å for MM2/3 with maximum errors of 0.067 Å and 0.029 Å, respectively. Conformational energies can be in error by more than 16 kJ/mol. The magnitude of errors for main group elements is even larger than for organic compounds [787].

7.5 Combining Quantum Mechanics and Force Fields – Embedding

Systems of interest in materials science are quite often solids. They gain their special properties by defects, which are usually very low in concentration. Therefore, simulations of these materials need to be able to handle large models of a material, models too large for electronic structure methods alone, but where certain electronic effects need to be studied.

The availability of potentials which are based on quantum chemical calculations permits to take into account the environment by embedding the site of interest in a larger model where different parts are treated with different approximations. We can divide the system under study into two parts: the inner region which is treated quantum mechanically and the outer region which is dealt with using a potential. Thus, the inner region is embedded in the environment modeled by a potential, but electronic structure effects in the inner region can still be studied. There are three possibilities for embedding. If the electrostatic potential of the atoms of the outer region can act on the atoms in the inner region and polarize them, the embedding is said to be "electronic". If polarization of the inner region by the atoms of the outer region is not allowed we have "mechanical" embedding. For electronic embedding two cases need to be distinguished: (i) only the atoms of the outer region polarize the atoms of the inner region and (ii) the atoms of the inner region also polarize the atoms in the outer region [837]. In the latter case a polarizable force field is needed such as a shell model potential. There have been numerous implementations of embedding in the past [838–848]. Most of the methods embed a cluster in a set of point charges, which are supposed to mimic the solid or add additional terms to the Fock matrix for the cluster. This approach requires a quantum mechanical program which can handle additional terms and is limited to optimizing the cluster only (e. g., point charges mimicking the solid are fixed in space). A more detailed discussion of the limits and merits of these methods has been provided by Sauer et al. [849, 850]. Embedding molecules in solvents or studying clusters as part of an ionic or molecular crystal pose no problem with respect to the termination of the cluster. For systems which have largely covalent bonds the cluster needs to be cut out of the system. There are two possibilities to accomplish this: (i) some of the bonds are cut or (ii) the cut is performed through atoms [837]. Both approaches have advantages and disadvantages. If bonds are broken care has to be taken to avoid artificial electronic effects. If a bond is broken radicals would be formed. Therefore, the broken bonds are usually saturated by hydrogen (so-called link atoms), but the hydrogen atoms create a special problem since their influence on the inner region has to be eliminated. Embedding the cluster (the inner region and the link atoms), e. g., in point charges, does not take care of this problem.

If the cut is performed through atoms no link atoms are necessary, but the atoms where the cut occurs (so-called connection atoms) need to have special electronic properties to model the seam between the quantum mechanically described inner region and the outer region described by a force field. It is possible to use especially parameterized atoms in semi-empirical calculations to fulfill this condition [837], but this approach is difficult to generalize and requires additional effort for the parameterization, although it has been tried for ab initio and density functional calculations [851]. A purely mechanical embedding scheme proposed by Eichler and co-workers [196] allows to combine any standard force field and quantum mechanical program, eliminating the need for software specifically written for embedding.

7.5.1   Mechanical Embedding

The energy of the whole system, E_S, in this scheme is calculated as the sum of the quantum mechanical energy of the inner region, E_I^{qm}, the force field energy of the outer region, E_O^{ff}, and an interaction term, E_{IO}

$$E_S = E_I^{qm} + E_O^{ff} + E_{IO} \tag{7.240}$$

There are several possibilities to calculate the interaction term. Parts of it can be calculated quantum mechanically, so that

$$E_{IO} = E_{IO}^{qm} + E_{IO}^{qm-ff} \tag{7.241}$$

where E_{IO}^{qm-ff} contains all interaction terms of the force field which are not included in the quantum mechanical calculation. E_{IO}^{qm} is the electrostatic interaction of the charge distribution of the inner region with the point charges of the force field of the outer region. However, this simple approach fails to account for polarization effects in the outer region, which would counteract. Therefore it might describe the interactions worse than complete neglect of this term. Thus, the energy of the whole system becomes

$$E_S = E_I^{qm} + E_O^{ff} + E_{IO}^{ff} \tag{7.242}$$

The evaluation of E_I^{qm} is a straightforward quantum mechanical calculation. $E_O^{ff} + E_{IO}^{ff}$ can be computed by subtracting the force field energy for the inner region from the force field energy of the whole system

$$E_O^{ff} + E_{IO}^{ff} = E_S^{ff} - E_I^{ff} \tag{7.243}$$

This approach makes it easy to account for the linking between the inner and the outer region. If link atoms are present since bonds have been cut their effect has to be eliminated. The energy of the whole system can be approximated as follows

$$E_S = E_C^{qm} + E_O^{ff} + E_{IO}^{ff} - E_L^{qm} - E_{IL}^{qm} \tag{7.244}$$

where the index C refers to the inner region plus the link region (the quantum mechanical cluster) and the index L to the link region. The last two terms form the correction of the energy of the total system considering that the link region is not part of the real system. The force field energy of the whole system can also be written down similarly

$$E_S^{ff} = E_C^{ff} + E_O^{ff} + E_{IO}^{ff} - E_L^{ff} - E_{IL}^{ff} \tag{7.245}$$

Thus $E_O^{ff} + E_{IO}^{ff}$ in Equation (7.244) can be substituted using force field energies

$$E_S = E_C^{qm} + E_S^{ff} - E_C^{ff} + \Delta \tag{7.246}$$

with

$$\Delta = E_L^{ff} - E_L^{qm} + E_{IL}^{ff} - E_{IL}^{qm} \tag{7.247}$$

Δ will be zero if the force field gives the same energies as the quantum mechanical method. In reality a close approximation can be reached, which makes it possible to neglect Δ. Equation (7.247) shows why it was important to develop a force field which mimics the quantum mechanical calculations. The final energy expression is

$$E_S = E_C^{qm} + E_S^{ff} - E_C^{ff} \tag{7.248}$$

If there are no link atoms (as, e. g., in a molecule embedded in solvent molecules) the cluster is equal to the inner region and the index C in the above equation can be replaced by I. To perform geometry optimizations forces have to be calculated. Since forces are the first derivatives of the energy with respect to atomic coordinates an analogous expression to Equation (7.248) can be obtained for the forces

$$f_S = f_C^{qm} + f_S^{ff} - f_C^{ff} \qquad \text{for atoms in the inner region} \tag{7.249}$$
$$f_S = f_S^{ff} \qquad \text{for atoms in the outer region} \tag{7.250}$$

The positions of the link atoms are not optimized. The link atom is always put on the bond axis between an atom of the inner region and an atom of the outer region. Its distance from the atom of the inner region is kept constant [196].

7.5.2 Electronic Embedding

Electronic embedding can be accomplished by including the charges of the atoms in the outer region in the one-electron potential of the Hamiltonian for the inner region. Since the atoms in the outer region move according to the forces acting on them, the limitation of using fixed point charges can be overcome. Problems arise, however, from the fact that if only the inner region is polarized by charges in the outer region, the atoms at the border between the inner and the outer region will be polarized much more compared to the real system where polarization of the outer region would counteract. Therefore, if a polarizable force field is available the atoms in the inner region should be allowed to polarize the atoms in the outer region as well. Alternatively, the atoms at the border between inner and outer region are often excluded from polarization [837].

7.5.3 Modeling Reactions with Embedding

Embedding can also be used to overcome one of the most serious shortcomings of force field based methods — the inability to deal with reactions. In reactions bonds are broken or formed. Molecular mechanics force fields rely on the definition of bonds before the calculation is started since they use bond terms. Changing the bonding during the calculation would make the potential energy surface discontinuous and therefore the forces undefined. However, quantum chemical methods are capable of modeling the making or breaking of bonds. If the embedding scheme as described above is used, the part of the molecule or solid where the bond is made or broken – the reaction center – can be put into the inner region which is treated quantum chemically and the effect of the environment can be included in the calculation through the force field part.

One problem remains to be solved for this approach. Since the energy of the entire system is to be calculated using the force field (see Eq. (7.244)) there is a need to define a force field energy for the reaction center. One way to solve this problem is to use the empirical valence bond (EVB) method suggested by Warshel [852, 853]. A smooth connection between two states described by interatomic potential functions such as a reactant and a product can be made by using an expression like

$$E = \frac{1}{2}(V_1 + V_2) - \frac{1}{2}\sqrt{(V_1 - V_2)^2 + 4V_{12}^2} \qquad (7.251)$$

where V_1 and V_2 describe the reactant and product, respectively as single minimum interatomic potential functions. The coupling element V_{12} needs to be determined such that it vanishes at the reactant and product states, that it results in the right barrier height, and that it decays smoothly in between. Warshel has proposed a simple exponential expression of the form

$$V_{12} = A \exp(-\mu(r - r_0)) \qquad (7.252)$$

where r and r_0 are the actual value and the reference value of a chosen internal coordinate and A and μ are fitted to reproduce the barrier height and width determined experimentally. A general derivation of the coupling element has been given by Chang and Miller [854] resulting in

$$V_{12}^2 = A \exp(\vec{B}^T \Delta \vec{q} - \frac{1}{2}(\Delta \vec{q})^T C \Delta \vec{q}) \qquad (7.253)$$

Here $\Delta\vec{q} = \vec{q} - \vec{q}_0$ with $\vec{q}$ being the internal coordinates of the actual structure and $\vec{q}_0$ the internal coordinates of the reference structure. The constant A, the vector $\vec{B}$, and the matrix C are chosen so that the EVB potential exactly reproduces the ab initio structure, energy, and force constant matrix of a given transition structure $\vec{q}_0$. Since Eq. (7.253) requires the full set of internal coordinates it is difficult to use for extended systems and is characterized by poor transferability. Sierka and Sauer observed that only very few atoms take part in a reaction and that it is therefore possible to reduce the set of internal coordinates included in Eq. (7.253) to the ones which change most during the reaction [855]. Thus, they arrive at a method where the parameters $\vec{q}_0$, $\vec{B}$, and C can be determined from ab initio data for small cluster models representing the transition state of a reaction. These parameters are also transferable to larger systems.

7.6 Monte Carlo Calculations

All the previous sections of this chapter dealt with methods to calculate energies and find equilibrium structures where the energy is a minimum. It is, however, often of interest to study either the behavior of a molecule at a non-equilibrium configuration, of a larger ensemble of molecules, or both. In particular, the prediction of macroscopic properties requires to take rather large numbers of molecules into account, since, e. g., pressure or temperature are only defined in the macroscopic world. The methods described so far do not allow this. We have to use statistical mechanics for this purpose, which will be the focus of the remainder of this chapter.

In an ensemble of N molecules we can calculate a macroscopic property, such as a pressure, for any instantaneous configuration of the molecules. The instantaneous value of a property, A, depends on the position, $\vec{r}^N(t)$, and the momenta, $\vec{p}^N(t)$, of each molecule. Over time the value of this property will fluctuate as the molecules move due to their interaction and the measured value of the property is the average of the instantaneous values over the time of the measurement. Thus, we can write the time average as

$$A_{average} = \lim_{\tau \to \infty} \frac{1}{\tau} \int_{t=0}^{\tau} A(\vec{p}^N(t), \vec{r}^N(t)) \mathrm{d}t \qquad (7.254)$$

which will approach the "true" value of the property as time approaches infinity. Calculating such an average seems to be straightforward. We have seen previously how to calculate interaction energies and forces between molecules. All we would have to do is to use Newton's law to calculate accelerations from the forces and use them to update positions and momenta. In reality, however, simulating a system of reasonable size would require dealing with "macroscopic" numbers of molecules (around 10^{23}), which is clearly out of reach for even the most powerful computer.

Boltzmann and Gibbs have proposed a method to replace the time average of Eq. (7.254) by an ensemble average

$$\langle A \rangle = \int \cdots \int A(\vec{p}^N, \vec{r}^N) \rho(\vec{p}^N, \vec{r}^N) d\vec{p}^N d\vec{r}^N \qquad (7.255)$$

where instead of the evolution of a single system in time a large number of replications of the system is considered simultaneously. The integration in Eq. (7.255) has to be done over all $6N$ positions and momenta. $\rho(\vec{p}^N(t), \vec{r}^N(t))$ is the probability density of the ensemble, e. g., the probability of finding a configuration with the moments $\vec{p}^N$ and the positions $\vec{r}^N$. This probability density is the Boltzmann distribution under the condition of constant number of particles, volume, and temperature (in the canonical ensemble often abbreviated as NVT)

$$\rho(\vec{p}^N, \vec{r}^N) = \frac{1}{Q} \exp\left(\frac{-E(\vec{p}^N, \vec{r}^N)}{k_B T}\right) \qquad (7.256)$$

Here Q is the partition function, k_B the Boltzmann constant, and T the temperature. The partition function for N identical particles can generally be written using a Hamilton function

$$Q_{NVT} = \frac{1}{N! h^{3N}} \int\!\!\int \exp\left[-\frac{H(\vec{p}^N, \vec{r}^N)}{k_B T}\right] d\vec{p}^N d\vec{r}^N \qquad (7.257)$$

(replacing the $6N$ integrals over all particle momenta and positions with only two integral signs). The Hamilton function can be written as sum of the kinetic and potential energy of the system

$$H(\vec{p}^N, \vec{r}^N) = \sum_i^N \frac{|\vec{p}_i|^2}{2m} + V(\vec{r}^N) \qquad (7.258)$$

which enables us to separate the integral in Eq. (7.257)

$$Q_{NVT} = \frac{1}{N! h^{3N}} \int \exp\left[-\frac{|\vec{p}|^2}{2m k_B T}\right] d\vec{p}^N \int \exp\left[-\frac{V(\vec{r}^N)}{k_B T}\right] d\vec{r}^N \quad (7.259)$$

assuming that the potential energy function does not depend on the velocities. Carrying out the integration over the momenta results in

$$Q_{NVT} = \frac{1}{N!} \left(\frac{2\pi m k_B T}{h^2}\right)^{\frac{3N}{2}} \underbrace{\int \exp\left(-\frac{V(\vec{r}^N)}{k_B T}\right) d\vec{r}^N}_{Z_{NVT}} \qquad (7.260)$$

where the term labeled Z_{NVT} is called the configurational integral. For ideal gases there is no interaction between particles and the potential energy function is therefore zero. Thus, the configurational integral for an ideal gas is equal to the volume, v^N, and the canonical partition function for an ideal gas becomes

$$Q_{NVT} = \frac{v^N}{N!} \left(\frac{2\pi mk_B T}{h^2} \right)^{\frac{3N}{2}} = \frac{v^N}{N!\Lambda^{3N}} \tag{7.261}$$

where

$$\Lambda = \sqrt{\frac{h^2}{2\pi mk_B T}} \tag{7.262}$$

is the de Broglie thermal wavelength.

A partition function for a real system can now be written as the product of the partition function for the ideal system and an "excess" partition function

$$Q_{NVT}^{excess} = \frac{1}{v^N} \int \exp \left(-\frac{V(\vec{r}^N)}{k_B T} \right) d\vec{r}^N \tag{7.263}$$

and therefore thermodynamic properties can be written in terms of an ideal gas value and an excess value. All the deviations from the ideal behavior are due to the presence of interactions between the particles, which can be calculated solely from the potential energy function. Since the potential energy function does not depend on the momenta of the particles it is sufficient to consider only the particle positions to calculate the excess contributions to thermodynamic functions.

To calculate a property as an ensemble average we would have to evaluate the integral in Eq. (7.255) where the probability now is

$$\rho(\vec{r}^N) = \frac{1}{Z} \underbrace{\exp \left[-\frac{V(\vec{r}^N)}{k_B T} \right]}_{\text{Boltzmann factor}} \tag{7.264}$$

with Z being the configurational integral. In general it is not possible to obtain analytical solutions for these integrals. However, it is possible to use numerical methods to evaluate them. A very common method for this is the use of a Monte Carlo algorithm. A naive approach to calculate, e. g., the average potential energy, would be to perform the following steps

1. Randomly generate $3N$ Cartesian coordinates and assign them to the particles.

2. Calculate the potential energy for these coordinates and the Boltzmann factor.

3. Add the Boltzmann factor and the potential energy to corresponding sums and return to step 1.

4. After a number of trial iterations calculate the average of the potential energy from these sums.

However, using completely random configurations makes this approach unfeasible since there will be a large number of configurations with large potential energies and very small Boltzmann factors, which do not contribute significantly to the sums. It is therefore important to sample only configurations with reasonable potential energies and, consequently, large Boltzmann factors. This *importance sampling* can be achieved by using Metropolis Monte Carlo [856], which biases the generation of configurations accordingly[5].

Metropolis Monte Carlo creates a Markov chain of states which is characterized by two conditions. First, the outcome of each trial depends only on the preceding trial and, second, each trial belongs to a finite set of possible outcomes. Markov chains can be described using state transition probabilities. In practice a Metropolis Monte Carlo calculations proceeds as follows

1. Generate an initial configuration with a reasonable low energy, e. g., by energy minimization.

2. Generate a move by displacing one particle in the current configuration by a random amount limited to a maximum distortion and calculate the energy of this new configuration.

3. If the new configuration is lower in energy than the preceding one it is kept (the move is accepted) and the simulation proceeds to step 2.

4. If the new configuration is higher in energy than its predecessor its Boltzmann factor is compared with a random number between zero and one. If the Boltzmann factor is greater than the random number the new configuration is kept (the move is accepted) and the simulation proceeds to step 2. Otherwise the preceding configuration is retained (the move is rejected) for the next move and the simulation proceeds to step 2.

5. After a given number of trials has been performed the integrals are evaluated from the accumulated sums of Boltzmann factors and property values for all accepted configurations.

A critical parameter in this scheme is the maximum distortion. This is an adjustable parameter which should be chosen so that around 50 % of the moves are accepted. If too many moves are accepted the sampling of the phase space will be slow.

Monte Carlo calculations can not only be performed in the canonical ensemble, but also in the isothermal-isobaric and grand canonical ensembles. In the

---

[5]Metropolis Monte Carlo is so ubiquitous in theoretical chemistry that referring to Monte Carlo usually means Metropolis Monte Carlo.

isothermal-isobaric ensemble (constant number of particles, constant temperature, constant pressure) a way to change the volume of the simulation cell is required. In the grand canonical ensemble (constant chemical potential, constant temperature, constant volume) the number of particles is a variable. Therefore, in addition to the moves described above random particle insertion and destruction attempts are made. The insertion or destruction attempts are accepted following the same strategy as for the moves: lower energy configurations are always accepted while for higher energy configurations the Boltzmann factor is compared with a random number. Grand canonical Monte Carlo simulations are able to predict, e. g., adsorption isotherms.

7.7 Molecular Dynamics Calculations

7.7.1 Basics

In the previous section we have seen how Monte Carlo calculations can be used to gain knowledge about thermodynamical properties of "real" systems without having to consider the momenta explicitly. Sometimes it is of value to study the development of a system in time, e. g., for investigating transport phenomena. For this purpose another method, called *molecular dynamics*, can be used. In molecular dynamics simulations the Newtonian equations of motion are integrated for all particles in the system. We know from classical physics (Newton's second law) that

$$\vec{f}_i = m_i \vec{a}_i \tag{7.265}$$

where $\vec{f}_i$ is the force acting on a particle, m_i the mass, and $\vec{a}_i$ the acceleration of a particle. The acceleration is related to the velocity, $\vec{v}_i$, and the position, $\vec{r}_i$, by

$$\vec{a}_i = \frac{\vec{f}_i}{m_i} = \frac{\mathrm{d}\vec{v}_i}{\mathrm{d}t} = \frac{\mathrm{d}^2\vec{r}_i}{\mathrm{d}t^2} \tag{7.266}$$

Since it is possible to calculate the forces on all particles in a system we can obtain their positions as function of time by integrating Eq. (7.266). Unfortunately, an analytical solution for the resulting system of equations is not possible in general since the motion of one particle is affected by all other particles. It is, however, possible to assume that for a reasonably short period of time, the time step Δt, any given particle will experience a constant force. In this case we can obtain the position of a particle at time $t + \Delta t$ from a Taylor series expansion

$$\vec{r}_i(t + \Delta t) = \vec{r}_i(t) + \vec{v}_i \Delta t + \frac{\vec{a}_i}{2}(\Delta t)^2 + \cdots \tag{7.267}$$

To predict the positions at time $t + \Delta t$ we need to know the positions at times t and $t - \Delta t$. Then we can write

$$\vec{r}_i(t + \Delta t) = \vec{r}_i(t) + \vec{v}_i \Delta t + \frac{\vec{a}_i}{2}(\Delta t)^2 \qquad (7.268)$$

$$\vec{r}_i(t - \Delta t) = \vec{r}_i(t) - \vec{v}_i \Delta t + \frac{\vec{a}_i}{2}(\Delta t)^2$$

By adding both equations we obtain our goal

$$\vec{r}_i(t + \Delta t) = 2\vec{r}_i(t) - \vec{r}_i(t - \Delta t) + \vec{a}_i(\Delta t)^2 \qquad (7.269)$$

This algorithm is known as the Verlet algorithm [857]. It is probably the most widely used algorithm for integrating the equations of motion, but it has some disadvantages. First, it is obviously not self-starting. We need to know the positions at time $t - \Delta t$ even at the beginning of the simulation. Therefore a different method has to be used to start the simulation. Second, this algorithm is not very stable numerically since it takes the difference of two large numbers and adds a small number. Third, velocities do not appear in the algorithm, but are required to calculate the kinetic energy and temperature of the system. They can be obtained in a number of ways, e. g., from

$$\vec{v}(t) = \frac{\vec{r}(t + \Delta t) - \vec{r}(t - \Delta t)}{2\Delta t} \qquad (7.270)$$

but they only become available after the next step. A better algorithm is the velocity Verlet algorithm [858], which uses

$$\vec{r}(t + \Delta t) = \vec{r}(t) + \vec{v}(t)\Delta t + \frac{\vec{a}(t)}{2}(\Delta t)^2 \qquad (7.271)$$

$$\vec{v}(t + \Delta t) = \vec{v}(t) + \frac{1}{2}[\vec{a}(t) + \vec{a}(t + \Delta t)]\Delta t$$

and obtains positions, velocities, and accelerations at the same time without loss of numerical precision. Other methods of integrating the equations of motions are the leap-frog algorithm [859], the Beeman algorithm [860], and predictor–corrector based methods according to Gear [861].

In a practical simulation the starting positions are obtained from, e. g., a geometry optimization. A set of initial velocities is selected randomly from a Maxwell–Boltzmann distribution for a given temperature T

$$p(v_{ix}) = \left(\frac{m_i}{2\pi k_B T}\right)^{1/2} \exp\left(-\frac{m_i v_{ix}^2}{2k_B T}\right) \qquad (7.272)$$

After that a trajectory is calculated by applying one of the integration algorithms. The system is usually allowed to equilibrate for some time to allow various parameters (e. g., temperature or pressure) to become stable. Then the data collection phase of the simulation can be started. Care has to be taken to choose an appropriate time step or the simulation will either be very slow or not stable. For fully flexible molecules 0.1 to 1 fs are usually appropriate. For more rigid systems the time step can be increased somewhat (cf., e. g., [686]).

7.7.2   Ensembles

Molecular dynamics simulations can be performed by keeping certain thermodynamic properties constant. If one of the integration algorithms mentioned above is applied without any further constraints the simulation is said to be performed in the NVE or microcanonical ensemble. The letters NVE stand for the properties which are kept constant. N stands for the number of particles in the system, which cannot change in a normal molecular dynamics simulation (but cf. Section 7.8). V denotes the volume and E the energy, which is constant since there is no way the system could gain or lose energy.

In practice one is often more interested in keeping the temperature constant since temperature control is a much more common way of performing experiments. Constant temperature molecular dynamics can be performed in the NVT (canonical) or NPT (isobaric-isothermic) ensembles. The number of particles has to be constant in both cases and in addition to the temperature the volume or the pressure are kept constant. To keep the temperature or the pressure constant additional constraints have to be introduced into the simulation.

The temperature is related to the kinetic energy of the system by

$$\langle E_{kin} \rangle_{NVT} = \frac{3}{2} N k_B T \tag{7.273}$$

The kinetic energy depends on the velocities of the particles. Therefore, the easiest way to control temperature in a molecular dynamics simulation is velocity-scaling [862]. It can be shown [686] that the velocities have to be multiplied by

$$\lambda = \sqrt{T_{desired}/T_{current}} \tag{7.274}$$

to obtain the desired temperature $T_{desired}$ if the temperature calculated from the kinetic energy of the current time step is $T_{current}$. Alternatively, the system can be coupled to a heat bath [863], which is fixed at the desired temperature. The velocities are scaled at each time step, such that the rate of change of temperature is proportional to the difference in temperature between the bath and the system

$$\frac{dT(t)}{dt} = \frac{1}{\tau}(T_{bath} - T(t)) \tag{7.275}$$

where τ is a coupling parameter whose magnitude determines how tightly the system is coupled to the bath. The scaling factor can be derived [686] to be

$$\lambda = \sqrt{1 + \frac{\Delta t}{\tau} \left(\frac{T_{bath}}{T(t)} - 1 \right)} \tag{7.276}$$

If the coupling parameter τ equals the time step Δt the velocity scaling algorithm is obtained. A coupling parameter of 0.4 ps has been recommended for time steps of 1 fs. Using this method allows the system to fluctuate about the desired temperature.

Both methods shown above do not generate rigorous canonical averages. Temperature differences between components of the system are artificially prolonged. In order to generate rigorous canonical ensembles either stochastic collisions or extended systems methods have to be used. In the stochastic collisions method a particle is randomly chosen at intervals and its velocity is reassigned by random selection from the Maxwell-Boltzmann distribution [864], which is equivalent to the system being in contact with a heat bath which emits "thermal particles" which collide with atoms in the system. Between each collision the system is simulated at constant energy. Unfortunately, this method does not generate a smooth trajectory.

Extended system methods [865, 866] consider the thermal reservoir to be an integral part of the system. The reservoir is represented by an additional degree of freedom, s. The reservoir has potential energy $(f + 1)k_B T \ln s$ where f is the number of degrees of freedom in the physical system and T is the desired temperature. The reservoir also has kinetic energy $(Q/2)(ds/dt)^2$ where Q can be considered as the fictitious mass of the extra degree of freedom. The magnitude of Q determines the coupling between the reservoir and the real system. Each state of the extended system that is generated by the molecular dynamics simulation corresponds to a unique state of the real system. The velocities of the atoms in the real system are given by

$$\vec{v}_i = s \frac{d\vec{r}_i}{dt} \tag{7.277}$$

where $\vec{r}_i$ is the velocity of the ith particle at position $\vec{r}_i$ in the real system. The time step $\Delta t'$ of the extended system is related to the time step Δt of the real system by

$$\Delta t = s \Delta t' \tag{7.278}$$

The value of the additional degree of freedom, s, can change and so the time step in real time can fluctuate. Therefore, the trajectory of the real system can be unevenly spaced in time although the extended system is simulated using regular time intervals. The fictitious mass Q controls the flow of energy between the system and the reservoir. Nosé has suggested that Q should be proportional to $f k_B T$. The constant of proportionality can be obtained by performing a series of trial simulations for a test system and observing how well the system maintains the desired temperature.

For molecular dynamics simulations of solids another important ensemble is the isothermal/isobaric or NPT ensemble, since most experiments are performed under the conditions of constant temperature and pressure. For a solid the unit cell can change during the simulation and the system therefore does not maintain a constant volume. In NPT ensemble the number of particles, the pressure, and the temperature are kept constant. Much like a macroscopic system maintains constant pressure by changing its volume constant pressure simulations are carried out by changing the volume of the simulation cell.

A volume change in an isobaric simulation can either be achieved by changing the volume in all directions simultaneously or in just one direction. Many of the methods used for pressure control are analogous to those used for temperature control. The easiest method for pressure control is simply scaling of the volume. Berendsen et al. have suggested the use of a "pressure bath" [863] in analogy to the temperature bath. The rate of change of pressure is given by

$$\frac{\mathrm{d}p(t)}{\mathrm{d}t} = \frac{1}{\tau_p}(p_{desired} - p(t)) \tag{7.279}$$

where τ_p is the coupling constant, $p_{desired}$ is the pressure desired and $p(t)$ the current pressure at time step t. The volume of the simulation box is scaled by a factor λ

$$\lambda = 1 - \kappa \frac{\Delta t}{\tau_p}(p - p_{desired}) \tag{7.280}$$

where κ is the isothermal compressibility, which is defined as

$$\kappa = -\frac{1}{V}\left(\frac{\partial V}{\partial p}\right)_T \tag{7.281}$$

The new positions are given by

$$\vec{r_i}' = \lambda^{1/3}\vec{r_i} \tag{7.282}$$

The scaling factor can be applied equally to all three directions (isotropically) or a scaling factor can be calculated and applied for each direction separately (anisotropically). The latter method is usually considered better since it allows the simulation box to change shape.

7.7.3 Analysis of Molecular Dynamics Trajectories

The evolution of a system in time during a molecular dynamics calculation is called a trajectory. Such a trajectory can be obtained by saving the coordinates and velocities of all atoms in the simulated system at regular intervals. This creates a massive amount of data, which is not easily understood. Trajectories can be visualized using molecular graphics programs and a first visual impression of the behavior of the system can be obtained this way. However, it is often desirable to perform an analysis of the trajectory to focus attention on specific features of the system.

Correlation Functions

It is often of interest to study whether certain properties of the system are correlated with each other. One way to analyze a molecular dynamics trajectory in search for correlations is the use of correlation functions. A correlation function is any function of the form

$$C_{xy} = \frac{1}{M} \sum_i^M x_i y_i \equiv \langle x_i y_i \rangle \tag{7.283}$$

where x_i and y_i are the points of two data sets where the existence of a correlation is to be determined and M is the number of points in each of the data sets. Usually, such a correlation function is normalized by dividing by the root-mean-square values of x and y

$$c_{xy} = \frac{\frac{1}{M} \sum_i^M x_i y_i}{\sqrt{\left(\frac{1}{M} \sum_i^M x_i^2\right) \left(\frac{1}{M} \sum_i^M y_i^2\right)}} = \frac{\langle x_i y_i \rangle}{\langle x_i^2 \rangle \langle y_i^2 \rangle} \tag{7.284}$$

A value of 0 than indicates no correlation and an absolute value of 1 indicates a high degree of correlation (either positive or negative). If the quantity of interest does not fluctuate about zero it is common to consider only the fluctuating part

$$c_{xy} = \frac{\frac{1}{M} \sum_i^M (x_i - \langle x \rangle)(y_i - \langle y \rangle)}{\sqrt{\left(\frac{1}{M} \sum_i^M (x_i - \langle x \rangle)^2\right) \left(\frac{1}{M} \sum_i^M (y_i - \langle y \rangle)^2\right)}} \tag{7.285}$$

If the value of some property at one time during a molecular dynamics simulation correlates with the value of the same or another property at a later time a time correlation is found. The time correlation function is written as

$$C_{xy}(t) = \langle x(t) y(0) \rangle \tag{7.286}$$

Some properties, such as the velocity of an atom, can be obtained for every atom; other properties, such as the dipole moment, can only be obtained for the entire system. In the former case the correlation function can be calculated by averaging over all atoms in the system

$$c_{vv}(t) = \frac{1}{N} \sum_{i}^{N} \frac{\langle \vec{v}_i(t) \cdot \vec{v}_i(0) \rangle}{\langle \vec{v}_i(0) \cdot \vec{v}_i(0) \rangle} \tag{7.287}$$

If the properties x and y are different a cross-correlation function is obtained, while if they are the same the correlation function is called an auto-correlation function. An auto-correlation function indicates to what extent a system retains a "memory" of its previous state. For example, the velocity auto-correlation function (Eq. 7.287) provides a measure of how much the velocity at time t is correlated with the velocity at time 0. Therefore, at the beginning of a simulation the velocity auto-correlation will have a value of one, whereas at long times it should become very small and systematically approach zero.

The time it takes a system to lose correlation is called the correlation time or the relaxation time. The length of the molecular dynamics simulation needs to be significantly greater than the correlation time. In that case, one can extract many sets of data from the trajectory and thus reduce the uncertainty in the calculation of the correlation function. If P steps of molecular dynamics are required for complete relaxation and the simulation has been run for a total of Q steps, $Q - P$ different sets of values can be used to calculate a value of the correlation function. The first set would run from steps 1 through N, the second from step 2 through $N + 1$, and so on. With M time origins, t_j, the velocity auto-correlation function is given by

$$C_{vv}(t) = \frac{1}{MN} \sum_{j}^{M} \sum_{i}^{N} \vec{v}_i(t_j) \cdot \vec{v}_i(t_j + t) \tag{7.288}$$

Therefore, properties with small correlation times can be predicted more accurately than properties with large correlation times. No property with a correlation time longer than the length of the molecular dynamics simulation can be determined accurately.

Whereas the velocity auto-correlation function is an example of a single particle correlation function which is not only averaged over time, but also over all the atoms there are also correlation functions which can only be calculated for the entire system. One such correlation function is the dipole auto-correlation function

$$c_{dipole}(t) = \frac{\langle \vec{\mu}(t) \cdot \vec{\mu}(0) \rangle}{\langle \vec{\mu}(0) \cdot \vec{\mu}(0) \rangle} \tag{7.289}$$

where

$$\vec{\mu}(t) = \sum_{i}^{N} \vec{\mu}_i(t) \tag{7.290}$$

is the vector sum of the dipole moments of the molecules forming the system. This correlation function is related to the adsorption spectrum of the system. To obtain the adsorption spectrum from the correlation function it is necessary to perform a Fourier transformation on the dipole auto-correlation function

$$c_{dipole}(\nu) = \int_{-\infty}^{\infty} c_{dipole}(t) \exp(-2\pi i \nu t) dt \qquad (7.291)$$

which yields the common representation of intensity as function of the vibrational frequency.

Transport

Since molecular dynamics simulations predict the evolution of a system in time they can be applied to study transport phenomena. Transport of matter is usually characterized by diffusion coefficients. Molecular dynamics trajectories can be used to calculate diffusion coefficients. The relation between the diffusion coefficient, D, and the mean square displacement is given by the Einstein equation

$$D = \lim_{t \to \infty} \frac{\langle |\vec{r}(t) - \vec{r}(0)|^2 \rangle}{6t} \qquad (7.292)$$

The mean square displacements can be calculated from the trajectory.

Alternatively, the diffusion coefficient can be calculated from an auto-correlation function using the Green-Kubo formula

$$\int_{0}^{\infty} \langle \vec{v}(\tau) \cdot \vec{v}(0) \rangle d\tau = \lim_{t \to \infty} \frac{\langle |\vec{r}(t) - \vec{r}(0)|^2 \rangle}{2t} = 3D \qquad (7.293)$$

It is clear from this formula that the long time tail of the auto-correlation function is important since it can contribute significantly to the integral. In practice the integral is evaluated numerically and the tail is usually fitted to a function which is then used to integrate to infinity.

7.8 Grand Canonical Molecular Dynamics

In the previous two sections we have seen that Monte Carlo calculations can be used to determine properties under a given set of conditions (temperature, pressure, chemical potential). We have also seen that molecular dynamics calculations can predict the dynamical properties of a system under a given set of conditions (temperature, pressure). Unlike Monte Carlo calculations the number of particles in the system has always to be constant for molecular dynamics simulations.

Therefore, it seems molecular dynamics simulations cannot be performed at constant chemical potential. This is, however, not true. It is possible to add creation/destruction attempts as performed in Monte Carlo calculations to molecular dynamics. Since these simulations are then performed at constant chemical potential in the grand canonical ensemble they are called grand canonical molecular dynamics simulations [867].

A grand canonical molecular dynamics simulation starts out just as an ordinary molecular dynamics simulation. After an equilibration period creation/destruction attempts are performed at given time steps. The creation/destruction attempts are successful based on the same criteria as in grand canonical Monte Carlo calculations. If a new particle is created it is assigned a random velocity drawn from a Maxwell-Boltzmann distribution. It then participates in the normal molecular dynamics propagation through time. If a destruction attempt is successful the particle is removed from the system. Care has to be taken not to disturb the molecular dynamics trajectories too much. Therefore, creation/destruction attempts can usually not be performed at every time step. To obtain a good sampling of phase space the number of creation/destruction attempts needs to be the same as in conventional Monte Carlo calculations. This results in very long simulations.

However, grand canonical molecular dynamics simulations can be used to predict properties which are otherwise impossible to calculate, such as adsorption isotherms with flexible hosts (cf. p. 96) [373]. Grand canonical molecular dynamics simulations have so far been mostly used in the simulation of fluids [868–874], to study the swelling of clays [875], and to investigate the diffusion of small gas molecules through zeolite membranes [364].

Appendix

Common Abbreviations in Computational Chemistry

ACM   Adiabatic Connection Method – a density functional

ACPF   Averaged Coupled Pair Functional – a method to include electron correlation

AM1   Austin Model 1 – a semi-empirical method (the second parameterization of MNDO)

AMBER   Assisted Model Building with Energy Refinement – a force field

ANO   Atomic Natural Orbital

AO   Atomic Orbital

APW   Augmented Plane Wave

ASA   Atomic Sphere Approximation

ASW   Augmented Spherical Wave – a method to calculate the electronic structure of closed-packed solids

au   atomic unit (see p. 259)

aug-cc-pVDZ   the augmented, correlation consistent, polarized

(aug-cc-pVTZ, aug-cc-pVQZ)   valence double- (triple-, quadruple)ζ basis set

B3LYP   the combination of Becke's three parameter nonlocal exchange functional [703] and Lee–Yang–Parr's nonlocal correlation functional [505, 876, 877] in DFT

B3P   the combination of Becke's three parameter nonlocal exchange functional [703] and Perdew's nonlocal correlation functional [698] in DFT

B3PW91   the combination of Becke's three parameter nonlocal exchange functional [703] and Perdew's and Wang's nonlocal correlation functional [701] in DFT

BLYP   the combination of Becke's nonlocal exchange functional [504] and Lee–Yang–Parr's nonlocal correlation functional [505, 876, 877] in DFT

BP86	the combination of Becke's nonlocal exchange functional [504] and Perdew's nonlocal correlation functional from 1986 [698] in DFT
BPW91	the combination of Becke's nonlocal exchange functional [504] and Perdew–Wang's nonlocal correlation functional from 1991 [697, 700] in DFT
CASPT2	Complete Active Space Perturbation Theory of second order
CASSCF	Complete Active Space Self Consistent Field
CC (CCSD, CCSDT)	Coupled Clusters – a method to include electron correlation (usually augmented with the type of excitations included, S – singles, D – doubles, T – triples)
CCSD(T)	Coupled Clusters with singles and double excitations and a perturbative treatment of triple excitations
CEPA	Coupled Electron Pair Approximation
CFF	Consistent Force Field
CHARMM	Chemistry at Harvard Molecular Mechanics – a force field and a molecular mechanics program
CI	Configuration Interaction – a method to include electron correlation
CNDO	Complete Neglect of Differential Overlap – a semi-empirical method
COMPASS	Condensed phase Optimized Molecular Potentials for Atomic Simulation Studies – a force field
COSMO	Conductor-like Screening Model – a method to account for solvation effects
CPF	Coupled Pair Functional
CPHF	Coupled Perturbed Hartree Fock
CVFF	Consistent Valence Force Field
DFPT	Density Functional Perturbation Theory
DFT	Density Functional Theory
DN (DNP)	a double-numerical basis set used in DFT (augmented with polarization functions)
DOS	Density of states
DZ (DZP)	a basis set of double-ζ quality (augmented with polarization functions)
EAM	Embedded Atom Method
E_h	Hartree – the atomic unit of energy (see p. 259)
EHT	Extended Hückel Theory
EVB	Empirical Valence Bond method (see p. 240)
FCI	Full Configuration Interaction – a method to include electron correlation

FLAPW	Full Potential Linearized Augmented Plane Waves
GCMC	Grand Canonical Monte Carlo
GCMD	Grand Canonical Molecular Dynamics
GGA	Generalized Gradient Approximation
GIAO	Gauge Including Atomic Orbitals
HF	Hartree–Fock
IGLO	Individual Gauge for Localized Orbitals
LAPW	Linearized Augmented Plane Waves
LCAO	Linear Combination of Atomic Orbitals
LDA	Local Density Approximation
LMTO	Linear Combination of Muffin-Tin Orbitals
LSD, LSDA	Local Spin Density Approximation
MBPT	Many-Body Perturbation Theory
MC	Monte Carlo
MCTDH	Multiconfigurational time dependent Hartree
MD	Molecular Dynamics
MM	Molecular Mechanics
MM1, MM2,	Molecular Mechanics force fields developed by
MM3, MM4	Norman Allinger
MNDO	Modified Neglect of Differential Overlap – a semi-empirical method
MO	Molecular Orbital
MP2, MP3,	Møller–Plesset perturbation theory of order 2, 3, 4,
MP4, . . .	. . . – a method to include electron correlation
MR-CI	Multireference configuration interaction
μVT	the grand-canonical ensemble (constant chemical potential, constant volume, constant temperature)
NPT	the isobaric-isothermic ensemble (constant number of particles, constant pressure, constant temperature)
NVE	the microcanonical ensemble (constant number of particles, constant volume, constant energy)
NVT	the canonical ensemble (constant number of particles, constant volume, constant temperature)
OPW	orthogonalized plane wave
PCM	Polarizable Continuum Model – a method for modeling solvation
PM3	a semi-empirical method (the third parameterization of MNDO)
QCISD	Quadratic Configuration Interaction with Singles and Doubles substitution
RHF	Restricted Hartree–Fock
SAC	Symmetry adapted cluster

SAM1	Semi-Ab Initio Method 1 – a semi-empirical method
SAO	Symmetrized Atomic Orbital
SCF	Self Consistent Field – the basic method of ab initio calculations (also referred to as Hartree–Fock)
SCIPCM	Self-consistent isodensity polarized continuum model
SOS	Sum Over States
SV	a split-valence basis set
TZ (TZP)	a basis set of triple-ζ quality (augmented with polarization functions)
UFF	Universal Force Field
UHF	Unrestricted Hartree–Fock
VWN	the Vosko–Wilk–Nusair local density functional [644]
ZORA	Zero-Order regular approximation – a method to account for relativistic effects

Basis Set Naming Conventions

Basis sets quoted in papers on quantum mechanical calculations usually follow some naming conventions. Basis sets developed by Pople and co-workers [676, 677] use a notation of a-bc$\cdots$G where a, b, c, etc. are some integers. a denotes the number of primitives per basis function for all core shells. b, c, etc. denote the number of primitives for each of the basis functions in the valence shell. The G stands for Gaussian denoting the use of Gaussian functions. Since there are more than one basis function in the valence shell, such a basis set is called a split-valence basis set. This basic notation is augmented by a number of asterisks and pluses. A single star added to the name of the basis, such as 6-31G*, means that there is a single set of polarization functions for all atoms except hydrogen and helium. Two stars, such as 6-31G**, mean that there are also polarization functions for hydrogen and helium. Plus signs are used to denote diffuse functions. More than one set of polarization functions is denoted by the number and angular momentum of the polarization functions in parentheses again split for hydrogen/helium and all other elements. A basis set named 6-311G(3df,3pd) is a basis set which has six primitives contracted to one function for all core shells and three functions for the valence shell where the first function consists of three primitives while the other two contain only one primitive. This basis set is augmented with a set of three d and one f polarization functions on all elements except hydrogen and helium and with three p and one f polarization functions on hydrogen and helium.

Other basis functions are usually classified according to the number of basis functions used per shell. A basis set which uses two basis functions per shell (including all shells in the atom core) is called a double-ζ (DZ) basis set. Three basis functions per shell constitutes a triple-ζ basis set, etc. If the larger number of basis functions is limited to the valence shell the basis set is called, e. g., a valence double-ζ basis set (Pople's 6-31G basis set is therefore a valence double-ζ basis set). Since this classification does not describe how many primitives are used per basis function this information is often provided in the following form: 11s7p1d/6s4p1d where the numbers in front of the slash denote the number of primitives for each angular momentum, while the numbers after the slash denote the number of contracted functions for each angular momentum.

There are also basis sets which have been developed specifically for use in correlated calculations. These basis sets are named cc-pVxZ where x can be D, T, Q, 5, $\cdots$. The abbreviation stands for correlation consistent polarized valence double (triple, quadruple, pentuple, $\cdots$) ζ.

Atomic Units

Quantum mechanical programs often use atomic units throughout since their use simplifies the equations which have to be programmed. If a quantity is expressed in atomic units it is expressed with respect to fundamental constants. According to the basic quantities of the international system of units atomic units can be written using

- the mass of the electron, $m_e = 9.109\,389\,7(54) \cdot 10^{-31}$ kg

- the charge of the electron, $e = 1.602\,177\,33(49) \cdot 10^{-19}$ As

- Planck's constant, $\hbar = 1.054\,572\,66(63) \cdot 10^{-34}$ Js, and

- the permittivity of vacuum, $4\pi\epsilon_0 = 1.112\,650\,055\ldots \cdot 10^{-10}$ F m^{-1} [878, 879].

The most common derived quantities are therefore

length	$1\ a_0$	$= \frac{4\pi\epsilon_0 \hbar^2}{m_e e^2}$	$= 5.291\,772\,49(24) \cdot 10^{-11}$ m
energy	$1\ E_h$	$= \frac{e^2}{4\pi\epsilon_0 a_0}$	$= 4.359\,748\,2(26) \cdot 10^{-18}$ J
time	1 au	$= \frac{\hbar}{E_h}$	$= 2.418\,884\,33 \cdot 10^{-17}$ s
velocity	1 au	$= \frac{\hbar}{m_e a_0}$	$= 2.187\,691\,41 \cdot 10^{6}$ m s^{-1}
frequency	1 au	$= \frac{E_h}{\hbar}$	$= 4.134\,137\,33 \cdot 10^{16}$ s^{-1}
dipole moment	1 au	$= 1\ a_0 e$	$= 8.478\,357\,92 \cdot 10^{-30}$ Asm

polarizability   1 au $= 4\pi\epsilon_0 a_0^3$ $= 1.648\,777\,63 \cdot 10^{-41}\ \mathrm{C^2m^2J^{-1}}$

force   1 au $= 1\ \frac{E_h}{a_0}$ $= 8.238\,729\,48 \cdot 10^{-8}\ \mathrm{N}$

Harmonic force constants

stretch/stretch   1 au $= 1\ \frac{E_h}{a_0^2}$ $= 1.556\,894\,12 \cdot 10^3\ \mathrm{N\ m^{-1}}$

stretch/bend   1 au $= 1\ \frac{E_h}{a_0 \mathrm{rad}}$ $= 8.238\,729\,48 \cdot 10^{-8}\ \mathrm{N\ rad^{-1}}$

bend/bend   1 au $= 1\ \frac{E_h}{\mathrm{rad}^2}$ $= 4.359\,748\,2(26) \cdot 10^{-18}\ \mathrm{J\ rad^{-2}}$

The atomic unit of length, a_0, is called Bohr, while the atomic unit of energy, E_h, is called Hartree.

Bibliography

[1] R. D. Groot and P. B. Warren, *J. Chem. Phys.*, **1997**, *107*.

[2] J. G. E. M. Fraaije, *J. Chem. Phys.*, **1993**, *99*.

[3] W. Heitler, *Elementare Wellenmechanik;* Friedr. Vieweg & Sohn, Braunschweig, 1961.

[4] M. Born and J. R. Oppenheimer, *Ann. Physik*, **1927**, *84*, 457.

[5] J. Sauer, *Chem. Rev.*, **1989**, *89*, 199.

[6] A. Imamura and H. Fujita, *J. Chem. Phys.*, **1974**, *61*.

[7] R. Dovesi, C. Pissani, C. Roetti, J. M. Ricart, and F. Illas, *Surf. Sci.*, **1984**, *148*.

[8] J.-Y. Saillard and R. Hoffmann, *J. Am. Chem. Soc.*, **1984**, *106*.

[9] S.-S. Sung and R. Hoffmann, *J. Am. Chem. Soc.*, **1985**, *107*.

[10] L. Subramanian and R. Hoffmann, *Inorg. Chem.*, **1992**, *31*.

[11] S. R. Boorse, P. Alemany, J. M. Burlitch, and R. Hoffmann, *Chem. Mater.*, **1993**, *5*.

[12] G. J. Miller, H. Deng, and R. Hoffmann, *Inorg. Chem.*, **1994**, *33*.

[13] R. Hoffmann, *J. Chem. Phys.*, **1963**, *39*.

[14] J. C. Phillips, *Phys. Rev.*, **1958**, *112*.

[15] M. T. Yin and M. L. Cohen, *Phys. Rev. B*, **1982**, *25*.

[16] B. G. Dick, Jr. and A. W. Overhauser, *Phys. Rev.*, **1958**, *112*, 90.

[17] W. Cochran, *Phys. Rev. Lett.*, **1959**, *2*.

[18] W. Cochran, *Crit. Rev. Solid State Sci.*, **1971**, *2*.

[19] M. Wilson, P. Madden, and B. J. Costa-Cabral, *J. Phys. Chem.*, **1996**, *100*.

[20] P. W. Fowler and P. A. Madden, *Mol. Phys.*, **1983**, *49*.

[21] P. W. Fowler and P. A. Madden, *Phys. Rev. B*, **1984**, *29*.

[22] M. A. Wilson, P. A. Madden, N. C. Pyper, and J. H. Harding, *J. Chem. Phys.*, **1996**, *104*.

[23] Neil W. Aschcroft and N. David Mermin, *Solid State Physics;* Saunders College Publishing, Fort Worth, 1976.

[24] M. S. Daw and M. I. Baskes, *Phys. Rev. Lett.*, **1983**, *50*.

[25] M. S. Daw and M. I. Baskes, *Phys. Rev. B*, **1984**, *29*, 6443.

[26] J. K. Nørskov and N. D. Lang, *Phys. Rev. B*, **1980**, *21*.

[27] R. A. Jackson, *Mol. Simul.*, **1987**, *2*.

[28] A. J. Stone and C.-S. Tong, *J. Comput. Chem.*, **1994**, *15*, 1377.

[29] M. W. Finnis, *Surf. Sci.*, **1991**, *241*.

[30] J. A. Pople, H. B. Schlegel, R. Krishnan, D. J. De Frees, J. S. Binkley, M. J. Frisch, R. A. Whiteside, R. F. Hout, and W. J. Hehre, *Int. J. Quantum Chem., Quantum Chem. Symp.*, **1981**, *15*, 269.

[31] R. F. Hout, Jr., B. A. Levi, and W. J. Hehre, *J. Comput. Chem.*, **1982**, *3*, 234.

[32] Computer physics communications (CPC): http://www.cpc.cs.qub.ac.uk/cpc.

[33] Computational chemistry list (CCL): http://ccl.osc.edu/ccl/welcome.html, http://ccl.osc.edu/ccl/srs.html.

[34] Collaborative computational projects (CCP): http://gserv1.dl.ac.uk/ccp/main.html.

[35] http://www.vrex.com/.

[36] http://www.accelrys.com/.

[37] http://www.scienomics.com/.

[38] http://www.emsl.pnl.gov/docs/nwchem/nwchem.html.

[39] http://www.materialsdesign.com/index.html.

[40] http://www.gaussian.com/.

[41] http://www.schrodinger.com/.

[42] http://www.turbomole.com/.

[43] http://www.ch.ic.ac.uk/gale/research/gulp.html.

[44] http://viewmol.sourceforge.net/.

[45] http://yaehmop.sourceforge.net/.

[46] W. Koch and R. Hertwig, In *Encyclopedia of Computational Chemistry*, P. v. R. Schleyer, Ed.; John Wiley & Sons, Chichester, 1998; Vol. 1; pages 689 – 700.

[47] W. Koch and M. C. Holthausen, *A Chemist's Guide to Density Functional Theory: An Introduction;* John Wiley & Sons, New York, 1999.

[48] P. Nortier, A. P. Borosy, and M. Allavena, *J. Phys. Chem. B*, **1997**, *101*, 1347.

[49] S. F. Vyboishchikov and J. Sauer, *J. Phys. Chem. A*, **2000**, *104*, 10913.

[50] J. D. Gale and A. K. Cheetham, *Zeolites*, **1992**, *12*, 674.

[51] K. Sohlberg, S. J. Pennycook, and S. T. Pantelides, *J. Am. Chem. Soc.*, **1999**, *121*, 7493.

[52] M. M. Thackeray, A. de Kock, and W. I. F. David, *Mat. Res. Bull.*, **1993**, *28*, 1041.

[53] K. M. Glassford and J. R. Chelikowsky, *Phys. Rev. B*, **1992**, *46*, 1284.

[54] K. M. Glassford and J. R. Chelikowsky, *Phys. Rev. B*, **1993**, *47*, 1732.

[55] M. M. Thackeray, *J. Electrochem. Soc.*, **1995**, *142*, 2558.

[56] X.-G. Wang, W. Weiss, S. K. Shaikhutdinov, M. Ritter, M. Petersen, F. Wagner, R. Schlögl, and M. Scheffler, *Phys. Rev. Lett.*, **1998**, *81*, 1038.

[57] V. I. Anisimov, I. S. Elfimov, N. Hamada, and K. Terakura, *Phys. Rev. B*, **1996**, *54*, 4387.

[58] J. A. Rodriguez, *Theor. Chem. Acc.*, **2002**, *107*, 117.

[59] B. Wessler, V. Jéhanno, W. Rossner, W.-F. Maier, J.-R. Hill, J. Tucker, G. Löwenhauser, and J.-U. Grabow; Combinatorial methods for the development of microwave dielectrics; In Werkstoffwoche-Partnerschaft GbR, Ed., *Materials Week 2002 – Proceedings*, Frankfurt, 2003. Werkstoff-Informationsgesellschaft mbH.

[60] J. L. Brito, J. Laine, and K. C. Pratt, *J. Mater. Sci.*, **1989**, *24*, 425.

[61] J. Miller, A. G. Sault, N. B. Jackson, L. Evans, and M. M. Gonzalez, *Catal. Lett.*, **1999**, *58*, 147.

[62] J. A. Rodriguez, J. C. Hanson, S. Chaturvedi, A. Maiti, and J. L. Brito, *J. Phys. Chem. B*, **2000**, *104*, 8145.

[63] J. A. Rodriguez, J. C. Hanson, S. Chaturvedi, A. Maiti, and J. L. Brito, *J. Chem. Phys.*, **2000**, *112*, 935.

[64] M. Nygren, L. G. M. Pettersson, Z. Barandiarián, and L. Seijo, *J. Chem. Phys.*, **1994**, *100*, 2010.

[65] G. Pacchioni, K. M. Neyman, and N. Rösch, *J. Electron Spectrosc. Relat. Phenom.*, **1994**, *69*, 13.

[66] J. A. Mejías, A. M. Márquez, J. Fernández-Sanz, M. Fernández-Garcia, J. M. Ricart, C. Sousa, and F. Illas, *Surf. Sci.*, **1995**, *327*, 59.

[67] I. V. Yudanov, V. A. Nasłuzov, K. M. Neyman, and N. Rösch, *Int. J. Quantum Chem.*, **1997**, *65*, 975.

[68] M. A. Nygren and L. G. M. Pettersson, *J. Chem. Phys.*, **1996**, *105*, 9339.

[69] L. Chen, R. Wu, N. Kioussis, and Q. Zhang, *Chem. Phys. Lett.*, **1998**, *290*, 255.

[70] J. A. Snyder, D. R. Alfonso, J. E. Jaffe, Z. Lin, A. C. Hess, and M. Gutowski, *J. Phys. Chem. B*, **2000**, *104*, 4717.

[71] J. A. Rodriguez, J. Hrbek, J. Dvorak, T. Jirsak, and A. Maiti, *Chem. Phys. Lett.*, **2001**, *336*, 377.

[72] S. Furuyama, H. Fujii, M. Kawamura, and T. Morimoto, *J. Phys. Chem.*, **1978**, *82*, 1028.

[73] R. Wichtendahl, M. Rodriguez-Rodrigo, U. Härtel, H. Kuhlenbeck, and H. J. Freund, *Phys. Status Solidi A*, **1999**, *173*, 93.

[74] G. Pacchioni, A. M. Ferrari, and P. S. Bagus, *Surf. Sci.*, **1996**, *350*, 159.

[75] M. Casarin, C. Maccato, and A. Vittadini, *J. Phys. Chem. B*, **1998**, *102*, 10745.

[76] D. C. Sorescu and J. T. Yates, Jr., *J. Phys. Chem. B*, **1998**, *102*, 4556.

[77] Z. Yang, R. Wu, Q. Zhang, and D. W. Goodman, *Phys. Rev. B*, **2001**, *63*, 045419.

[78] H. J. Freund, *Faraday Discuss.*, **1999**, *114*, 1.

[79] D. C. Sorescu, C. N. Rusu, and J. T. Yates, Jr., *J. Phys. Chem. B*, **2000**, *104*, 4408.

[80] J. A. Rodriguez, T. Jirsak, G. Liu, J. Hrbek, J. Dvorak, and A. Maiti, *J. Am. Chem. Soc.*, **2001**, *123*, 9597.

[81] S. Chaturvedi, J. A. Rodriguez, T. Jirsak, and J. Hrbek, *J. Phys. Chem. B*, **1998**, *102*, 7033.

[82] J. A. Rodriguez and A. Maiti, *J. Phys. Chem. B*, **2000**, *104*, 3630.

[83] J. A. Rodriguez, T. Jirsak, S. Chaturvedi, and J. Hrbek, *Surf. Sci.*, **1998**, *407*, 171.

[84] T. Jirsak, J. Dvorak, and J. A. Rodriguez, *J. Phys. Chem. B*, **1999**, *103*, 5550.

[85] J. A. Rodriguez, T. Jirsak, and J. Hrbek, *J. Phys. Chem. B*, **1999**, *103*, 1966.

[86] J. A. Rodriguez, T. Jirsak, J. Z. Larese, and A. Maiti, *Chem. Phys. Lett.*, **2000**, *330*, 475.

[87] J. A. Rodriguez, T. Jirsak, S. Sambasivan, D. Fischer, and A. Maiti, *J. Chem. Phys.*, **2000**, *112*, 9929.

[88] J. A. Rodriguez, T. Jirsak, M. Pérez, S. Chaturvedi, M. Kuhn, L. González, and A. Maiti, *J. Am. Chem. Soc.*, **2000**, *122*, 12362.

[89] J. A. Rodriguez, T. Jirsak, M. Pérez, L. González, and A. Maiti, *J. Chem. Phys.*, **2001**, *114*, 4186.

[90] J. Rodriguez, M. Pérez, T. Jirsak, L. González, A. Maiti, and J. Z. Larese, *J. Phys. Chem. B*, **2001**, *105*, 5497.

[91] J. A. Rodriguez, T. Jirsak, L. González, J. Evans, M. Pérez, and A. Maiti, *J. Chem. Phys.*, **2001**, *115*, 10914.

[92] G. D. Mahan, *Solid State Ionics*, **1980**, *1*, 29.

[93] G. V. Lewis and C. R. A. Catlow, *J. Phys. C: Solid State Phys.*, **1985**, *18*, 1149.

[94] M. J. L. Sangster and A. M. Stoneham, *Phil. Mag.*, **1980**, *43*, 597.

[95] C. R. A. Catlow, *Proc. R. Soc. A*, **1977**, *333*, 533.

[96] http://www.ri.ac.uk/Potentials/.

[97] S. M. Woodley, P. D. Battle, J. D. Gale, and C. R. A. Catlow, *Phys. Chem. Chem. Phys.*, **1999**, *1*, 2535.

[98] B. Ammundsen, G. R. Burns, M. S. Islam, H. Kanoh, and J. Rozière, *J. Phys. Chem. B*, **1999**, *103*, 5175.

[99] M. S. Islam and B. Ammundsen, In *Materials for Lithium-ion Batteries*, C. Julien and Z. Stoynov, Eds.; Kluwer Academic Publishers, Netherlands, 2000; pages 293 – 307.

[100] A. M. Stoneham, *J. Am. Ceram. Soc.*, **1981**, *64*, 54.

[101] D. J. Binks, R. W. Grimes, A. L. Rohl, and D. H. Gay, *J. Mater. Science*, **1996**, *31*, 1151.

[102] P. W. Tasker, *J. Phys. C*, **1979**, *12*, 4977.

[103] A. M. Stoneham and P. W. Tasker, In *Surface and near-surface chemistry of oxide materials*, J. Nowotny and L.-C. Dufour, Eds.; Elsevier Science Publishers B. V., Amsterdam, 1988; page 1.

[104] P. W. Tasker, *Adv. in Ceram.*, **1984**, *10*, 176.

[105] M. R. Welton-Cook and W. Berndt, *J. Phys. C*, **1982**, *15*, 5691.

[106] P. W. Tasker, In *Computer simulation in physical metallurgy*, G. Jacucci, Ed.; Reidel, Dordrecht, 1986; page 21.

[107] P. de Sainte Claire, K. C. Hass, W. F. Schneider, and W. L. Hase, *J. Chem. Phys.*, **1997**, *106*, 7331.

[108] S. Jiang, R. Frazier, E. S. Yamaguchi, M. Blanco, S. Dasgupta, Y. Zhou, T. Cagin, Y. Tang, and W. A. Goddard, III, *J. Phys. Chem. B*, **1997**, *101*, 7702.

[109] S. Vyas, R. W. Grimes, D. H. Gay, and A. L. Rohl, *J. Chem. Soc., Faraday Trans.*, **1998**, *94*, 427.

[110] D. C. Sayle, D. H. Gay, A. L. Rohl, C. R. A. Catlow, J. H. Harding, M. A. Perrin, and P. Nortier, *J. Mater. Chem.*, **1996**, *6*, 653.

[111] D. C. Sayle, C. R. A. Catlow, M.-A. Perrin, and P. Nortier, *J. Phys. Chem.*, **1996**, *100*, 8940.

[112] D. C. Sayle, A. R. George, C. R. A. Catlow, M. A. Perrin, and P. Nortier, *Revue de l'institut français du pétrole*, **1996**, *51*, 43.

[113] P. M. Oliver, S. C. Parker, and W. C. Mackrodt, *Modell. Simul. Mater. Sci. Eng.*, **1993**, *1*, 755.

[114] P. M. Oliver, G. W. Watson, and S. C. Parker, *Phys. Rev. B*, **1995**, *52*, 5323.

[115] P. A. Mulheran and J. H. Harding, *Modell. Simul. Mater. Sci. Eng.*, **1992**, *1*, 39.

[116] E. D. Skouras, V. N. Burganos, and A. C. Payatakes, *J. Chem. Phys.*, **1999**, *110*, 9244.

[117] E. D. Skouras, V. N. Burganos, and A. C. Payatakes, *J. Chem. Phys.*, **2001**, *114*, 545.

[118] M. Leslie and M. J. Gillan, *J. Phys. C: Solid State Phys.*, **1985**, *18*, 973.

[119] N. F. Mott and M. J. Littleton, *Trans. Faraday Soc.*, **1938**, *34*, 485.

[120] C. R. A. Catlow, *Ann. Rev. Mater. Sci.*, **1986**, *16*, 517.

[121] J. H. Harding, In *Defects and Disorder in Crystalline and Amorphous Solids*, C. R. A. Catlow, Ed.; Kluwer Academic Publishers, Amsterdam, 1994; pages 315 – 339.

[122] J. Corish, In *Defects and Disorder in Crystalline and Amorphous Solids*, C. R. A. Catlow, Ed.; Kluwer Academic Publishers, Amsterdam, 1994; pages 413 – 434.

[123] C. R. A. Catlow, In *Computer Simulation of Solids*, C. R. A. Catlow and W. C. Mackrodt, Eds.; Springer-Verlag, Berlin, 1982; page 130; Vol 166 of Lecture Notes in Physics.

[124] S. M. Tomlinson, C. M. Freeman, C. R. A. Catlow, H. Donnerberg, and M. Leslie, *J. Chem. Soc., Faraday Trans. 2*, **1989**, *85*, 367.

[125] M. S. D. Read, M. S. Islam, G. W. Watson, F. King, and F. E. Hancock, *J. Mater. Chem.*, **2000**, *10*, 2298.

[126] J.-R. Hill, C. M. Freeman, and M. H. Rossouw, *J. Solid State Chem.*, **2004**, *177*, 165.

[127] F. Beniere and C. R. A. Catlow, Eds., *Mass Transport in Solids;* Plenum, New York, 1983.

[128] G. E. Murch and A. S. Nowick, Eds., *Diffusion in Crystalline Solids;* Academic, New York, 1984.

[129] A. B. Lidiard, In *Handbuch der Physik*, S. Flugge, Ed.; Springer, Berlin, 1957; Vol. 20.

[130] C. R. A. Catlow, *Proc. R. Soc. London, Ser. A*, **1978**, *364*, 473.

[131] B. Vessal, M. Amini, D. Fincham, and C. R. A. Catlow, *Philos. Mag. B.*, **1989**, *60*, 753.

[132] A. D. Murray, G. E. Murch, and C. R. .A. Catlow, *Solid State Ionics*, **1986**, *18*, 196.

[133] M. S. Islam, *J. Mater. Chem.*, **2000**, *10*, 1027.

[134] R. A. Davies, M. S. Islam, and J. D. Gale, *Solid State Ionics*, **1999**, *126*, 323.

[135] R. A. Davies, M. S. Islam, A. V. Chadwick, and G. E. Rush, *Solid State Ionics*, **2000**, *130*, 115.

[136] G. Balducci, J. Kašpar, P. Fornasiero, and M. Graziani, *J. Phys. Chem. B*, **1997**, *101*, 1750.

[137] J. Sauer, *J. Phys. Chem.*, **1987**, *91*, 2315.

[138] A. C. Hess, P. F. McMillan, and M. O'Keeffe, *J. Phys. Chem.*, **1988**, *92*, 1785.

[139] W. B. De Almeida and P. J. O'Malley, *Chem. Phys. Lett.*, **1991**, *178*, 483.

[140] W. B. De Almeida and P. J. O'Malley, *J. Mol. Structure*, **1991**, *246*, 179.

[141] G. J. Kramer, A. J. M. de Man, and R. A. van Santen, *J. Am. Chem. Soc.*, **1991**, *113*, 6435.

[142] J. Sauer, *J. Mol. Catal.*, **1989**, *54*, 312.

[143] J.-R. Hill and J. Sauer, *J. Phys. Chem.*, **1995**, *99*, 9536.

[144] J.-R. Hill and J. Sauer, *J. Phys. Chem.*, **1994**, *98*, 1238.

[145] E. L. Uzunova and G. S. Nikolov, *J. Phys. Chem. A*, **2000**, *104*, 5302.

[146] E. L. Uzunova and G. St. Nikolov, *J. Phys. Chem. B*, **2000**, *104*, 7299.

[147] D. J. M. Burkhard, B. H. W. S. de Jong, A. J. H. M. Meyer, and J. H. van Lenthe, *Geochimica et Cosmochimica Acta*, **1991**, *55*, 3453.

[148] J.-R. Hill, *Entwicklung eines auf quantenchemischen Rechnungen basierenden Molekülmechanik-Potentials zur Simulation von Zeolithstrukturen;* PhD thesis, Humboldt-Universität zu Berlin, **1992**.

[149] J. Sauer and J.-R. Hill, *Chem. Phys. Lett.*, **1994**, *218*, 333.

[150] M. J. Rice, A. K. Chakraborty, and A. T. Bell, *J. Phys. Chem. B*, **2000**, *104*, 9987.

[151] L. A. M. M. Barbosa, G. M. Zhidomirov, and R. A. van Santen, *Phys. Chem. Chem. Phys.*, **2000**, *2*, 3909.

[152] D. Zhou, D. Ma, X. Liu, and X. Bao, *J. Chem. Phys.*, **2001**, *114*, 9125.

[153] G. N. Vayssilov and N. Rösch, *J. Phys. Chem. B*, **2001**, *105*, 4277.

[154] J. E. Šponer, Z. Sobalík, J. Leszczynski, and B. Wichterlová, *J. Phys. Chem. B*, **2001**, *105*, 8285.

[155] T. A. Wesolowski, A. Goursot, and J. Weber, *J. Chem. Phys.*, **2001**, *115*, 4791.

[156] D. Berthomieu, S. Krishnamurty, B. Coq, G. Delahay, and A. Goursot, *J. Phys. Chem. B*, **2001**, *105*, 1149.

[157] D. L. Bhering, A. Ramírez-Solís, and C. J. A. Mota, *J. Phys. Chem. B*, **2003**, *107*, 4342.

[158] J.-R. Hill, C. M. Freeman, and B. Delley, *J. Phys. Chem. A*, **1999**, *103*, 3772.

[159] J.-R. Hill, J. A. Stuart, A. R. Minihan, E. Wimmer, and C. J. Adams, *Phys. Chem. Chem. Phys.*, **2000**, *2*, 4249.

[160] T. Demuth, J. Hafner, L. Benco, and H. Toulhoat, *J. Phys. Chem. B*, **2000**, *104*, 4593.

[161] M. Catti, B. Civalleri, and P. Ugliengo, *J. Phys. Chem. B*, **2000**, *104*, 7259.

[162] L. A. M. M. Barbosa, R. A. van Santen, and J. Hafner, *J. Am. Chem. Soc.*, **2001**, *123*, 4530.

[163] T. Bucko, L. Benco, T. Demuth, and J. Hafner, *J. Chem. Phys.*, **2002**, *117*, 7295.

[164] F. Pascale, P. Ugliengo, B. Civalleri, R. Orlando, P. D'Arco, and R. Dovesi, *J. Chem. Phys.*, **2002**, *117*, 5337.

[165] T. Bucko, L. Benco, and J. Hafner, *J. Chem. Phys.*, **2003**, *118*, 8437.

[166] L. A. M. M. Barbosa and R. A. van Santen, *J. Phys. Chem. B*, **2003**, *107*, 4532.

[167] A. V. Larin, D. N. Trubnikov, and D. P. Vercauteren, *Int. J. Quantum Chem.*, **2003**, *92*, 71.

[168] A. Damin, S. Bordiga, A. Zecchina, K. Doll, and C. Lamberti, *J. Chem. Phys.*, **2003**, *118*, 10183.

[169] X. Rozanska, R. A. van Santen, F. Hutschka, and J. Hafner, *J. Am. Chem. Soc.*, **2001**, *123*, 7655.

[170] X. Rozanska, R. A. van Santen, and F. Hutschka, *J. Phys. Chem. B*, **2002**, *106*, 4652.

[171] X. Rozanska, T. Demuth, F. Hutschka, J. Hafner, and R. A. van Santen, *J. Phys. Chem. B*, **2002**, *106*, 3248.

[172] X. Rozanska, R. A. van Santen, T. Demuth, F. Hutschka, and J. Hafner, *J. Phys. Chem. B*, **2003**, *107*, 1309.

[173] E. Fois, A. Gamba, G. Tabacchi, S. Quartieri, and G. Vezzalini, *J. Phys. Chem. B*, **2001**, *105*, 3012.

[174] T. Demuth, L. Benco, J. Hafner, and H. Toulhoat, *Int. J. Quantum Chem.*, **2001**, *84*, 110.

[175] A. T. Hagler and C. S. Ewig, *Comput. Phys. Commun.*, **1994**, *84*, 131.

[176] S. Tsuneyuki, M. Tsukada, H. Aoki, and Y. Matsui, *Phys. Rev. Lett.*, **1988**, *61*, 869.

[177] S. Tsuneyuki, Y. Matsui, H. Aoki, and M. Tsukada, *Nature*, **1989**, *339*, 209.

[178] D. E. Akporiaye and G. D. Price, *Zeolites*, **1989**, *9*, 321.

[179] B. W. H. van Beest, G. J. Kramer, and R. A. van Santen, *Phys. Rev. Lett.*, **1990**, *64*, 1955.

[180] A. C. Lasaga and G. V. Gibbs, *Phys. Chem. Minerals*, **1987**, *14*, 107.

[181] K.-P. Schröder and J. Sauer, *J. Phys. Chem.*, **1996**, *100*, 11043.

[182] M. Sierka and J. Sauer, *Faraday Discuss.*, **1997**, *106*, 41.

[183] R. G. Della Valle and H. C. Andersen, *J. Chem. Phys.*, **1991**, *94*, 5056.

[184] G. J. Kramer, N. P. Farragher, B. W. H. van Beest, and R. A. van Santen, *Phys. Rev. B*, **1991**, *43*, 5068.

[185] A. J. M. de Man, B. W. H. van Beest, M. Leslie, and R. A. van Santen, *J. Phys. Chem.*, **1990**, *94*, 2524.

[186] A. J. M. de Man, H. Küppers, and R. A. van Santen, *J. Phys. Chem.*, **1992**, *96*, 2092.

[187] J. S. Tse and D. D. Klug, *J. Chem. Phys.*, **1991**, *95*, 9176.

[188] M. Mabilia, R. A. Pearlstein, and A. J. Hopfinger, *J. Am. Chem. Soc.*, **1987**, *109*, 7960.

[189] S. Grigoras and T. H. Lane, *J. Comput. Chem.*, **1988**, *9*, 25.

[190] J. B. Nicholas, A. J. Hopfinger, F. R. Trouw, and L. E. Iton, *J. Am. Chem. Soc.*, **1991**, *113*, 4792.

[191] J. R. Maple, M.-J. Hwang, T. P. Stockfisch, U. Dinur, M. Waldman, C. S. Ewig, and A. T. Hagler, *J. Comput. Chem.*, **1994**, *15*, 162.

[192] M.-J. Hwang, T. P. Stockfisch, and A. T. Hagler, *J. Am. Chem. Soc.*, **1994**, *116*, 2515.

[193] J. R. Maple, U. Dinur, and A. T. Hagler, *Proc. Natl. Acad. Sci. USA*, **1988**, *85*, 5350.

[194] D. W. Lewis, C. R. A. Catlow, and J. M. Thomas, *Faraday Discuss.*, **1997**, *106*, 451.

[195] C. Bussai, S. Hannongbua, S. Fritzsche, and R. Haberlandt, *Chem. Phys. Lett.*, **2002**, *354*, 310.

[196] U. Eichler, C. M. Kölmel, and J. Sauer, *J. Comput. Chem.*, **1996**, *18*, 463.

[197] J. A. Hriljac, M. M. Eddy, A. K. Cheetham, J. A. Donohue, and G. J. Ray, *J. Solid State Chem.*, **1993**, *106*, 66.

[198] U. Eichler, M. Brändle, and J. Sauer, *J. Phys. Chem. B*, **1997**, *101*, 10035.

[199] M. Brändle and J. Sauer, *J. Mol. Catalysis*, **1997**, *119*, 19.

[200] L. Rodriguez-Santiago, M. Sierka, V. Branchadell, M. Sodupe, and J. Sauer, *J. Am. Chem. Soc.*, **1998**, *120*, 1545.

[201] M. Brändle and J. Sauer, *J. Am. Chem. Soc.*, **1998**, *120*, 1556.

[202] D. Nachtigallová, P. Nachtigall, M. Sierka, and J. Sauer, *Phys. Chem. Chem. Phys.*, **1999**, *1*, 2019.

[203] P. Nachtigall, D. Nachtigallová, and J. Sauer, *J. Phys. Chem. B*, **2000**, *104*, 1738.

[204] M. Sierka and J. Sauer, *J. Phys. Chem. B*, **2001**, *105*, 1603.

[205] J. M. Martínez-Magadán, A. Cuán, and M. Castro, *Int. J. Quantum Chem.*, **2002**, *88*, 750.

[206] L. A. Clark, M. Sierka, and J. Sauer, *J. Am. Chem. Soc.*, **2003**, *125*, 2136.

[207] A. Damin, S. Bordiga, A. Zecchina, and C. Lamberti, *J. Chem. Phys.*, **2002**, *117*, 226.

[208] G. De Luca, A. Arbouznikov, A. Goursot, and P. Pullumbi, *J. Phys. Chem. B*, **2001**, *105*, 4663.

[209] H. Soscún, O. Castellano, J. Hernández, and A. Hinchliffe, *Int. J. Quantum Chem.*, **2001**, *82*, 143.

[210] J. Sauer and R. Ahlrichs, *J. Chem. Phys.*, **1990**, *93*, 2575.

[211] J. Sauer, In *Advances in Molecular Electronic Structure Theory;* JAI Press Inc., 1994; Vol. 2; page 111.

[212] W. J. Mortier, J. Sauer, J. A. Lercher, and H. Noller, *J. Phys. Chem.*, **1984**, *88*, 905.

[213] I. N. Senchenya, V. B. Kazansky, and S. Beran, *J. Phys. Chem.*, **1986**, *90*, 4857.

[214] J. Limtrakul and D. Tantanak, *Chem. Phys.*, **1996**, *208*, 331.

[215] S. F. Boys and F. Bernardi, *Mol. Phys.*, **1970**, *19*, 553.

[216] H. B. Shore, J. H. Rose, and E. Zaremba, *Phys. Rev. B*, **1977**, *15*, 2858.

[217] K. Schwarz, *J. Phys. B: Atom. Molec. Phys.*, **1978**, *11*, 1339.

[218] K. Schwarz, *Chem. Phys. Lett.*, **1978**, *57*, 605.

[219] J. P. Perdew, *Chem. Phys. Lett.*, **1979**, *64*, 127.

[220] A. Zunger, J. P. Perdew, and G. L. Oliver, *Solid State Comm.*, **1980**, *34*, 933.

[221] L. A. Cole and J. P. Perdew, *Phys. Rev. A*, **1982**, *25*, 1265.

[222] J. M. Galbraith and H. F. Schaefer III, *J. Chem. Phys.*, **1996**, *105*, 862.

[223] N. Rösch and S. B. Trickey, *J. Chem. Phys.*, **1997**, *106*, 8940.

[224] J. A. Ryder, A. K. Chakraborty, and A. T. Bell, *J. Phys. Chem. B*, **2000**, *104*, 6998.

[225] S. P. Yuan, J. G. Wang, Y. W. Li, and H. Jiao, *J. Phys. Chem. A*, **2002**, *106*, 8167.

[226] J. M. Vollmer and T. N. Truong, *J. Phys. Chem. B*, **2000**, *104*, 6308.

[227] G. N. Vayssilov, J. A. Lercher, and N. Rösch, *J. Phys. Chem. B*, **2000**, *104*, 8614.

[228] I. P. Zaragoza, J. M. Martínez-Magadán, R. Santamaria, D. Dixon, and M. Castro, *Int. J. Quantum Chem.*, **2000**, *80*, 125.

[229] J. T. Fermann, C. Blanco, and S. Auerbach, *J. Chem. Phys.*, **2000**, *112*, 6779.

[230] J. T. Fermann and S. Auerbach, *J. Chem. Phys.*, **2000**, *112*, 6787.

[231] J. T. Fermann and S. M. Auerbach, *J. Phys. Chem. A*, **2001**, *105*, 2879.

[232] S. A. Zygmunt, L. A. Curtiss, P. Zapol, and L. E. Iton, *J. Phys. Chem. B*, **2000**, *104*, 1944.

[233] L. Benco, T. Demuth, J. Hafner, F. Hutschka, and H. Toulhoat, *J. Chem. Phys.*, **2001**, *114*, 6327.

[234] M. Boronat, P. Viruela, and A. Corma, *J. Phys. Chem. B*, **1999**, *103*, 7809.

[235] M. Boronat, P. Viruela, and A. Corma, *Phys. Chem. Chem. Phys.*, **2000**, *2*, 3327.

[236] M. Boronat, C. M. Zicovich-Wilson, P. Viruela, and A. Corma, *J. Phys. Chem. B*, **2001**, *105*, 11169.

[237] R. J. Correa and C. J. A. Mota, *Phys. Chem. Chem. Phys.*, **2002**, *4*, 375.

[238] J. Limtrakul, T. Nanok, S. Jungsuttiwong, P. Khongpracha, and T. N. Truong, *Chem. Phys. Lett.*, **2001**, *349*, 161.

[239] I. Milas and M. A. C. Nascimento, *Chem. Phys. Lett.*, **2001**, *338*, 67.

[240] T. Demuth, L. Benco, J. Hafner, H. Toulhoat, and F. Hutschka, *J. Chem. Phys.*, **2001**, *114*, 3703.

[241] B. Arstad, S. Kolboe, and O. Swang, *J. Phys. Chem. B*, **2002**, *106*, 12722.

[242] V. V. Mihaleva, R. A. van Santen, and A. P. J. Jansen, *J. Phys. Chem. B*, **2001**, *105*, 6874.

[243] J. A. Ryder, A. K. Chakraborty, and A. T. Bell, *J. Phys. Chem. B*, **2002**, *106*, 7059.

[244] F. Tielens, W. Langenaeker, A. R. Ocakoglu, and P. Geerlings, *J. Comp. Chem.*, **2000**, *21*, 909.

[245] D. P. Tieleman, D. van der Spoel, and H. J. C. Berendsen, *J. Phys. Chem. B*, **2000**, *104*, 6380.

[246] K.-P. Schröder and J. Sauer, *J. Phys. Chem.*, **1993**, *97*, 6579.

[247] W. J. Hehre, L. Radom, P. v. R. Schleyer, and J. A. Pople, *Ab Initio Molecular Orbital Theory;* Wiley Interscience, New York, 1986.

[248] G. D. Price, I. G. Wood, and D. E. Akporiaye; In C. R. A. Catlow, Ed., *Modelling of Structure and Reactivity in Zeolites*, page 19. Academic Press, London, San Diego, New York, Boston, Sydney, Tokyo, Toronto, 1992.

[249] J. V. Smith and C. S. Blackwell, *Nature*, **1983**, *302*, 223.

[250] G. Engelhardt and R. Radeglia, *Chem. Phys. Lett.*, **1984**, *271*, 108.

[251] C. A. Fyfe, H. Strobl, G. T. Kokotailo, C. T. Pasztor, G. E. Barlow, and S. Bradley, *Zeolites*, **1988**, *8*, 132.

[252] V. Moravetski, J.-R. Hill, U. Eichler, A. K. Cheetham, and J. Sauer, *J. Am. Chem. Soc.*, **1996**, *118*, 13015.

[253] D. Sykes, J. D. Kubicki, and T. C. Farrar, *J. Phys. Chem. A*, **1997**, *101*, 2715.

[254] B. Bussemer, K.-P. Schröder, and J. Sauer, *J. Solid State Nucl. Magn. Reson.*, **1997**, *9*, 145.

[255] L. M. Bull, B. Bussemer, T. Anupõld, A. Reinhold, A. Samoson, J. Sauer, A. K. Cheetham, and R. Dupree, *J. Am. Chem. Soc.*, **2000**, *122*, 4948.

[256] J. O. Ehresmann, W. Wang, B. Herreros, D.-P. Luigi, T. N. Venkatraman, W. Song, J. B. Nicholas, and J. F. Haw, *J. Am. Chem. Soc.*, **2002**, *124*, 10868.

[257] F. Mauri, B. G. Pfrommer, and S. G. Louie, *Phys. Rev. Lett.*, **1996**, *77*, 5300.

[258] F. Haase, J. Sauer, and J. Hutter, *Chem. Phys. Lett.*, **1997**, *266*, 397.

[259] I. Štich, J. D. Gale, K. Terakura, and M. C. Payne, *Chem. Phys. Lett.*, **1998**, *283*, 402.

[260] B. L. Trout, B. H. Suits, R. J. Gorte, and D. White, *J. Phys. Chem. B*, **2000**, *104*, 11734.

[261] S. Iarlori, D. Ceresoli, M. Bernasconi, D. Donadio, and M. Parrinello, *J. Phys. Chem. B*, **2001**, *105*, 8007.

[262] G. Poulet, P. Sautet, and A. Tuel, *J. Phys. Chem. B*, **2002**, *106*, 8599.

[263] E. Dempsey, G. H. Kühl, and D. H. Olson, *J. Phys. Chem.*, **1969**, *73*, 387.

[264] L. Levien, C. T. Prewitt, and D. J. Weidner, *Am. Mineral.*, **1980**, *65*, 920.

[265] K.-P. Schröder; Personal communication, **1992**.

[266] R. A. Jackson and C. R. A. Catlow, *Mol. Simul.*, **1988**, *1*, 207.

[267] E. de Vos Burchart, V. A. Verheij, H. van Bekkum, and B. van de Graaf, *Zeolites*, **1992**, *12*, 183.

[268] E. de Vos Burchart, H. van Bekkum, and B. van de Graaf, *J. Chem. Soc., Faraday Trans.*, **1992**, *88*, 1161.

[269] J. Sefcik, E. Demiralp, T. Cagin, and W. A. Goddard III, *J. Comp. Chem.*, **2002**, *23*, 1507.

[270] D. R. Peacor, *Z. Kristallogr.*, **1973**, *138*, 274.

[271] L. Liu, W. A. Bassett, and T. Takahashi, *J. Geophys. Res.*, **1974**, *79*, 1160.

[272] L. Levien and C. T. Prewitt, *Am. Mineral.*, **1981**, *66*, 324.

[273] J. W. Richardson, Jr., J. J. Pluth, J. V. Smith, W. J. Dytrych, and D. M. Bibby, *J. Phys. Chem.*, **1988**, *92*, 243.

[274] A. Alberti, P. Davoli, and G. Vezzalini, *Z. Kristallogr.*, **1986**, *175*, 249.

[275] D. H. Olson, G. T. Kokotailo, S. L. Lawton, and W. M. Meier, *J. Phys. Chem.*, **1981**, *85*, 2238.

[276] R. G. Bell, R. A. Jackson, and C. R. A. Catlow, *J. Chem. Soc., Chem. Commun.*, **1990**, *10*, 782.

[277] R. M. Dessau, J. L. Schlenker, and J. B. Higgins, *Zeolites*, **1990**, *10*, 522.

[278] A. J. M. de Man, R. A. van Santen, and E. T. C. Vogt, *J. Phys. Chem.*, **1992**, *96*, 10460.

[279] E. de Vos Burchart, H. van Bekkum, B. van de Graaf, and E. T. C. Vogt, *J. Chem. Soc., Faraday Trans.*, **1992**, *88*, 2761.

[280] M. J. Annen, D. Young, M. E. Davis, O. B. Cavin, and C. R. Hubbard, *J. Phys. Chem.*, **1991**, *95*, 1380.

[281] R. W. G. Wyckoff, *Crystal Structures;* Wiley, New York, 1963.

[282] M. E. Striefler and G. R. Barsch, *Phys. Rev. B*, **1975**, *12*, 4553.

[283] G. Sastre, V. Fornes, and A. Corma, *J. Phys. Chem. B*, **2002**, *106*, 701.

[284] P. Demontis, G. B. Suffritti, S. Quartieri, E. S. Fois, and A. Gamba, *J. Phys. Chem.*, **1988**, *92*, 867.

[285] P. Demontis, G. B. Suffritti, S. Quartieri, A. Gamba, and E. S. Fois, *J. Chem. Soc., Faraday Trans.*, **1991**, *87*, 1657.

[286] M. K. Song, J. M. Shin, H. Chon, and M. S. Jhon, *J. Phys. Chem.*, **1989**, *93*, 6463.

[287] V. A. Ermoshin, K. S. Smirnov, and D. Bougeard, *Chem. Phys.*, **1996**, *209*, 41.

[288] P. Bornhauser and D. Bougeard, *J. Phys. Chem. B*, **2001**, *105*, 36.

[289] E. Martínez Morales, C. M. Zicovich-Wilson ad J. E. Sánchez Sánchez, and L. J. Alvarez, *Chem. Phys. Lett.*, **2000**, *327*, 224.

[290] E. Martínez Morales, J. J. Fripiat, and L. J. Alvarez, *Chem. Phys. Lett.*, **2001**, *349*, 286.

[291] A. Alavi, L. J. Alvarez, S. R. Elliot, and I. R. McDonald, *Phil. Mag. B*, **1992**, *65*, 489.

[292] L. Huang and J. Kieffer, *J. Chem. Phys.*, **2003**, *118*, 1487.

[293] H. Jobic, K. S. Smirnov, and D. Bougeard, *Chem. Phys. Lett.*, **2001**, *344*, 147.

[294] P. A. Jacobs, J. A. Martens, J. Weitkamp, and H. K. Beyer, *Faraday Discuss. Chem. Soc.*, **1981**, *72*, 353.

[295] M. W. Anderson and J. Klinowski, *Zeolites*, **1986**, *6*, 455.

[296] R. L. Stevenson, *J. Catal.*, **1971**, *21*, 113.

[297] D. Freude, J. Klinowski, and H. Hamdan, *Chem. Phys. Lett.*, **1988**, *149*, 355.

[298] D. Fenzke, M. Hunger, and H. Pfeifer, *J. Magn. Reson.*, **1991**, *95*, 477.

[299] D. H. Olson and E. Dempsey, *J. Catal.*, **1969**, *13*, 221.

[300] J. W. Ward, *J. Phys. Chem.*, **1969**, *73*, 2086.

[301] P. A. Jacobs and J. B. Uytterhoeven, *J. Chem. Soc., Faraday Trans. 1*, **1973**, *69*, 359.

[302] W. J. Mortier, J. J. Pluth, and J. V. Smith, *J. Catal.*, **1976**, *45*, 367.

[303] Z. Jirák, S. Vratislav, J. Zajicek, and V. Bosáček, *J. Catal.*, **1977**, *49*, 115.

[304] V. Bosáček, S. Beran, and Z. Jirák, *J. Phys. Chem.*, **1981**, *85*, 3856.

[305] M. Czjzek, H. Jobic, A. N. Fitch, and T. Vogt, *J. Phys. Chem.*, **1992**, *96*, 1535.

[306] K.-P. Schröder, J. Sauer, M. Leslie, C. R. A. Catlow, and J. M. Thomas, *Chem. Phys. Lett.*, **1992**, *188*, 320.

[307] K. Beck, H. Pfeifer, and B. Staudte, *Microporous Mater.*, **1993**, *2*, 1.

[308] K.-P. Schröder, J. Sauer, M. Leslie, and C. R. A. Catlow, *Zeolites*, **1992**, *12*, 20.

[309] R. Grau-Crespo, A. G. Peralta, A. R. Ruiz-Salvador, A. Gómez, and R. López-Cordero, *Phys. Chem. Chem. Phys.*, **2000**, *2*, 5716.

[310] B. R. Gelin and M. Karplus, *Biochemistry*, **1979**, *18*, 1256.

[311] U. Burkert and N. L. Allinger, *Molecular Mechanics;* ACS Monograph No. 177. American Chemical Society, Washington, D.C., 1982.

[312] S. J. Weiner, P. A. Kollman, D. A. Case, U. C. Singh, C. Ghio, G. Alagona, S. Profeta, Jr., and P. Weiner, *J. Am. Chem. Soc.*, **1984**, *106*, 765.

[313] W. L. Jorgensen and J. Tirado-Rives, *J. Am. Chem. Soc.*, **1988**, *110*, 1657.

[314] N. L. Allinger, Y. H. Yuh, and J.-H. Lii, *J. Am. Chem. Soc.*, **1989**, *111*, 8551.

[315] J.-H. Lii and N. L. Allinger, *J. Am. Chem. Soc.*, **1989**, *111*, 8566.

[316] J.-H. Lii and N. L. Allinger, *J. Am. Chem. Soc.*, **1989**, *111*, 8576.

[317] J. R. Maple, M.-J. Hwang, T. P. Stockfisch, and A. T. Hagler, *Isr. J. Chem.*, **1994**, *34*, 195.

[318] J. P. Hoogenboom, H. L. Tepper, N. F. A. van der Vegt, and W. J. Briels, *J. Chem. Phys.*, **2000**, *113*, 6875.

[319] D. I. Kopelevich and H.-C. Chang, *J. Chem. Phys.*, **2001**, *115*, 9519.

[320] A. G. Bezus, A. V. Kiselev, A. A. Lopatkin, and P. Q. Du, *J. Chem. Soc., Faraday Trans. 2*, **1978**, *74*, 367.

[321] A. V. Kiselev and P. Q. Du, *J. Chem. Soc., Faraday Trans. 2*, **1981**, *77*, 1.

[322] A. V. Kiselev and P. Q. Du, *J. Chem. Soc., Faraday Trans. 2*, **1981**, *77*, 17.

[323] A. V. Kiselev, A. A. Lopatkin, and A. A. Shulga, *Zeolites*, **1985**, *5*, 261.

[324] P. A. Wright, J. M. Thomas, A. K. Cheetham, and A. K. Nowak, *Nature*, **1985**, *318*, 611.

[325] A. K. Nowak, A. K. Cheetham, S. D. Pickett, and S. Ramdas, *Mol. Simul.*, **1987**, *1*, 67.

[326] F. Vigné-Maeder, In *Modelling of Molecular Structures and Properties*, J.-L. Rivail, Ed.; Elsevier Science Publishers, Amsterdam, 1990; Vol. 71 of *Studies in Physical and Theoretical Chemistry;* pages 135 – 142.

[327] J. O. Titiloye, S. C. Parker, F. S. Stone, and C. R. A. Catlow, *J. Phys. Chem.*, **1991**, *95*, 4038.

[328] K. Makrodimitris, G. K. Papadopoulos, and D. N. Theodorou, *J. Phys. Chem. B*, **2001**, *105*, 777.

[329] G. Manos, L. J. Dunne, M. F. Chaplin, and Z. Du, *Chem. Phys. Lett.*, **2001**, *335*, 77.

[330] G. Maurin, P. Senet, S. Devautour, P. Gaveau, F. Henn, V. E. Van Doren, and J. C. Giuntini, *J. Phys. Chem. B*, **2001**, *105*, 9157.

[331] S. Buttefey, A. Boutin, C. Mellot-Draznieks, and A. H. Fuchs, *J. Phys. Chem. B*, **2001**, *105*, 9569.

[332] V. Lachet, S. Buttefey, A. Boutin, and A. H. Fuchs, *Phys. Chem. Chem. Phys.*, **2001**, *3*, 80.

[333] S. Yashonath, J. M. Thomas, A. K. Nowak, and A. K. Cheetham, *Nature*, **1988**, *331*, 601.

[334] B. Smit and C. J. J. den Ouden, *J. Phys. Chem.*, **1988**, *92*, 7169.

[335] E. G. Derouane, J.-M. André, L. Leherte, P. Galet, D. Vanderveken, D. P. Vercauteren, and J. G. Fripiat, In *Theoretical Aspects of Heterogenous Catalysis*, J. B. Moffat, Ed.; Van Nostrand Reinhold, New York, 1990; pages 1 – 51; Van Nostrand Reinhold Catalysis Series.

[336] T. J. Hou, L. L. Zhu, and X. J. Xu, *J. Phys. Chem. B*, **2000**, *104*, 9356.

[337] D. Paschek and R. Krishna, *Phys. Chem. Chem. Phys.*, **2000**, *2*, 2389.

[338] R. Krishna and D. Paschek, *Phys. Chem. Chem. Phys.*, **2001**, *3*, 453.

[339] D. Paschek and R. Krishna, *Chem. Phys. Lett.*, **2001**, *342*, 148.

[340] D. Paschek and R. Krishna, *Phys. Chem. Chem. Phys.*, **2001**, *3*, 3185.

[341] A. Goj, D. S. Sholl, E. D. Akten, and D. Kohen, *J. Phys. Chem. B*, **2002**, *106*, 8367.

[342] T. J. H. Vlugt and M. Schenk, *J. Phys. Chem. B*, **2002**, *106*, 12757.

[343] S. C. Turaga and S. M. Auerbach, *J. Chem. Phys.*, **2003**, *118*, 6512.

[344] J. B. Nicholas, F. R. Trouw, J. E. Mertz, L. E. Iton, and A. J. Hopfinger, *J. Phys. Chem.*, **1993**, *97*, 4149.

[345] S. Yashonath, P. Demontis, and M. L. Klein, *Chem. Phys. Lett.*, **1988**, *153*, 551.

[346] C. J. J. den Ouden, B. Smit, A. F. H. Wielers, R. A. Jackson, and A. K. Nowak, *Mol. Simul.*, **1989**, *4*, 121.

[347] E. Cohen de Lara, R. Kahn, and A. M. Goulay, *J. Chem. Phys.*, **1989**, *90*, 7482.

[348] P. Demontis, E. S. Fois, G. B. Suffritti, and S. Quartieri, *J. Phys. Chem.*, **1990**, *94*, 4329.

[349] R. L. June, A. T. Bell, and D. N. Theodorou, *J. Phys. Chem.*, **1990**, *94*, 8232.

[350] A. K. Nowak, C. J. J. den Ouden, S. D. Pickett, B. Smit, A. K. Cheetham, M. F. M. Post, and J. M. Thomas, *J. Phys. Chem.*, **1991**, *95*, 848.

[351] C. R. A. Catlow, C. M. Freeman, B. Vessal, S. M. Tomlinson, and M. Leslie, *J. Chem. Soc., Faraday Trans.*, **1991**, *87*, 1947.

[352] S. J. Goodbody, K. Watanabe, D. MacGowan, J. P. R. B. Walton, and N. Quirke, *J. Chem. Soc., Faraday Trans.*, **1991**, *87*, 1951.

[353] J. R. Hufton, *J. Phys. Chem.*, **1991**, *95*, 8836.

[354] P. Demontis, G. B. Suffritti, E. S. Fois, and S. Quartieri, *J. Phys. Chem.*, **1992**, *96*, 1482.

[355] P. Demontis, G. B. Suffritti, and P. Mura, *Chem. Phys. Lett.*, **1992**, *101*, 553.

[356] M. Kawano, B. Vessal, and C. R. A. Catlow, *J. Chem. Soc., Chem. Comm.*, **1992**, *12*, 879.

[357] P. Demontis and G. B. Suffritti, In *Proceedings from the ninth international zeolite conference, Montreal 1992, Vol. II*, R. von Ballmoos, J. B. Higgins, and M. M. J. Treacy, Eds.; Butterworth-Heinemann, Stoneham, 1993; page 137.

[358] E. J. Maginn, A. T. Bell, and D. N. Theodorou, *J. Phys. Chem.*, **1993**, *97*, 4173.

[359] D. Dumont and D. Bougeard, In *Zeolites and Related Microporous Materials: State of the Art 1994, Part C*, J. Weitkamp, H. G. Karge, H. Pfeifer, and W. Hölderich, Eds.; Elsevier, Amsterdam, 1994; pages 2131–2138.

[360] K. S. Smirnov, *Chem. Phys. Lett.*, **1994**, *229*, 250.

[361] D. Dumont and D. Bougeard, *Zeolites*, **1995**, *15*, 650.

[362] K. T. Thomson, A. V. McCormick, and H. T. Davis, *J. Chem. Phys.*, **2000**, *112*, 3345.

[363] R. Chitra, A. V. Anil Kumar, and S. Yashonath, *J. Chem. Phys.*, **2001**, *114*, 11.

[364] M. G. Martin, A. P. Thompson, and T. M. Nenoff, *J. Chem. Phys.*, **2001**, *114*, 7174.

[365] S. Y. Bhide and S. Yashonath, *J. Chem. Phys.*, **2002**, *116*, 2175.

[366] A. I. Skoulidas and D. S. Sholl, *J. Phys. Chem. B*, **2002**, *106*, 5058.

[367] L. Leherte, G. C. Lie, K. N. Swamy, E. Clementi, E. G. Derouane, and J.-M. André, *Chem. Phys. Lett.*, **1988**, *145*, 237.

[368] L. Leherte, J.-M. André, D. P. Vercauteren, and E. G. Derouane, *J. Mol. Catal.*, **1989**, *54*, 426.

[369] L. Leherte, J.-M. André, E. G. Derouane, and D. P. Vercauteren, *Catalysis Today*, **1991**, *10*, 177.

[370] L. Leherte, J.-M. André, E. G. Derouane, and D. P. Vercauteren, *J. Chem. Soc., Faraday Trans.*, **1991**, *87*, 1959.

[371] L. Leherte, J.-M. André, E. G. Derouane, and D. P. Vercauteren, *Comput. Chem.*, **1991**, *15*, 273.

[372] L. Leherte, J.-M. André, E. G. Derouane, and D. P. Vercauteren, *Int. J. Quantum Chem.*, **1992**, *42*, 1291.

[373] J.-R. Hill, A. R. Minihan, E. Wimmer, and C. J. Adams, *Phys. Chem. Chem. Phys.*, **2000**, *2*, 4255.

[374] P. Cicu, P. Demontis, S. Spanu, G. B. Suffritti, and A. Tilocca, *J. Chem. Phys.*, **2000**, *112*, 8267.

[375] P. Demontis, G. Stara, and G. B. Suffritti, *J. Phys. Chem. B*, **2003**, *107*, 4426.

[376] S. Yashonath, P. Demontis, and M. L. Klein, *J. Phys. Chem.*, **1989**, *93*, 5016.

[377] T. Inui and Y. Nakazaki, *Zeolites*, **1991**, *11*, 434.

[378] H. Klein, H. Fuess, and G. Schrimpf, *J. Phys. Chem.*, **1996**, *100*, 11101.

[379] T. Mosell, G. Schrimpf, and J. Brickmann, *J. Phys. Chem. B*, **1997**, *101*, 9476.

[380] T. Mosell, G. Schrimpf, and J. Brickmann, *J. Phys. Chem. B*, **1997**, *101*, 9485.

[381] F. Jousse, S. M. Auerbach, and D. P. Vercauteren, *J. Phys. Chem. B*, **2000**, *104*, 2360.

[382] S. E. Jalili, L. J. Dunne, G. Manos, and A. Khettar, *Chem. Phys. Lett.*, **2003**, *367*, 324.

[383] K. S. Smirnov, *J. Phys. Chem. B*, **2001**, *105*, 7405.

[384] E. Jaramillo, C. P. Grey, and S. M. Auerbach, *J. Phys. Chem. B*, **2001**, *105*, 12319.

[385] K. H. Lim, F. Jousse, S. M. Auerbach, and C. P. Grey, *J. Phys. Chem. B*, **2001**, *105*, 9918.

[386] R. L. June, A. T. Bell, and D. N. Theodorou, *J. Phys. Chem.*, **1992**, *96*, 1051.

[387] S. P. Bates, W. J. M. van Well, R. A. van Santen, and B. Smit, *J. Am. Chem. Soc.*, **1996**, *118*, 6753.

[388] G. Sastre, N. Raj, C. R. A. Catlow, R. Roque-Malherbe, and A. Corma, *J. Phys. Chem. B*, **1998**, *102*, 3198.

[389] G. Sastre, C. R. A. Catlow, A. Chica, and A. Corma, *J. Phys. Chem. B*, **2000**, *104*, 416.

[390] D. Schuring, A. P. J. Jansen, and R. A. van Santen, *J. Phys. Chem. B*, **2000**, *104*, 941.

[391] L. N. Gergidis, D. N. Theodorou, and H. Jobic, *J. Phys. Chem. B*, **2000**, *104*, 5541.

[392] T. J. H. Vlugt, C. Dellago, and B. Smit, *J. Chem. Phys.*, **2000**, *113*, 8791.

[393] P. Demontis, J. Gulín González, G. B. Suffritti, and A. Tilocca, *J. Am. Chem. Soc.*, **2001**, *123*, 5069.

[394] G. Sastre, C. R. A. Catlow, and A. Corma, *J. Phys. Chem. B*, **2002**, *106*, 956.

[395] A. Sayeed, S. Mitra, A. V. Anil Kumar, R. Mukhopadhyay, S. Yashonath, and S. L. Chaplot, *J. Phys. Chem. B*, **2003**, *107*, 527.

[396] S. El Amrami and M. Kolb, *J. Chem. Phys.*, **1993**, *98*, 1509.

[397] M. Gosh, S. Yashonath, G. Ananthakrishna, P. Demontis, and G. B. Suffritti, *J. Phys. Chem.*, **1994**, *98*, 9354.

[398] D. I. Kopelevich and H.-C. Chang, *J. Chem. Phys.*, **2001**, *114*, 3776.

[399] S. Yashonath and S. Bandyopadhyay, *Chem. Phys. Lett.*, **1994**, *228*, 284.

[400] S. Bandyopadhyay and S. Yashonath, *J. Phys. Chem.*, **1995**, *99*, 4286.

[401] G. Schrimpf, M. Schlenkrich, J. Brickmann, and P. Bopp, *J. Phys. Chem.*, **1992**, *96*, 7404.

[402] F. Leroy, B. Rousseau, and A. H. Fuchs, *Phys. Chem. Chem. Phys.*, **2004**, *6*, 775.

[403] P. Demontis, G. B. Suffritti, A. Alberti, S. Quartieri, E. S. Fois, and A. Gamba, *Gazz. Chim. Ital.*, **1986**, *116*, 459.

[404] P. Demontis, G. B. Suffritti, E. S. Fois, A. Gamba, and G. Morosi, *Mater. Chem. Phys.*, **1991**, *29*, 357.

[405] E. E. Dil'mukhambetov, Z. K. Zhamasheva, V. I. Lygin, and E. G. Chadiarov, *Zh. Fiz. Khim.*, **1992**, *66*, 1942.

[406] J. M. Shin, K. T. No, and M. S. Jhon, *J. Phys. Chem.*, **1988**, *92*, 4533.

[407] G. K. Moon, S. G. Choi, H. S. Kim, and S. H. Lee, *Bull. Korean Chem. Soc.*, **1992**, *13*, 317.

[408] G. K. Moon, S. G. Choi, H. S. Kim, and S. H. Lee, *Bull. Korean Chem. Soc.*, **1993**, *14*, 356.

[409] S. H. Lee, G. K. Moon, S. G. Choi, and H. S. Kim, *J. Phys. Chem.*, **1994**, *98*, 1561.

[410] F. M. Higgins, N. H. de Leeuw, and S. C. Parker, *J. Mater. Chem.*, **2002**, *12*, 124.

[411] P. Demontis and G. B. Suffritti, *Chem. Rev.*, **1997**, *97*, 2845.

[412] S. L. Meisel, J. P. McCullogh, C. H. Lechthaler, and P. B. Weisz, *Chem. Tech.*, **1976**, *6*, 86.

[413] J. D. Gale, C. R. A. Catlow, and J. R. Carruthers, *Chem. Phys. Lett.*, **1993**, *216*, 155.

[414] F. Haase and J. Sauer, *J. Am. Chem. Soc.*, **1995**, *117*, 3780.

[415] E. Nusterer, P. E. Blöchl, and K. Schwarz, *Angew. Chem., Intl. Ed. Engl.*, **1996**, *35*, 175.

[416] E. Nusterer, P. E. Blöchl, and K. Schwarz, *Chem. Phys. Lett.*, **1996**, *253*, 448.

[417] K. Schwarz, E. Nusterer, P. Margl, and P. E. Blöchl, *Int. J. Quantum Chem.*, **1997**, *61*, 369.

[418] R. Shah, M. C. Payne, and J. D. Gale, *Int. J. Quantum Chem.*, **1997**, *61*, 393.

[419] J. Andzelm, N. Govind, G. Fitzgerald, and A. Maiti, *Int. J. Quantum Chem.*, **2003**, *91*, 467.

[420] S. H. Garofalini, *J. Chem. Phys.*, **1982**, *76*, 3189.

[421] L. V. Woodcock, C. A. Angell, and P. Cheeseman, *J. Chem. Phys.*, **1976**, *65*, 1565.

[422] S. H. Garofalini, *J. Chem. Phys.*, **1983**, *78*, 2069.

[423] S. H. Garofalini and S. Conover, *J. Non-Cryst. Solids*, **1985**, *74*, 171.

[424] S. H. Garofalini and S. M. Levine, *J. Am. Ceramic Soc.*, **1985**, *68*, 376.

[425] B. P. Feuston and S. H. Garofalini, *J. Chem. Phys.*, **1988**, *89*, 5818.

[426] B. P. Feuston and S. H. Garofalini, *J. Chem. Phys.*, **1989**, *91*, 564.

[427] A. B. Rosenthal and S. H. Garofalini, *J. Am. Ceramic Soc.*, **1987**, *70*, 821.

[428] B. Vessal, *J. Non-Cryst. Solids*, **1994**, *177*, 103.

[429] T. F. Soules, *J. Chem. Phys.*, **1979**, *71*, 4570.

[430] D. M. Zirl and S. H. Garofalini, *J. Am. Ceramic Soc.*, **1990**, *73*, 2848.

[431] M. Misawa, D. L. Price, and K. Suzuki, *J. Non-Cryst. Solids*, **1980**, *37*, 85.

[432] D. A. Litton and S. H. Garofalini, *J. Non-Cryst. Solids*, **1997**, *217*, 250.

[433] M. Benoit, S. Ispas, P. Jund, and R. Jullien, *Eur. Phys. J. B*, **2000**, *13*, 631.

[434] R. Car and M. Parrinello, *Phys. Rev. Lett.*, **1985**, *55*, 2471.

[435] W. Smith, G. N. Greaves, and M. J. Gillan, *J. Non-Cryst. Solids*, **1995**, *192*, 267.

[436] B. Vessal, A. C. Wright, and A. C. Hannon, *J. Non-Cryst. Solids*, **1996**, *196*, 233.

[437] N. M. Vedishcheva, B. A. Shakhmatkin, M. M. Shultz, B. Vessal, A. C. Wright, B. Bachra, A. G. Clare, A. C. Hannon, and R. N. Sinclair, *J. Non-Cryst. Solids*, **1995**, *192 & 193*, 292.

[438] C. Huang and A. N. Cormack, *J. Chem. Phys.*, **1990**, *93*, 8180.

[439] B. Smit, *J. Phys. Chem.*, **1995**, *99*, 5597.

[440] B. Vessal, G. N. Greaves, P. T. Marten, A. V. Chadwick, R. Mole, and S. Houde-Walter, *Nature*, **1992**, *356*, 504.

[441] W. H. Green, D. Jayatilaka, A. Willets, R. D. Amos, and N. C. Handy, *J. Chem. Phys.*, **1990**, *93*, 4965.

[442] A. Corma, C. R. A. Catlow, and G. Sastre, *J. Phys. Chem. B*, **1998**, *102*, 7085.

[443] S. H. Garofalini, *J. Non-Cryst. Solids*, **1984**, *67*, 133.

[444] T. Uchino and T. Yoko, *J. Phys. Chem. B*, **1998**, *102*, 8372.

[445] D. Timpel, K. Scheerschmidt, and S. H. Garofalini, *J. Non-Cryst. Solids*, **1997**, *221*, 187.

[446] G. J. Ackland, G. Tichy, V. Vitek, and M. W. Finnis, *Philos. Mag. A*, **1987**, *56*, 735.

[447] D. Timpel and K. Scheerschmidt, *J. Non-Cryst. Solids*, **1998**, *232*, 245.

[448] S. Rossano, A. Y. Ramos, and J.-M. Delaye, *J. Non-Cryst. Solids*, **2000**, *273*, 48.

[449] D. K. Belashchenko, *J. Non-Cryst. Solids*, **1996**, *212*, 205.

[450] A. Takada, C. R. A. Catlow, and G. D. Price, *J. Phys. Condens. Matter*, **1995**, *7*, 8659.

[451] A. Takada, C. R. A. Catlow, and G. D. Price, *J. Phys. Condens. Matter*, **1995**, *7*, 8693.

[452] A. H. Verhoef and H. W. den Hartog, *J. Non-Cryst. Solids*, **1992**, *146*, 267.

[453] T. F. Soules and A. K. Varshneya, *J. Am. Ceram. Soc.*, **1981**, *64*, 145.

[454] B. Delley, *J. Phys. Chem.*, **1996**, *100*, 6107.

[455] M. Takahashi, H. Toyuki, N. Umesaki, K. Kawamura, M. Tatsumisago, and T. Minami, *J. Non-Cryst. Solids*, **1992**, *150*, 103.

[456] G. Cormier, T. Peres, and J. A. Capobianco, *J. Non-Cryst. Solids*, **1996**, *195*, 125.

[457] S. Gruenhut, M. Amini, D. R. MacFarlane, and P. Meakin, *Mol. Simul.*, **1997**, *19*, 139.

[458] S. Gruenhut and D. R. MacFarlane, *J. Non-Cryst. Solids*, **1995**, *184*, 356.

[459] BIOSYM Technologies, *Catalysis User Guide 2.3.0;* BIOSYM Technologies, San Diego, 1993.

[460] J. Horbach, W. Kob, K. Binder, and C. A. Angell, *Phys. Rev. E*, **1996**, *54*, 5897.

[461] S. H. Garofalini and D. M. Zirl, *J. Vac. Sci. Technol.*, **1988**, *6*, 975.

[462] E. B. Webb and S. H.Garofalini, *Surf. Sci.*, **1994**, *319*, 381.

[463] J. K. Norskov, *Phys. Rev. B*, **1982**, *26*, 2875.

[464] J. D. Ferry, *Viscoelastic Properties of Polymers;* Wiley Interscience, New York, 1986.

[465] I. Akiyama K. Takase and N. Ohtori, *Materials Transaction JIM*, **1999**, *40*, 1258.

[466] P. Ewald, *Ann. Phys*, **1921**, *64*, 253.

[467] P. J. D. Lindan G. V. Paolini and J. H. Harding, *J. Chem. Phys.*, **1997**, *106*, 3681.

[468] P. J. D. Lindan and M. J. Gillan, *J. Phys.: Condens. Matter*, **1991**, *3*, 3929.

[469] W. T. Ashurst and W. G. Hoover, *Phys. Rev. A*, **1975**, *11*, 658.

[470] R. Khare, *J. Chem. Phys.*, **1997**, *107*, 2589.

[471] S. A. Gupta, *J. Chem. Phys.*, **1997**, *107*, 10316.

[472] S. A. Gupta, *J. Chem. Phys.*, **1997**, *107*, 10326.

[473] S. A. Gupta, *J. Chem. Phys.*, **1997**, *107*, 10335.

[474] N. W. Ashcroft and N. D. Mermin, *Solid State Physics;* Saunders College Publishing, 1976.

[475] R. Hoffmann, *Solids and Surfaces: A Chemist's View of Bonding in Extended Structures;* VCH Press, 1988.

[476] P. A. Cox, *The Electronic Structure and Chemistry of Solids;* Oxford University Press, 1987.

[477] C. M. Wolfe, N. Holonyak Jr., and G. E. Stillman, *Physical Properties of Semiconductors, Chapter 4;* Prentice Hall, Englewood Cliffs, New Jersey, 1989.

[478] M. C. Payne, M. P. Teter, D. C. Allan, T. A. Arias, and J. D. Joannopoulos, *Rev. Mod. Phys.*, **1992**, *64*, 1045.

[479] H. J. Monkhorst and J. D. Pack, *Phys. Rev. B*, **1976**, *13*, 5188.

[480] S. Baroni, S. de Gironcoli, A. Dal Corso, and P. Giannozzi, *Rev. Mod. Phys.*, **2001**, *73*, 515.

[481] Molecular Simulations Inc.; Program CASTEP; San Diego, **1998**.

[482] http://www.accelrys.com/mstudio/ms_modeling/castep.html.

[483] R. C. G. Leckey and J. D. Riley, *Critical Rev. Solid State and Mat. Sci.*, **1992**, *17*, 307.

[484] N. Troullier and J. L. Martins, *Phys. Rev. B*, **1991**, *43*, 1993.

[485] D. Vanderbilt, *Phys. Rev. B*, **1990**, *41*, 7892.

[486] H. P. Maruska and J. J. Tietjen, *Appl. Phys. Lett.*, **1969**, *15*, 327.

[487] A. F. Wright and J. S. Nelson, *Phys. Rev. B*, **1994**, *50*, 2159.

[488] T. C. Koopman, *Physica*, **1933**, *1*, 104.

[489] V. Fock, *Z. Phys.*, **1930**, *61*, 126.

[490] J. C. Slater, *Phys. Rev.*, **1930**, *35*, 210.

[491] W. von der Linden and P. Horsch, *Phys. Rev. B*, **1988**, *37*, 8351.

[492] F. Aryasetiawan, *Phys. Rev. B*, **1992**, *46*, 13051.

[493] P. E. Maslen, N. C. Handy, R. D. Amos, and D. Jayatilaka, *J. Chem. Phys.*, **1992**, *97*, 4233.

[494] A. Seidl, A. Görling, P. Vogl, J. A. Majewski, and M. Levy, *Phys. Rev. B*, **1996**, *53*, 3764.

[495] S. Picozzi, A. Continenza, R. Asahi, W. Mannstadt, A. J. Freeman, W. Wolf, E. Wimmer, and C. B. Geller, *Phys. Rev. B*, **2000**, *61*, 4677.

[496] E. Wimmer, H. Krakauer, M. Weinert, and A. J. Freeman, *Phys. Rev. B*, **1981**, *24*, 864.

[497] H. J. F. Jansen and A. J. Freeman, *Phys. Rev. B*, **1984**, *30*, 561.

[498] M. L. Hitchman and K. F. Jensen, *Chemical Vapor Deposition : Principles and Applications;* Academic Press, 1993.

[499] N. Imaishi, T. Sato, M. Kimura, and Y. Akiyama, *J. Crystal Growth*, **1997**, *180*, 680.

[500] http://www.reactiondesign.com/products/open/chemkin.html.

[501] P. Ho, M. E. Coltrin, J. S. Binkley, and C. F. Melius, *J. Phys. Chem.*, **1985**, *89*, 4647.

[502] D. J. DeFrees, B. A. Levi, S. K. Pollack, W. J. Hehre, J. S. Binkley, and J. A. Pople, *J. Am. Chem. Soc.*, **1979**, *101*, 4085.

[503] P. J. Hay, *J. Phys. Chem.*, **1996**, *100*, 5.

[504] A. D. Becke, *Phys. Rev. A*, **1988**, *38*, 3098.

[505] C. Lee, W. Yang, and R. G. Parr, *Phys. Rev. B*, **1988**, *37*, 785.

[506] P. Ho, M. E. Coltrin, J. S. Binkley, and C. F. Melius, *J. Phys. Chem.*, **1986**, *90*, 3399.

[507] K. Sato, H. Haruta, and Y. Kumashiro, *Phys. Rev. B*, **1997**, *55*, 15467.

[508] E. Kaxiras, *Phys. Rev. Lett.*, **1990**, *64*, 551.

[509] K. Shiraishi, T. Ito, and T. Ohno, *Solid State Electron.*, **1994**, *37*, 601.

[510] K. Shiraishi and T. Ito, *J. Cryst. Growth*, **1995**, *150*, 158.

[511] G. Brocks, P. J. Kelly, and R. Car, *Phys. Rev. Lett.*, **1991**, *66*, 1729.

[512] Z. Y. Zhang, Y. T. Lu, and H. Metiu, *Surf. Sci. Lett.*, **1991**, *248*, L250.

[513] F. H. Stillinger and T. A. Weber, *Phys. Rev. B*, **1985**, *31*, 5262.

[514] Z. Zhang, F. Wu, H. J. W. Zandvliet, B. Poelsema, H. Metiu, and M. G. Lagally, *Phys. Rev. Lett.*, **1995**, *74*, 3644.

[515] T. A. Arias and J. D. Joannopoulos, *Phys. Rev. B*, **1994**, *49*, 1425.

[516] L. J. Clarke, I. Stich, and M. C. Payne, *Comp. Phys. Commun.*, **1992**, *72*, 14.

[517] T. Starkloff and J. D. Joannopoulos, *Phys. Rev. B*, **1977**, *16*, 5212.

[518] D. R. Hamann, M. Schluter, and C. Chiang, *Phys. Rev. Lett.*, **1979**, *43*, 1494.

[519] A. Maiti, M. F. Chisholm, S. J. Pennycook, and S. T. Pantelides, *Phys. Rev. Lett.*, **1996**, *77*, 1306.

[520] M. F. Chisholm, A. Maiti, S. J. Pennycook, and S. T. Pantelides, *Phys. Rev. Lett.*, **1998**, *81*, 132.

[521] A. Maiti, T. Kaplan, M. Mostoller, M. F. Chisholm, S. J. Pennycook, and S. T. Pantelides, *Appl. Phys. Lett.*, **1997**, *70*, 336.

[522] R. Jones, A. Umerski, P. Sitch, M. I. Heggie, and S. Oberg, *Phys. Stat. Sol.*, **1993**, *137*, 389.

[523] T. P. Sheahan, *Introduction to High-Temperature Superconductivity;* Plenum Press, 1994.

[524] R. M. Hazen, *The Breakthrough: The Race for the Superconductor;* Summit Books, 1989.

[525] C. W. Chu, *Proc. Natl. Acad. of Sciences*, **1987**, *84*, 4681.

[526] R. J. Cava, *Science*, **1990**, *247*, 656.

[527] N. L. Allan and W. Mackrodt, *Phil. Mag. Lett.*, **1989**, *60*, 183.

[528] N. L. Allan and W. Mackrodt, *J. Chem. Soc. Faraday Trans. 2*, **1990**, *86*, 1227.

[529] Y. Tokura, H. Tagaki, and S. Uchida, *Nature*, **1989**, *337*, 3457.

[530] M. Evain, M.-H. Whangbo, M. A. Beno, U. Geiser, and J. M. Williams, *J. Am. Chem. Soc.*, **1987**, *109*, 7917.

[531] Y. Q. Jia, *J. Phys. Condens. Matter*, **1991**, *3*, 2833.

[532] M. S. Islam, *Supercond. Sci. Technol.*, **1990**, *3*, 531.

[533] S. J. Rothman, J. L. Roubort, U. Welp, and J. E. Baker, *Phys. Rev. B*, **1991**, *44*, 2326.

[534] X. Zhang and C. R. A. Catlow, *Molec. Simul.*, **1994**, *12*, 115.

[535] M. Ratner and D. Ratner, *Nanotechnology - A Gentle Introduction to the Next Big Idea;* Prentice-Hall, Englewood Cliffs, NJ, 2002.

[536] M. Gross, *Travels to the Nanoworld: Miniature Machinery in Nature and Technology;* Perseus Publishing, Cambridge, MA, 2001.

[537] M. Wilson, K. Kannangara, G. Smith, M. Simmons, and C. Crane, *Nanotechnology: Basic Science and Emerging Technologies;* CRC Press, Boca Raton, FL, 2002.

[538] G. Timp, *Nanotechnology;* Springer-Verlag, New York, London, Heidelberg, 1999.

[539] K. E. Drexler, *Nanosystems: Molecular Machinery, Manufacturing, and Computation;* John Wiley & Sons, New York, NY, 1992.

[540] P. J. F. Harris, *Carbon Nanotubes and Related Structures;* Cambridge University Press, Cambridge, UK, 2002.

[541] M. S. Dresselhaus, G. Dresselhaus, and P. Avouris, *Carbon Nanotubes: Synthesis, Structure, Properties, and Application;* Springer-Verlag, New York, London, Heidelberg, 2001.

[542] B. C. Satishkumar, A. Govindaraj, E. M. Vogl, L. Basumallick, and C. N. R. Rao, *J. Mater. Res.*, **1997**, *12*, 604.

[543] Z. L. Wang, *Nanowires and Nanobelts: Materials, Properties, and Devices;* Kluwer Academic Publishers, Netherlands, 2003.

[544] S. Bandyopadhyay and H. S. Nalwa, *Quantum Dots and Nanowire;* American Scientific Publishers, Stevenson Ranch, CA, 2002.

[545] D. Koruga, *Fullerene C60: History, Physics, Nanobiology, Nanotechnology;* Elsevier, Netherlands, 1993.

[546] G. Schmid, *Nanoparticles: From Theory to Application;* John Wiley & Sons, New York, NY, 2003.

[547] http://www.nanospectra.com/.

[548] P. Harrison, *Quantum Wells, Wires and Dots: Theoretical and Computational Physics;* John Wiley & Sons, New York, NY, 2000.

[549] D. Bimberg, M. Grundmann, and N. N. Ledentsov, *Quantum Dot Heterostructures;* John Wiley & Sons, New York, NY, 1999.

[550] Y. Masumoto and T. Takagahara, *Semiconductor Quantum Dots;* Springer-Verlag, New York, London, Heidelberg, 2002.

[551] http://www.qdots.com/.

[552] C. Joachim and S. Roth, *Atomic and Molecular Wires;* Kluwer Academic Press, Netherlands, 1997.

[553] C. P. Collier, G. Mattersteig, E. W. Wong, Y. Luo, K. Beverly, J. Sampaio, F. M. Raymo, J. F. Stoddart, and J. R. Heath, *Science*, **2000**, *289*, 1172.

[554] W. R. Salaneck, S. Stafstrom, and J. L. Brédas, *Conjugated Polymer Surfaces and Interfaces: Electronic and Chemical Structure of Interfaces for Polymer Light Emitting Devices;* Cambridge University Press, Cambridge, UK, 1996.

[555] http://www.csixty.com/learn.htm.

[556] L. R. Rudnick and P. A. Schweitzer, *Lubricant Additives: Chemistry and Applications;* Marcel Dekker, New York, NY, 2003.

[557] J. M. J. Frechet and D. A. Tomalia, *Dendrimers and Other Dendritic Polymers;* John Wiley & Sons, New York, NY, 2001.

[558] D. T. Bong, T. D. Clark, J. R. Granja, and M. R. Ghadiri, *Angew. Chem.*, **2001**, *40*, 988.

[559] M. Schena, Ed., *DNA Microarrays: A Practical Approach;* Oxford University Press, New York, 1999.

[560] G. Hardiman, *Pharmacogenomics*, **2002**, *3*, 293.

[561] A. Talapatra, R. Rouse, and G. Hardiman, *Pharmacogenomics*, **2002**, *3*, 527.

[562] A. K. Dillow, A. M. Lowman, and K. A. Hudgins, Eds., *Biomimetic Materials and Design: Biointerfacial Strategies, Tissue Engineering, and Targeted Drug Delivery;* Marcel Dekker, New York, NY, 2002.

[563] R. H. Baughman, A. A. Zakhidov, and W. A. de Heer, *Science*, **2002**, *297*, 787.

[564] P. M. Ajayan and O. Zhou, In *Carbon Nanotubes: Synthesis, Structure, Properties and Applications*, M. S. Dresselhaus, G. Dresselhaus, and P. Avouris, Eds.; Springer-Verlag, New York, London, Heidelberg, 2001; pages 391–425.

[565] C. Sanchez, R. M. Laine, S. Yang, and C. J. Brinker, Eds., *Organic/Inorganic Hybrid Materials;* Vol. 726 of *MRS Symp. Ser.;* Warrendale, PA, 2002.

[566] D. Piner, J. Zhu, F. Xu, S. Hong, and C. A. Mirkin, *Science*, **1999**, *283*, 661.

[567] A. L. Barabasi, F. Liu, T. Pearsall, and M. Krishnamurthy, Eds., *Epitaxial Growth-Principles and Applications;* Materials Research Society, Warrendale, PA, 1999.

[568] D. M. Mattox, *Handbook of Physical Vapor Deposition (PVD) Processing;* Noyes Publications, New Jersey, 1998.

[569] H. O. Pierson, *Handbook of Chemical Vapor Deposition: Principles, Technology, and Applications;* William Andrew Publishing, New York, NY, 2 ed., 1999.

[570] A. M. Morales and C. M. Lieber, *Science*, **1998**, *279*, 208.

[571] J. Hu, T. W. Odom, and C. M. Lieber, *Acc. Chem. Res.*, **1999**, *32*, 435.

[572] C. A. Mirkin, R. L. Letsinger, R. C. Mucic, and J. J. Storhoff, *Nature*, **1996**, *382*, 607.

[573] C. A. Mirkin, *MRS Bull.*, **2000**, *25*, 43.

[574] C. A. Mirkin, *Inorg. Chem.*, **2000**, *39*, 2258.

[575] S.-J. Park, A. A. Lazarides, and C. A. Mirkin, *Angew. Chem., Int. Ed.*, **2001**, *40*, 2909.

[576] http://www.aeiveos.com/ bradbury/papers/pbaonp.html.

[577] R. G. Parr and W. Yang, *Density Functional Theory of Atoms and Molecules;* Oxford University Press, New York, NY, 1989.

[578] F. Jensen, *Introduction to Computational Chemistry;* John Wiley & Sons, Chichester, 1999.

[579] S. Datta, *Electronic Transport in Mesoscopic Systems;* Cambridge University Press, Cambridge, UK, 1997.

[580] P. E. Blochl, C. Joachim, and A. J. Fisher, Eds., *Computations for the Nano-Scale;* Kluwer Academic Press, Netherlands, 1993.

[581] M. Reith, *Nano-Engineering in Science and Technology: An Introduction to the World of Nano-Design;* Imperial College Press, London, UK, 2003.

[582] *Technical Proceedings of the 2003 Nanotechnology Conference and Trade Show;* Vol. 3; NSTI Publications, Cambridge, MA, 2003.

[583] *Technical Proceedings of the 2002 International Conference on Computational Nanoscience and Nanotechnology;* Vol. 2; NSTI Publications, Cambridge, MA, 2002.

[584] *Technical Proceedings of the 2001 International Conference on Computational Nanoscience and Nanotechnology;* Vol. 2; NSTI Publications, Cambridge, MA, 2001.

[585] L. Merhari, L. T. Wille, K. Gonsalves, M. F. Gyure, S. Matsui, and L. J. Whitman, Eds., *Materials Issues and Modeling for Device Nanofabrication;* Vol. 584 of *MRS Proceedings;* Warrendale, PA, 1999.

[586] E. J. Brandas, H. Adachi, M. Uda, and R. Sekine, Eds., *Advances in Quantum Chemistry: DV-Xa for Advanced Nano Materials and Other Interesting Topics in Materials Science;* Vol. 42; Academic Press, 2003.

[587] P. R. Wallace, *Phys. Rev.*, **1947**, *71*, 622.

[588] J. W. Mintmire, B. I. Dunlap, and C. T. White, *Phys. Rev. Lett.*, **1992**, *68*, 631.

[589] N. Hamada, S. Sawada, and A. Oshiyama, *Phys. Rev. Lett.*, **1992**, *68*, 1579.

[590] R. Saito, M. Fujita, G. Dresselhaus, and M. S. Dresselhaus, *Appl. Phys. Lett.*, **1992**, *60*, 2204.

[591] C. T. White, D. H. Robertson, and J. W. Mintmire, *Phys. Rev. B*, **1993**, *47*, 5485.

[592] R. A. Jishi, D. Inomata, K. Nakao, M. S. Dresselhaus, and G. Dresselhaus, *J. Phys. Soc. Jpn.*, **1994**, *63*, 2252.

[593] C. T. White, J. W. Mintmire, R. C. Mowrey, D. W. Brenner, D. H. Robertson, J. A. Harrison, and B. I. Dunlap; VCH Publishers, New York, 1993.

[594] S. Iijima, *Nature*, **1991**, *354*, 56.

[595] S. Iijima, T. Ichihashi, and Y. Ando, *Nature*, **1992**, *356*, 776.

[596] T. W. Ebbesen and P. M. Ajayan, *Nature*, **1992**, *358*, 220.

[597] D. S. Bethune, C. H. Klang, M. S. de Vries, G. Gorman, R. Savoy, J. Vazquez, and R. Beyers, *Nature*, **1993**, *363*, 605.

[598] S. Iijima and T. Ichihashi, *Nature*, **1993**, *363*, 603.

[599] H. Dai, In *Carbon Nanotubes*, M. Dresselhaus, G. Dresselhaus, and P. Avouris, Eds.; Springer Verlag, New York, London, Heidelberg, 2000.

[600] C. Journet, W. K. Maser, P. Bernier, A. Loiseau, M. Lamy de la Chapelle, S. Lefrant, P. Deniard, R. Lee, and J. E. Fischer, *Nature*, **1997**, *388*, 756.

[601] A. Thess, R. Lee, P. Nikolaev, H. Dai, P. Petit, J. Robert, C. Xu, Y. H. Lee, S. G. Kim, A. G. Rinzler, D. T. Colbert, G. Scuseria, D. Tománek, J. E. Fischer, and R. E. Smalley, *Science*, **1996**, *273*, 483.

[602] P. Nikolaev, M. J. Bronikowski, R. K. Bradley, F. Rohmund, D. T. Colbert, K. A. Smith, and R. E. Smalley, *Chem. Phys. Lett.*, **1999**, *313*, 91.

[603] *Physics World*, **2000**, *13*, 29–53.

[604] M. Terrones, *Annual Review of Materials Research*, **2003**, *33*, 419.

[605] W. A. de Heer, A. Chatelain, and D. Ugarte, *Science*, **1995**, *270*, 1179.

[606] A. G. Rinzler, J. H. Hafner, P. Nikolaev, L. Lou, S. G. Kim, D. Tománek, P. Nordlander, D. T. Colbert, and R. E. Smalley, *Science*, **1995**, *269*, 1550.

[607] P. G. Collins and A. Zettl, *Appl. Phys. Lett.*, **1996**, *69*, 1969.

[608] Y. V. Gulyaev, N. I. Sinitsyn, G. V. Torgashov, Sh. T. Mevlyut, A. I. Zhbanov, Yu. F. Zakharchenko, Z. Ya. Kosakovskaya, L. A. Chernozatonskii, O. E. Glukhova, and I. G. Torgashov, *J. Vac. Sci. Technol. B*, **1997**, *15*, 422.

[609] S. Fan, M. G. Chapline, N. R. Franklin, T. W. Tombler, A. M. Cassell, and H. Dai, *Science*, **1997**, *283*, 512.

[610] O. M. Küttel, O. Groenig, C. Emmenegger, and L. Schlapbach, *Appl. Phys. Lett.*, **1998**, *73*, 2113.

[611] M. J. Fransen, Th. L. van Rooy, and P. Kruit, *Appl. Surf. Sci.*, **1999**, *146*, 312.

[612] S. J. Tans, A. R. M. Verchueren, and C. Dekker, *Nature*, **1998**, *393*, 49.

[613] W. A. de Heer and R. Martel, *Physics World*, **2000**, page 49.

[614] C. Liu, Y. Y. Fan, M. Liu, H. T. Cong, H. M. Cheng, and M. S. Dresselhaus, *Science*, **1999**, *286*, 1127.

[615] E. W. Wong, P. E. Sheehan, and C. M. Lieber, *Science*, **1997**, *277*, 1971.

[616] H. Dai, *Physics World*, **2000**, page 43.

[617] J. Kong, N. R. Franklin, and H. Dai, *Science*, **2000**, *287*, 622.

[618] P. G. Collins, K. Bradley, and A. Zettl, *Science*, **2000**, *287*, 1801.

[619] T. W. Tombler, C. Zhou, L. Alexseyev, J. Kong, H. Dai, L. Liu, C. S. Jayanthi, M. Tang, and S. Y. Wu, *Nature*, **2000**, *405*, 769.

[620] M. Guthold, M. R. Falvo, W. G. Matthews, S. Paulson, S. Washburn, D. Erie, R. Superfine, F. P. Brooks, and R. M. Taylor, *IEEE Asme Transactions on Mechatronics*, **2000**, *5*, 189.

[621] P. Kim and C. M. Lieber, *Science*, **1999**, *286*, 2148.

[622] J. Bernholc, D. Brenner, M. B. Nardelli, V. Meunier, and C. Roland, *Annual Review of Materials Research*, **2002**, *32*, 347.

[623] D. Tománek and R. Enbody, Eds., *Science and Applications of Nanotubes;* Kluwer Academic Publishers, Netherlands, 2000.

[624] *Physica B: Condensed Matter*, **2002**, *323*.

[625] Z. F. Ren, Z. P. Huang, J. W. Xu, J. H. Wang, P. Bush, M. P. Siegal, and P. N. Provencio, *Science*, **1998**, *282*, 1105.

[626] S. Fan, *Science*, **1999**, *283*, 512.

[627] Q. H. Wang, A. A. Setlur, J. M. Lauerhaas, J. Y. Dai, E. W. Seelig, and R. P. H. Chang, *Appl. Phys. Lett.*, **1998**, *72*, 2912.

[628] Y. Saito, S. Uemura, and K. Hamaguchi, *Jpn. J. Appl. Phys. Part 2*, **1998**, *37*, L346.

[629] Y. B. Choi, D. S. Chung, J. H. Kang, H. Y. Kim, Y. W. Jin, I. T. Han, Y. H. Lee, J. E. Jung, N. S. Lee, G. S. Park, and J. M. Kim, *Appl. Phys. Lett.*, **1999**, *75*, 3129.

[630] K. A. Dean, P. von Allmen, and B. R. Chalamala, *J. Vac. Sci. Technol. B*, **1999**, *17*, 1959.

[631] A. Maiti, C. J. Brabec, C. Roland, and J. Bernholc, *Phys. Rev. Lett.*, **1994**, *73*, 2468.

[632] B. Delley, *J. Chem. Phys.*, **1990**, *92*, 508.

[633] B. Delley, *Int. J. Quantum Chem*, **1998**, *69*, 423.

[634] B. Delley, *J. Chem. Phys.*, **2000**, *113*, 7756.

[635] http://www.accelrys.com/mstudio/dmol3.html.

[636] A. Maiti, J. Andzelm, N. Tanpipat, and P. von Allmen, *Phys. Rev. Lett.*, **2001**, *87*, 155502.

[637] M. Grujicic, G. Cao, and B. Gersten, *Appl. Surf. Sci.*, **2003**, *206*, 167.

[638] A. Buldum and J. P. Lu; Vol. 3, page 297, 2003.

[639] L. Liu, C. S. Jayanthi, M. Tang, S. Y. Wu, T. W. Tombler, C. Zhou, L. Alexseyev, J. Kong, and H. Dai, *Phys. Rev. Lett.*, **2000**, *84*, 4950.

[640] A. K. Rappé, C. J. Casewit, K. S. Colwell, W. A. Goddard III, and W. M. Skiff, *J. Am. Chem. Soc.*, **1992**, *114*, 10024.

[641] N. Yao and V. Lordi, *J. Appl. Phys.*, **1998**, *84*, 1939.

[642] Z. Lin and J. Harris, *J. Phys.: Condens. Matter*, **1992**, *4*, 1055.

[643] X. P. Li, J. W. Andzelm, J. Harris, and A. M. Chaka; A fast density-functional method for chemistry; In B. B. Laird, R. B. Ross, and T. Ziegler, Eds., *Chemical Applications of Density Functional Theory*, Vol. 629 of *ACS Symposium Series*, page 388, 1996.

[644] S. H. Vosko, L. Wilk, and M. Nusair, *Can. J. Phys.*, **1980**, *58*, 1200.

[645] A. Maiti, *Phys. Stat. Sol. B*, **2001**, *226*, 87.

[646] A. Svizhenko, M. P. Anantram, T. R. Govindan, B. Biegel, and R. Venugopal, *J. Appl. Phys.*, **2002**, *91*, 2343.

[647] D. A. Papaconstantopoulos, M. J. Mehl, S. C. Erwin, and M. R. Pederson, *Tight-Binding Approach to Computational Materials Science;* Vol. 491 of *MRS Proceedings 491;* Materials Research Society, Warrendale, PA, 1998.

[648] D. Lohez and M. Lanoo, *Phys. Rev. B*, **1983**, *27*, 5007.

[649] A. Maiti, A. Svizhenko, and M. P. Anantram, *Phys. Rev. Lett.*, **2002**, *88*, 126805.

[650] C. L. Kane and E. J. Mele, *Phys. Rev. Lett.*, **1997**, *78*, 1932.

[651] R. Heyd, A. Charlier, and E. McRae, *Phys. Rev. B*, **1997**, *55*, 6820.

[652] L. Yang, M. P. Anantram, J. Han, and J. P. Lu, *Phys. Rev. B*, **1999**, *60*, 13874.

[653] L. Yang and J. Han, *Phys. Rev. Lett.*, **2000**, *85*, 154.

[654] A. Kleiner and S. Eggert, *Phys. Rev. B*, **2001**, *63*, 073408.

[655] P. E. Lammert, P. Zhang, and V. H. Crespi, *Phys. Rev. Lett.*, **2000**, *84*, 2453.

[656] J.-Q. Lu, J. Wu, W. Duan, F. Liu, B. F. Zhu, and B. L. Gu, *Phys. Rev. Lett.*, **2003**, *90*, 156601.

[657] E. D. Minot, Y. Yaish, V. Sazonova, J.-Y. Park, M. Brink, and P. L. McEuen, *Phys. Rev. Lett.*, **2003**, *90*, 156401.

[658] J. Cao, Q. Wang, and H. Dai, *Phys. Rev. Lett.*, **2003**, *90*, 157601.

[659] A. Maiti, *Nature Materials*, **2003**, *2*, 440.

[660] Z. R. Dai, Z. W. Pan, and Z. L. Wang, *Solid State Commun.*, **2001**, *118*, 351.

[661] M. Huang, Y. Wu, H. Feick, N. Tran, E. Weber, and P. Yang, *Adv. Mater.*, **2001**, *13*, 113.

[662] M. Huang, S. Mao, H. Feick, H. Yan, Y. Wu, H. Kind, E. Weber, R. Russo, and P. Yang, *Science*, **2001**, *292*, 1897.

[663] Y. Cui, Q. Wei, H. Park, and C. M. Lieber, *Science*, **2001**, *293*, 1289.

[664] E. Comini, G. Faglia, G. Sberveglieri, Z. Pan, and Z. L. Wang, *Appl. Phys. Lett.*, **2002**, *81*, 1869.

[665] F. Favier, E. C. Walter, M. P. Zach, T. Benter, and R. M. Penner, *Science*, **2001**, *293*, 2227.

[666] C. G. Founstadt and R. H. Rediker, *J. Appl. Phys.*, **1971**, *42*, 2911.

[667] M. Law, H. Kind, F. Kim, B. Messer, and P. Yang, *Angew. Chem., Int. Ed.*, **2002**, *41*, 2405.

[668] A. Szabo and N. S. Ostlund, *Modern Quantum Chemistry;* Dover, New York, 1996.

[669] A. Maiti, J. Rodriguez, M. Law, P. Kung, J. McKinney, and P. Yang, *Nano Letters*, **2003**, *3*, 1025.

[670] J. A. Rodriguez, T. Jirsak, G. Liu, J. Hrbek, J. Dvorak, and A. Maiti, *J. Am. Chem. Soc.*, **2001**, *123*, 3597.

[671] H. Kind, H. Yan, M. Law, B. Messer, and P. Yang, *Adv. Mater.*, **2002**, *14*, 158.

[672] W. Pauli, *Z. Physik*, **1925**, *31*, 765.

[673] E. A. Hylleraas, *Z. Physik*, **1930**, *65*, 209.

[674] C. C. J. Roothaan, *Rev. Mod. Phys.*, **1951**, *23*, 161.

[675] S. F. Boys, *Proc. Roy. Soc.*, **1950**, *A200*, 542.

[676] M. J. Frisch, J. A. Pople, and J. S. Binkley, *J. Chem. Phys.*, **1984**, *80*, 3265.

[677] R. Poirier, R. Kari, and I. G. Csizmadia, *Handbook of Gaussian Basis Sets;* Physical Sciences Data, 24. Elsevier, Amsterdam, 1985.

[678] S. Huzinaga, *Computer Physics Reports*, **1985**, *2*, 279.

[679] J. Almlöf, K. Faegri, Jr., and K. Korsell, *J. Comput. Chem.*, **1982**, *3*, 385.

[680] M. Häser and R. Ahlrichs, *J. Comput. Chem.*, **1989**, *10*, 104.

[681] R. P. Feynman, *Phys. Rev.*, **1939**, *56*, 340.

[682] P. Pulay; In H. F. Schaefer III, Ed., *Application of Electronic Structure Theory*, page 153. Plenum, New York, 1977.

[683] P. Pulay; In K. P. Lawley, Ed., *Ab Initio Methods in Quantum Chemistry II*, page 241. John Wiley & Sons Ltd., Chichester, New York, Brisbane, Toronto, Singapore, 1987.

[684] J. A. Pople, R. Krishnan, H. B. Schlegel, and J. S. Binkley, *Int. J. Quantum Chem.: Quantum Chemistry Symposium*, **1979**, *13*, 225.

[685] J. F. Gaw and N. C. Handy; The Royal Society of Chemistry, London, 1984; page 291.

[686] A. R. Leach, *Molecular Modeling – Principles and Applications;* Addison Wesley Longman Ltd., Singapore, 1997.

[687] L. H. Thomas, *Proc. Camb. Phil. Soc.*, **1927**, *23*, 524.

[688] E. Fermi, *Z. Phys.*, **1928**, *48*, 73.

[689] E. Fermi, *Rend. Accad. Lincei*, **1928**, *7*, 342.

[690] R. G. Parr and W. Yang, *Density-functional Theory of Atoms and Molecules;* Vol. 16 of *The International Series of Monographs on Chemistry;* Oxford University Press, New York, 1989.

[691] E. Teller, *Rev. Mod. Phys.*, **1962**, *34*, 627.

[692] P. Hohenberg and W. Kohn, *Phys. Rev. B*, **1964**, *136*, 864.

[693] W. Kohn and L. J. Sham, *Phys. Rev. A*, **1965**, *140*, 1133.

[694] J. C. Slater, *Phys. Rev.*, **1951**, *81*, 385.

[695] J. P. Perdew, *Phys. Rev. Lett.*, **1985**, *55*, 1665.

[696] J. P. Perdew and Y. Wang, *Phys. Rev. B*, **1986**, *33*, 8800.

[697] J. P. Perdew, In *Electronic structure of solids*, P. Ziesche and H. Eschrig, Eds.; Akademie Verlag, Berlin, Germany, 1991.

[698] J. P. Perdew, *Phys. Rev. B*, **1986**, *33*, 8822.

[699] B. Miehlich, A. Savin, H. Stoll, and H. Preuss, *Chem. Phys. Lett.*, **1989**, *157*, 200.

[700] J. P. Perdew, *Physica*, **1991**, *B172*, 1.

[701] J. P. Perdew and Y. Wang, *Phys. Rev. B*, **1992**, *45*, 13244.

[702] A. D. Becke, *J. Chem. Phys.*, **1992**, *98*, 1372.

[703] A. D. Becke, *J. Chem. Phys.*, **1993**, *98*, 5648.

[704] A. D. Becke, *J. Chem. Phys.*, **1993**, *98*, 1372.

[705] H. Sambe and R. H. Felton, *J. Chem. Phys.*, **1975**, *62*, 1122.

[706] B. I. Dunlap, J. W. D. Connolly, and J. R. Sabin, *J. Chem. Phys.*, **1979**, *71*, 3386.

[707] B. I. Dunlap, J. W. D. Connolly, and J. R. Sabin, *J. Chem. Phys.*, **1979**, *71*, 4993.

[708] R. Jones and A. Sayyash, *J. Phys.*, **1986**, *C19*, L653.

[709] F. Bloch, *Z. Phys.*, **1928**, *52*, 555.

[710] E. Wimmer; Materials science with density functional theory; In *DFT: A Bridge between Physics and Chemistry*, Vrije Universiteit Brussels, 1998.

[711] J. C. Phillips and L. Kleinmann, *Phys. Rev.*, **1959**, *116*, 287.

[712] E. Antoncik, *J. Phys. Chem. Solids*, **1959**, *10*, 314.

[713] V. Heine, In *Solid State Physics, Vol. 24*, H. Ehrenreich, F. Seitz, and D. Turnbull, Eds.; Academic Press, Orlando, 1970; page 1.

[714] J. Kübler and V. Eyert, In *Electronic and Magnetic Properties of Metals and Ceramics, Part I*, R. W. Cahn, P. Haasen, and E. J. Kramer, Eds.; VCH, Weinheim, 1992; page 1.

[715] D. D. Koelling and G. O. Arbman, *J. Phys. F*, **1975**, *5*, 2041.

[716] O. K. Andersen, *Phys. Rev. B*, **1975**, *12*, 3060.

[717] J. A. Pople and G. A. Segal, *J. Chem. Phys.*, **1965**, *43*, S136.

[718] W. Thiel and A. A. Voityuk, *J. Phys. Chem.*, **1996**, *100*, 616.

[719] W. Thiel, *Adv. Chem. Phys.*, **1996**, *93*, 703.

[720] M. J. S. Dewar, C. Jie, and J. Yu, *Tetrahedron*, **1993**, *49*, 5003.

[721] W. Kutzelnigg, *Isr. J. Chem.*, **1980**, *19*, 193.

[722] M. Schindler and W. Kutzelnigg, *J. Chem. Phys.*, **1982**, *76*, 1919.

[723] F. London, *J. Phys. Radium*, **1937**, *8*, 397.

[724] K. Wolinski, J. F. Hinton, and P. Pulay, *J. Am. Chem. Soc.*, **1990**, *112*, 8251.

[725] T. Helgaker and P. Jørgensen, *J. Chem. Phys.*, **1991**, *95*, 2595.

[726] J. Gauss, *Chem. Phys. Lett.*, **1992**, *191*, 614.

[727] J. Gauss, *J. Chem. Phys.*, **1993**, *99*, 3629.

[728] K. Ruud, T. Helgaker, R. Kobayashi, P. Jørgensen, K. L. Bak, and H. J. Aa. Jensen, *J. Chem. Phys.*, **1994**, *100*, 8178.

[729] J. Gauss, *Chem. Phys. Lett.*, **1994**, *229*, 198.

[730] J. Gauss and J. F. Stanton, *J. Chem. Phys.*, **1995**, *103*, 3561.

[731] J. Gauss and J. F. Stanton, *J. Chem. Phys.*, **1995**, *102*, 251.

[732] J. Gauss, *Ber. Bunsenges. Phys. Chem.*, **1995**, *99*, 1001.

[733] D. F. McIntosh and K. H. Michaelian, *Canadian J. Spectros.*, **1979**, *24*, 1.

[734] D. F. McIntosh and K. H. Michaelian, *Canadian J. Spectros.*, **1979**, *24*, 35.

[735] D. F. McIntosh and K. H. Michaelian, *Canadian J. Spectros.*, **1979**, *24*, 65.

[736] E. Madelung, *Physik. Z.*, **1910**, *11*, 898.

[737] M. Born, *Atomtheorie des festen Zustandes;* Teubner, Leipzig, 1923.

[738] A. J. Dekker, *Solid State Physics;* Prentice-Hall, Inc., N.J., 1957.

[739] A. J. Pertsin and A. I. Kitaigorodski, *The Atom-Atom Potential Method;* Springer, Berlin, 1987.

[740] D. E. Williams, *J. Chem. Phys.*, **1965**, *45*, 3770.

[741] D. E. Williams, *J. Chem. Phys.*, **1967**, *47*, 4680.

[742] R. J.-M. Pellenq and D. Nicholson, In *Proceedings of the Fourth International Conference on Fundamentals of Adsorption, Kyoto*, M. Suzuki, Ed.; Kodansha, Tokyo, 1993; page 515.

[743] R. J.-M. Pellenq and D. Nicholson, *J. Phys. Chem.*, **1994**, *98*, 13339.

[744] D. Nicholson, A. Boutin, and R. J.-M. Pellenq, *Mol. Simul.*, **1996**, *17*, 217.

[745] B. van de Graaf, S. L. Njo, and K. S. Smirnov, In *Reviews in Computational Chemistry*, K. B. Lipkowitz and D. B. Boyd, Eds.; Wiley – VCH, New York, 1999; Vol. 14; pages 137–223.

[746] R. Taylor, *Physica B*, **1985**, *131*, 103.

[747] M. W. Finnis and J. E. Sinclair, *Phil. Mag. A*, **1984**, *50*, 45.

[748] B. M. Axilrod and E. Teller, *J. Chem. Phys.*, **1943**, *11*, 299.

[749] A. E. Howard, U. Chandra Singh, M. Billeter, and P. A. Kollman, *J. Am. Chem. Soc.*, **1988**, *110*, 6984.

[750] C. R. A. Catlow and A. N. Cormack, *Int. Rev. Phys. Chem.*, **1987**, *6*, 227.

[751] M. J. Sanders, M. Leslie, and C. R. A. Catlow, *J. Chem. Soc., Chem. Commun.*, **1984**, page 1271.

[752] J. O. Titiloye, S. C. Parker, and S. Mann, *J. Cryst. Growth*, **1993**, *131*, 533.

[753] C. R. A. Catlow, C. M. Freeman, M. S. Islam, R. .A. Jackson, M. Leslie, and S. M. Tomlinson, *Phil. Mag. A*, **1988**, *58*, 123.

[754] A. M. Stoneham, *Physica B*, **1985**, *131*, 69.

[755] C. R. A. Catlow and A. M. Stoneham, *J. Phys. C.: Solid State Phys.*, **1983**, *15*, 4321.

[756] K. B. Wiberg and P. R. Rablen, *J. Comp. Chem.*, **1993**, *14*, 1504.

[757] J. Meister and W. H. E. Schwarz, *J. Phys. Chem.*, **1994**, *98*, 8245.

[758] S. L. Mayo, B. D. Olafson, and W. A. Goddard III, *J. Phys. Chem.*, **1990**, *94*, 8897.

[759] H. Sun, S. J. Mumby, J. R. Maple, and A. T. Hagler, *J. Am. Chem. Soc.*, **1994**, *116*, 2978.

[760] H. Sun, S. J. Mumby, J. R. Maple, and A. T. Hagler, *J. Phys. Chem.*, **1995**, *99*, 5873.

[761] H. Sun, *Macromolecules*, **1995**, *28*, 701.

[762] H. Sun, *Macromolecules*, **1993**, *26*, 5924.

[763] H. Sun, *J. Comput. Chem.*, **1994**, *15*, 752.

[764] E. de Vos Burchart; *Studies on Zeolites: Molecular Mechanics, Framework Stability and Crystal Growth;* PhD thesis, Technical University of Delft, **1992**.

[765] B. W. H. van Beest, G. J. Kramer, and R. A. van Santen, *Phys. Rev. Lett.*, **1995**, *64*, 1955.

[766] N. Karasawa and W. A. Goddard III, *Macromolecules*, **1992**, *25*, 7268.

[767] K. Watanabe, N. Austin, and M. R. Stapleton, *Molecular Simulation*, **1995**, *15*, 197.

[768] B. Feuston and S. Garofalini, *J. Phys. Chem.*, **1990**, *94*, 5351.

[769] G. V. Lewis, *Physica B*, **1985**, *131*, 114.

[770] J. R. Hart and A. K. Rappé, *J. Chem. Phys.*, **1992**, *97*, 1109.

[771] Molecular Simulations Inc.; Program Cerius^2, Version 3.9; 9685 Scranton Road, San Diego, **1999**.

[772] P. Gombás, *Die statistische Theorie des Atoms und ihre Anwendungen;* Springer-Verlag, Vienna, 1949.

[773] J. E. Lennard-Jones, *Proc. Phys. Soc.*, **1931**, *43*, 461.

[774] J. H. de Boer, *Faraday Trans.*, **1936**, *33*, 10.

[775] B. Vessal, M. Amini, M. Leslie, and C. R. .A. Catlow, *Molecular Simulations*, **1990**, *5*, 1.

[776] C. R. A. Catlow and W. C. Mackrodt, *Computer Simulation of Solids;* Lecture notes in physics, 166. Springer-Verlag, Berlin, 1982.

[777] W. L. Jorgensen, *J. Chem. Phys.*, **1982**, *77*, 4156.

[778] G. Brink and L. Glasser, *J. Comp. Chem.*, **1982**, *3*, 47.

[779] M. C. Wojcik and K. Hermansson, *Chem. Phys. Lett.*, **1998**, *289*, 211.

[780] C. R. A. Catlow, C. M. Freeman, and R. L. Royle, *Physica B*, **1985**, *131*, 1.

[781] N. L. Allinger, M. T. Tribble, M. A. Miller, and D. H. Wertz, *J. Am. Chem. Soc.*, **1971**, *93*, 1637.

[782] N. L. Allinger, In *Advances in Physical Organic Chemistry*, V. Gold and D. Bethell, Eds.; Academic Press, London, New York, San Francisco, 1976; Vol. 13; pages 1 – 82.

[783] N. L. Allinger, *J. Am. Chem. Soc.*, **1977**, *99*, 8127.

[784] H.-Q. Ding, N. Karasawa, and W. A. Goddard III, *J. Chem. Phys.*, **1992**, *97*, 4309.

[785] D. Wolf, P. Keblinski, S. R. Phillpot, and J. Eggebrecht, *J. Chem. Phys.*, **1999**, *110*, 8254.

[786] C. J. Casewit, K. S. Colwell, and A. K. Rappé, *J. Am. Chem. Soc.*, **1992**, *114*, 10035.

[787] C. J. Casewit, K. S. Colwell, and A. K. Rappé, *J. Am. Chem. Soc.*, **1992**, *114*, 10046.

[788] C. R. Landis, D. M. Root, and T. Cleveland, In *Reviews in Computational Chemistry*, K. B. Lipkowitz and D. B. Boyd, Eds.; VCH Publishers Inc., New York, 1995; Vol. 6; pages 73 – 148.

[789] T. S. Bush, J. D. Gale, C. R. A. Catlow, and P. D. Battle, *J. Mater. Chem.*, **1994**, *4*, 831.

[790] U. Dinur and A. T. Hagler, In *Reviews in Computational Chemistry*, K. B. Lipkowitz and D. B. Boyd, Eds.; VCH, New York, 1991; Vol. 2; pages 99 – 164.

[791] U. Dinur and A. T. Hagler, *J. Chem. Phys.*, **1989**, *91*, 2949.

[792] U. Dinur and A. T. Hagler, *J. Chem. Phys.*, **1989**, *91*, 2959.

[793] A. T. Hagler, E. Huler, and S. Lifson, *J. Am. Chem. Soc.*, **1974**, *96*, 5319.

[794] A. T. Hagler and S. Lifson, *J. Am. Chem. Soc.*, **1974**, *96*, 5327.

[795] A. T. Hagler and S. Lifson, *Acta Crystallogr. B*, **1974**, *30*, 1336.

[796] S. Lifson, A. T. Hagler, and P. Dauber, *J. Am. Chem. Soc.*, **1979**, *101*, 5111.

[797] A. T. Hagler, S. Lifson, and P. Dauber, *J. Am. Chem. Soc.*, **1979**, *101*, 5122.

[798] A. T. Hagler, S. Lifson, and P. Dauber, *J. Am. Chem. Soc.*, **1979**, *101*, 5131.

[799] S. M. Bachrach, In *Reviews in Computational Chemistry*, K. B. Lipkowitz and D. B. Boyd, Eds.; VCH, New York, 1994; Vol. 5; pages 171 – 227.

[800] D. E. Williams, *J. Comput. Chem.*, **1988**, *9*, 745.

[801] D. E. Williams, *Biopolymers*, **1990**, *29*, 1367.

[802] A. Warshel and S. Lifson, *J. Chem. Phys.*, **1970**, *53*, 582.

[803] F. A. Momany, L. M. Carruthers, R. F. McGuire, and H. A. Scheraga, *J. Phys. Chem.*, **1974**, *78*, 1595.

[804] J. L. M. Dillen, *J. Comput. Chem.*, **1990**, *11*, 1125.

[805] J.-R. Hill, *J. Comput. Chem.*, **1997**, *18*, 211.

[806] S. Kristyan and P. Pulay, *Chem. Phys. Lett.*, **1994**, *229*, 175.

[807] R. J. Bartlett and J. F. Stanton, In *Reviews in Computational Chemistry*, K. B. Lipkowitz and D. B. Boyd, Eds.; VCH, New York, 1994; Vol. 5; pages 65 – 169.

[808] H.-J. Böhm and R. Ahlrichs, *J. Chem. Phys.*, **1982**, *77*, 2028.

[809] H.-J. Böhm, R. Ahlrichs, P. Scharf, and H. Schiffer, *J. Chem. Phys.*, **1984**, *81*, 1389.

[810] H.-J. Böhm, C. Meissner, and R. Ahlrichs, *Mol. Phys.*, **1984**, *53*, 651.

[811] K. P. Sagarik, R. Ahlrichs, and S. Brode, *Mol. Phys.*, **1986**, *57*, 1247.

[812] K. P. Sagarik and R. Ahlrichs, *Chem. Phys. Lett.*, **1986**, *131*, 74.

[813] K. P. Sagarik and R. Ahlrichs, *J. Chem. Phys.*, **1987**, *86*, 5117.

[814] BIOSYM Technologies, Inc.; PROBE Manual; San Diego, **1989**.

[815] C. R. A. Catlow, *Proc. R. Soc. London, Ser. A*, **1977**, *353*, 533.

[816] H. Sun and D. Rigby, *Spectrochim. Acta*, **1997**, *A53*, 1301.

[817] M. Wilson and P. A. Madden, *Faraday Discuss.*, **1997**, *106*, 339.

[818] M. J. Gillan, *Information Quarterly for Computer Simulation of Condensed Phases*, **1993**, *36*, 16.

[819] J. D. Gale, C. R. A. Catlow, and W. C. Mackrodt, *Model. Simul. Mater. Sci. Eng.*, **1992**, *1*, 73.

[820] J. H. Harding, In *Molecular Simulation and Industrial Applications*, K. E. Gubbins and N. Quirke, Eds.; Gordon and Breach Science Publishers, Amsterdam B.V., 1996; page 71.

[821] M. Stoneham, J. Harding, and A. Harker, *MRS Bulletin*, **1996**, *February*, 29.

[822] N. J. Henson, A. K. Cheetham, and J. D. Gale, *Chem. Mater.*, **1994**, *6*, 1647.

[823] C. M. Freeman, C. R. .A. Catlow, J. M. Thomas, and S. Brode, *Chem. Phys. Lett.*, **1991**, *186*, 137.

[824] L. M. Bull, N. J. Henson, A. K. Cheetham, J. M. Newsam, and S. J. Heyes, *J. Phys. Chem.*, **1993**, *97*, 11776.

[825] J. A. Horsley, J. D. Fellmann, E. G. Derouane, and C. M. Freeman, *J. Catal.*, **1994**, *147*, 231.

[826] C. M. Kölmel, Y. S. Li, C. M. Freeman, S. M. Levine, M.-J. Hwang, J. R. Maple, and J. M. Newsam, *J. Phys. Chem.*, **1994**, *98*, 12911.

[827] C. M. Freeman, D. W. Lewis, T. V. Harris, A. K. Cheetham, N. Henson, P. A. Cox, A. M. Gorman, S. M. Levine, J. M. Newsam, E. Hernandez, and C. R. A. Catlow, In *Materials Modeling and Simulation (ACS Symp. Ser.);* American Chemical Society, Washington, DC, 1995; page 5.

[828] J. C. Shelley and D. R. Bérard, In *Reviews in Computational Chemistry*, K. B. Lipkowitz and D. B. Boyd, Eds.; Wiley – VCH, New York, 1998; Vol. 12; pages 137 – 205.

[829] M. Waldman and A. T. Hagler, *J. Comput. Chem.*, **1993**, *14*, 1077.

[830] B. Vessal; In *Catalysis and Sorption Consortium Meeting Minutes*, Biosym Technologies Inc., 1994.

[831] J.-R. Hill, C. M. Freeman, and L. Subramanian, In *Reviews in Computational Chemistry*, K. B. Lipkowitz and D. B. Boyd, Eds.; Wiley-VCH, John Wiley and Sons, Inc., New York, 2000; Vol. 16; pages 141 – 216.

[832] V. S. Allured, C. M. Kelly, and C. R. Landis, *J. Am. Chem. Soc.*, **1991**, *113*, 1.

[833] S. Barlow, A. L. Rohl, and D. O'Hare, *J. Chem. Soc., Chem. Commun.*, **1996**, page 257.

[834] S. Barlow, A. L. Rohl, S. G. Shi, C. M. Freeman, and D. O'Hare, *J. Am. Chem. Soc.*, **1996**, *118*, 7578.

[835] R. M. Badger, *J. Chem. Phys.*, **1934**, *2*, 128.

[836] R. M. Badger, *J. Chem. Phys.*, **1935**, *3*, 710.

[837] I. Antes and W. Thiel, *J. Phys. Chem. A*, **1999**, *103*, 9290.

[838] R. Vetrivel, C. R. A. Catlow, and E. A.Colbourn, *Proc. Royal Soc. Lond.*, **1988**, *A417*, 81.

[839] R. Vetrivel, C. R. A. Catlow, and E. A.Colbourn, *J. Phys. Chem.*, **1989**, *93*, 4597.

[840] E. H. Teunissen, C. Roetti, C. Pisani, A. J. M. de Man, A. P. J. Jansen, R. Orlando, R. A. van Santen, and R. Dovesi, *Modell. Simul. Mater. Sci. Eng.*, **1994**, *2*, 921.

[841] E. H. Teunissen, A. P. J. Jansen, R. A. van Santen, R. Orlando, and R. Dovesi, *J. Chem. Phys.*, **1994**, *101*, 5865.

[842] S. P. Greatbanks, P. Sherwood, and I. H. Hillier, *J. Phys. Chem.*, **1994**, *98*, 8134.

[843] E. H. Teunissen, A. P. J. Jansen, and R. A. van Santen, *J. Phys. Chem.*, **1995**, *99*, 1873.

[844] S. P. Greatbanks, I. H. Hillier P. Sherwood, R. J. Hall, N. A. Burton, and I. R. Gould, *Chem. Phys. Lett.*, **1995**, *234*, 367.

[845] A. Kyrlidis, S. J. Cook, A. K. Chakraborty, A. T. Bell, and D. N. Theodorou, *J. Phys. Chem.*, **1995**, *99*, 1505.

[846] J. Gao, In *Reviews in Computational Chemistry*, K. B. Lipkowitz and D. B. Boyd, Eds.; VCH, New York, 1995; Vol. 7; pages 119 – 185.

[847] D. Bakowies and W. Thiel, *J. Phys. Chem.*, **1996**, *100*, 10580.

[848] T. Matsubara, F. Maseras, N. Koga, and K. Morokuma, *J. Phys. Chem.*, **1996**, *100*, 2573.

[849] J. Sauer and M. Sierka, *J. Comp. Chem.*, **2000**, *21*, 1470.

[850] J. Sauer, P. Ugliengo, E. Garrone, and V. R. Saunders, *Chem. Rev.*, **1994**, *94*, 2095.

[851] Y. Zhang, T.-S. Lee, and W. Yang, *J. Chem. Phys.*, **1999**, *110*, 46.

[852] A. Warshel, Ed., *Computer Modeling of Chemical Reactions in Enzymes and in Solutions;* Wiley, New York, 1991.

[853] J. Åqvist and A. Warshel, *Chem. Rev.*, **1993**, *93*, 2523.

[854] Y.-T. Chang and W. H. Miller, *J. Phys. Chem.*, **1990**, *94*, 5884.

[855] M. Sierka and J. Sauer, *J. Chem. Phys.*, **2000**, *112*, 6983.

[856] N. Metropolis, A. W. Rosenbluth, M. N. Rosenbluth, A. H. Teller, and E. Teller, *J. Chem. Phys.*, **1953**, *21*, 1087.

[857] L. Verlet, *Phys. Rev.*, **1967**, *159*, 98.

[858] W. C. Swope, H. C. Anderson, P. H. Berens, and K. R. Wilson, *J. Chem. Phys.*, **1982**, *76*, 637.

[859] R. W. Hockney, *Methods in Comp. Phys.*, **1970**, *9*, 136.

[860] D. Beeman, *J. Comp. Phys.*, **1976**, *20*, 130.

[861] C. W. Gear, *Numerical Initial Value Problems in Ordinary Differential Equations;* Prentice-Hall, Englewood Cliffs, New Jersey, 1971.

[862] L. V. Woodcock, *Chem. Phys. Lett.*, **1971**, *10*, 257.

[863] H. J. C. Berendsen, J. P. M. Postma, W. F. van Gunsteren, A. Di Nola, and J. R. Haak, *J. Chem. Phys.*, **1984**, *81*, 3684.

[864] H. C. Andersen, *J. Chem. Phys.*, **1980**, *72*, 2384.

[865] S. Nosé, *Mol. Phys.*, **1984**, *53*, 255.

[866] W. G. Hoover, *Phys. Rev. A*, **1985**, *31*, 1695.

[867] M. M. Cielinski; Computer simulation of the Lennard–Jones fluid in the grand canonical ensemble; Master's thesis, University of Maine, Orono, **1985**.

[868] M. Lupkowski and F. van Swol, *J. Chem. Phys.*, **1991**, *95*, 1995.

[869] A. Papadopoulou, E. D. Becker, M. Lupkowski, and F. van Swol, *J. Chem. Phys.*, **1993**, *98*, 4897.

[870] S. Weerasinghe and B. M. Pettitt, *Mol. Phys.*, **1994**, *82*, 897.

[871] L. F. Vega, K. S. Shing, and L. F. Rull, *Mol. Phys.*, **1994**, *82*, 439.

[872] G. S. Heffelfinger and F. van Swol, *J. Chem. Phys.*, **1994**, *100*, 7548.

[873] C. M. Lo and B. Palmer, *J. Chem. Phys.*, **1995**, *102*, 925.

[874] R. F. Cracknell, D. Nicholson, and N. Quirke, *Phys. Rev. Lett.*, **1995**, *74*, 2463.

[875] R. M. Shroll and D. E. Smith, *J. Chem. Phys.*, **1999**, *111*, 9025.

[876] R. Colle and O. Salvetti, *Theor. Chim. Acta*, **1975**, *37*, 329.

[877] C. Lee and C. Sosa, *J. Chem. Phys.*, **1994**, *100*, 9018.

[878] E. R. Cohen and B. N. Taylor, *J. of Research of the National Bureau of Standards*, **1987**, *92*, 85.

[879] E. R. Cohen and B. N. Taylor, *Rev. Mod. Phys.*, **1987**, *59*, 1121.

Index

T - #0102 - 101024 - C0 - 234/156/18 [20] - CB - 9780824724191 - Gloss Lamination